W9-AFW-122
1775 404
the Coop
$49.95

Projects in Scientific Computation

Projects in Scientific Computation

Richard E. Crandall

Diskette included

Springer-Verlag

Richard E. Crandall
Reed College
Portland, Oregon
USA

Cover illustration: Hilbert fractal, a space-filling curve, prepared in *Mathematica*® program.

Publisher:	Allan M. Wylde
Publishing Associate:	Cindy Peterson
TELOS Production and ManufacturingManager:	Sue Purdy Pelosi
Administrative Assistant:	Kate McNally Young
Project Management:	Jan Benes, Black Hole Publishing Service
Cover Designer:	Iva Frank
Cover Production Artist:	Peter Altenberg
Copy Editor:	Paul Green

Camera ready copy was prepared by the author on a NeXT computer.

Published by TELOS, The Electronic Library of Science, Santa Clara, California
TELOS is an imprint of Springer-Verlag New York, Inc.

Cataloging-in-Publication data is available from the Library of Congress

Printed in the United States of America

9 8 7 6 5 4 3 2 1

ISBN 0-387-97808-9 Springer-Verlag New York Berlin Heidelberg
ISBN 3-540-97808-9 Springer-Verlag Berlin Heidelberg New York

TELOS, The Electronic Library of Science, is an imprint of Springer-Verlag New York with publishing facilities in Santa Clara, California. Its publishing program encompasses the natural and physical sciences, computer science, economics, mathematics, and engineering. All TELOS publications have a computational orientation to them, as TELOS' primary publishing strategy is to wed the traditional print medium with the emerging new electronic media in order to provide the reader with a truly interactive multimedia information environment. To achieve this, every TELOS publication delivered on paper has an associated electronic component. This can take the form of book/diskette combinations, book/CD-ROM packages, books delivered via networks, electronic journals, newsletters, plus a multitude of other exciting possibilities. Since TELOS is not committed to any one technology, any delivery medium can be considered.

The range of TELOS publications extends from research level reference works through textbook materials for the higher education audience, practical handbooks for working professionals, as well as more broadly accessible science, computer science, and high technology trade publications. Many TELOS publications are interdisciplinary in nature, and most are targeted for the individual buyer, which dictates that TELOS publications be priced accordingly.

Of the numerous definitions of the Greek word "telos," the one most representative of our publishing philosophy is "to turn," or "turning point." We perceive the establishment of the TELOS publishing program to be a significant step towards attaining a new plateau of high quality information packaging and dissemination in the interactive learning environment of the future. TELOS welcomes you to join us in the exploration and development of this frontier as a reader and user, an author, editor, consultant, strategic partner, or in whatever other capacity might be appropriate.

TELOS, The Electronic Library of Science
Springer-Verlag Publishers
3600 Pruneridge Avenue, Suite 200
Santa Clara, CA 95051

Preface

Projects in Scientific Computation

"[Mr. Isaac Asimov], in your short story "Saving Humanity," a character says all problems in the world are created by computers, which are making human beings unnecessary. Do you believe that?"

[*Reporter, Leaders Magazine,* 1989 *vol.* 12 *no.* 4]

"That's just what a character is saying. It's not my view. I am not responsible for the views of my characters."

[*Isaac Asimov*]

Likewise I am not responsible for the deep and fascinating problems that have come my way–from researchers, engineers, and students–over the last decade. This book is comprised of selected Projects ranging from the tutorial to the open-ended and, at times, to the utterly impossible. There are two primary institutions at which I collected these Projects. First, as Chief Scientist at NeXT, Inc. of Redwood City, California in the early 1990s, I collected research results and truly confounding computational dilemmas that were directed to myself or to colleagues at that corporation. The second source of the Projects is Reed College, Portland, Oregon where, primarily in the 1980s, I was fortunate to be able to interact with some of the sharpest students of science to be found in the liberal arts sector, or for that matter, in any sector. In large measure it has been the special questions and needs of such brilliant students (they know who they are), which needs dictated the design of the Projects. Much of the material herein has accordingly been used in my scientific computation lectures, and thus tested to some degree by the attending students.

Subject-Project book structure

The book is arranged by Chapter, Subject, Project according to the hierarchy:

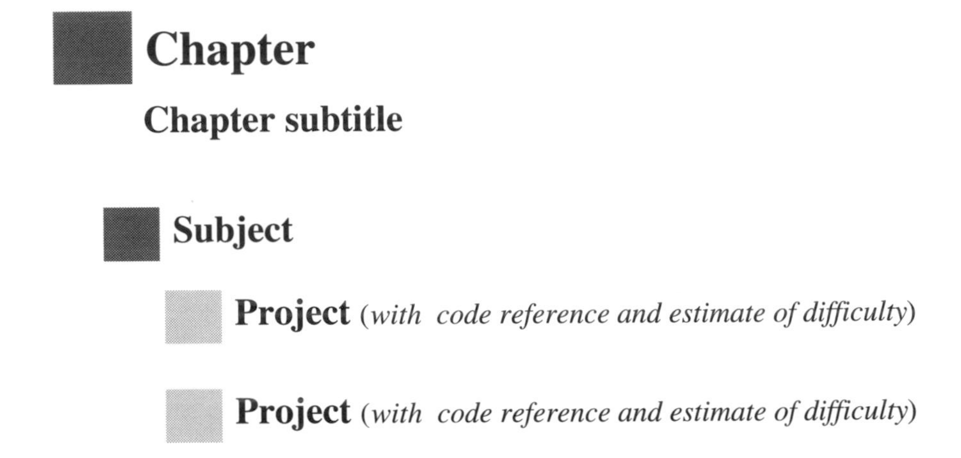

To navigate, one may use the Table of Contents, the Index, or the page headers to locate, respectively, Project titles, key words and phrases, or Subject names.

Project design

One fundamental belief to which I have come around over these years is that there is nothing wrong with letting the responsibility for creativity rest on the shoulders of the reader. Some readers will protest that they have not the time, which is understandable; but other readers, I hope, will be excited by how very far they can go within a Project. So here is how I design a typical Project in the book: I pretend a student or colleague genuinely wants to investigate a problem or field. I presume further that this person has been frustrated in part by past disappointments with standard homework-style problem sessions. For one thing, I like to try to arrange a Project's result to be as attractive as possible (fine identity, strong statistical statement, pretty graph, etc.) and in this way present a kind of "reward" for wading through such a junglc.

In many of the Projects the basic message is "do this or that hard, wide-ranging, or, at minimum, laborious task . . ." I have found it more productive, especially when addressing students, to convey the scope of the issue rather than engage in long tutorials; to explain, say, what an experienced colleague of mine has done with the problem. Thus a typical Project, though it will contain some indication of difficulty, may neither be obviously simple nor obviously difficult. After all, to paraphrase the eminent mathematician C. L. Siegel, one cannot know the true difficulty of a Project until it has been solved! What we *can* assume, I think, is that when Siegel's notion is applicable the burden of creativity does indeed rest with the reader. All these remarks amount to an admission that the burden is *intentional*, and, I hope, ultimately productive and rewarding.

Project difficulty

Attached to each Project is a statement of its principal focus: computation, theory, or graphics, or some combination thereof. Each Project is given an estimated difficulty level, though, as just explained, this assessment cannot be completely reliable. The difficulty levels in any case are defined:

Level 1: Introductory, elementary, tutorial;
Level 2: A level that should be accessible to a college major in the field;
Level 3: Difficult for most everyone, requiring creativity, labor;
Level 4: Exploratory research problem, perhaps intractable.

I trust it is clear why there would be any difficulty level 4 Project steps herein. Recall that many tough problems were sent to me by colleagues and students. Such problems, if they had not been intractable, would probably not have come to my attention in the first place.

One question a reader might ask is: "What good is it to label a Project 'difficulty level 1-3,' which is almost the whole difficulty range?" My answer is that such a span of levels indicates you can find at least one elementary exercise therein, and, if so moved, you can also find some challenging problem(s).

Support software

Also attached to most Projects is a list of support code. Almost all such code appears in the Appendix, with the exception of code that is just too voluminous on which to spend paper. Such code resides instead on a companion diskette.

The code was all developed under the NEXTSTEP™ operating system, and takes the form either of C or of *Mathematica*™ source. However, many of the author's own results, as well as observations within Projects from other collaborators, were obtained via NEXTSTEP object-oriented applications, using primarily Objective-C.

Most of the code is portable. When special libraries (such as sound features) are involved, the source comments reflect this. The reader should beware that many of the *Mathematica* sources are relatively ancient, corresponding to problems posed and solved in the late 1980s. Such sources are expected to run under *Mathematica* 1.0 or beyond, and often will have an "old-fashioned" flavor. I thought it best, when such a source actually worked well, to leave well enough alone and place it in the Appendix as something that is known to work. Here I must admit something. Whenever I give such an older source to a *Mathematica* expert (I am not one of these) the probability that the code will promptly be improved is $1-\epsilon$, and I hereby apologize for this.

I have been reminded by colleagues and reviewers that the software of the modern world is fairly grown up (or at least as grown up as it ever has been!), and that much public and commercial software packages exist which might work well with the Projects herein. For symbolic, numerical, and algebraic manipulation one may use *Maple*™ and *MatLab*™. I apologize to aficianados of these products that I do not explicitly use them herein. I certainly believe that the Projects will work harmoniously with those packages if that is the reader's desire. For public software, one may peruse ftp network sites, newsgroups, and announcements of all types to find a great deal of software that should also work well with this book. An example of when the reader can do well to look outside the book is the following. Though rudimentary Runge-Kutta differential equation solvers are outlined in Subject 1.2, I have been reminded that the true sophisticate should contemplate a Runge-Kutta-Felberg (say RKF-45) scheme with error estimation and variable step size. Thus, we find a suggestion in Project 1.2.3 that the reader may do well to look outside the book for truly sophisticated techniques. This notion, of pointing the interested reader to realms upon which I have not touched, or am not qualified to touch, is a theme that pervades the entire book.

How to use this book for teaching

The following remarks reflect what I have found to be the most effective way to use the Projects in a college (undergraduate) course context.

I have found it important to demand of every student who starts with this material that a language such as C or *Mathematica* be in hand *on day one of the course*. This demand has a twofold purpose. First, it ensures that failure to make progress will not merely be a disguised failure to grasp programming; and second, the demand removes *a priori* any expectation on the part of the student that a programming course is being taught. The course is a scientific computation course, and I am saying that a programming course (or solid outside experience) must precede it.

No student can reasonably be expected to solve a large fraction of the Projects. As I have intimated, the book is designed so that hard and impossibile Projects are not uncommon. Therefore I would recommend:

- That the conventional "final exam, percentile achievement score" paradigm be dropped, and replaced with a teacher's estimate of how far the student has progressed in terms of interest and ability to explore, without regard for the *number* or *fraction* of Project steps turned in;
- That students work on just *a very few* Projects, if good productive work on those few be deemed as important as equivalent integrated labor over many Projects;
- That students who pursue the difficulty level 3-4 problems be given liberal credit for partial work; indeed many of the Project steps at those difficulty levels exist, even in the current literature, as only partially solved dilemmas.

I have had some success in starting out a semester with a preliminary scouring of the Projects for difficulty level 1 steps. The opening Project of Chapter 2, on plane geometry, is a good starting point, especially for students who are a little intimidated by the harder parts of the book. What I have found most successful is to start a course with Chapters 1-2 all at once, so that students more comfortable with numerical methods can start turning in work from Chapter 1, whilst students who prefer the graphics/geometry approach can start in on Chapter 2. In this regard, yet another possibility for the teacher is:

- As an option, drop also the conventional serial-chapter course approach and allow students to work asynchronously if they can do well in such a mode.

Relevant to this last option, it has happened that I have had to struggle not to allow Chapter 4 (on the FFT) to consume a whole semester. Instead I have let certain students stay with it, and allowed others move elsewhere. If a teacher were, on the other hand, to allow the FFT chapter to dominate, I would welcome that choice as entirely consistent with the purpose of the book. To my mind the chapter chronology and emphasis employed by the instructor should depend only upon one thing: the teacher's immediate perception of the progress and interest of the students. If you believe me, then it is unlikely that such a course in successive years will turn out the same. Indeed, I have learned that with this material there is no telling where the course will lead. That can be a fulfilling experience on all sides.

Because graphics is so integral to many of the Projects, my advice on the matter is:

- Prior to the course, endeavor to supply truly decent graphics tools. *Mathematica* and other mathematical processors are self-contained of course. But consider color (e.g. for the last chapter's image processing) and color printing means. NEXTSTEP application building with clean PostScript™ and TIFF image handling is especially effective.

I feel that one final word of psychological advice is important. It is usually valuable to an institution, and certainly valuable to the students present, to ensure that no one subgroup feels abandoned. When chemistry and biology majors are taking the course, it is good at least to point out that Projects from their fields reside prominently within the book, and that the opening Projects are a kind of preparation for the professional hands-on science to come. This lends a legitimate sense of pride to such students, who otherwise are too often overwhelmed by highly computer-literate or mathematically inclined peers who may perform better on the opening fundamentals. It is important not to "lose" the interdisciplinary flavor by inadvertently "losing" students of various disciplines. Perhaps the strongest interdisciplinary message I have been lucky to convey to students is this: *Forget the paradigm of software categories*. Don't be upset just because you think you will never again in your career use an FFT, or test a prime number. The fundamental work in the book involves the type of *thinking* one can expect to encounter in professional scientific computation work. The Euler constant γ may have little to do with atomic orbitals, but if a student does solid work on that constant I can pretty much guarantee that certain mental pathways to computational chemical physics will be less congested and easier to navigate.

How to use this book for research

It is my hope that this book will stand as useful reference material for researchers, in the following sense. When you, the researcher, confront a computational problem, I hope you will recall some Subjects herein that *might* be related. If you turn to that Subject, there may be a literature reference or a similar problem posed.

It may be helpful to peruse the source code list at the start of the Appendix, for you will often be able to tell from the program's name the actual research problem being addressed.

The good news is that, as intimated in the opening of this Preface, the book is essentially built upon a foundation of research problems submitted to me or indirectly brought to my attention. So in a sense I have fashioned a reference book for myself. Thus, there is a chance that it can serve as a kind of reference for others.

Golf

In summary, this book was designed to support computational research and to convey to students a sense of how one's brain feels when engaging in that research. I am told (and, based on brief personal experience, I completely believe) that golf is a noble but supremely frustrating game. I think the same is true of the world of computation. So the reader, whether he or she be researcher, student, or teacher, will naturally experience some consternation, some CPU envy (there never will be enough available cycles), some sadly immortal bugs, and a humbling sense of awe at the vastness of the universe of impossible questions. If these frustrations do not sit squarely, side by side with any excitement and sense of reader accomplishment, I have not done my job.

Acknowledgements

Primary acknowledgement pertinent to the industrial sector has to start with S. Jobs, founder of NeXT, Inc., who not only has allowed scientific computation to proceed in his domain, but has provided perennial encouragement in regard to the importance of algorithms for public consumption. His wish, that certain computational science methods appear in one form or another within his products, has given rise to many of the techniques and results reported herein. I also thank S. Wolfram, who gave us the *Mathematica* language and gave me excellent advice on how to present studies in computation. Some of the research problems herein can be traced back to the days when *Mathematica* was being ported to the NEXTSTEP platform (some problems began as tests of the new software). I am grateful to G. Tribble, who over the years steadfastly supported algorithm development at NeXT, Inc.

Engineering colleagues whom I wish to thank include J. Doenias, S. Gillespie, J. Adams, C. Norrie, A. Tevanian, P. Graffagnino, J-M. Hullot, M. Paquette, L. Hourvitz, L. Boynton, B. Yamamoto, M. Minnick, J. Smith, M. Meyer, B. Pinkerton, G. Crow, R. Page, M. Byron, J. Newlin, K. McCurley, E. Winfree, B. Smith, T. Gayley, D. Slowinski, R. Silverman, R. Frye, T. Gray, J. Keiper, S. Russell, D. Moss, C. Hallstrom, A. Laird, W. Hunter, J. Seamons, J. Coursey, M. DeMoney. Expert programmer T. Matteson provided tools and insight for the organization of the source code Appendix. A. Bittner has, over the years of scientific development at NeXT, provided expert administrative assistance.

Primary acknowledgement for the aid I received from the academic sector must be directed to J. P. Buhler. His mathematical knowledge and algorithmic advice have found their way into many facets of this work. In fact, whole sections of the book arose either from our collaboration, or from his encyclopedic grasp of what is currently solved and what is not.

Academic colleagues whom I also wish to thank include M. Levich, A. Odlyzko, J. Powell, J. Delord, G. Schlickeiser, R. Mayer, D. Griffiths, D. Perkinson, S. Arch, R. Kaplan, G. Gwilliam, M. McClellan, P Russell, N. Wheeler, T. Wieting, A. Jones, G. Elliott, B. McNamara, M. Bedau, T. Dunne, A. Prinz, S. Christenson, B. Fagin, I. Vardi, S. Wagstaff, V. Miller, R. Ormond, G. Butz, M. Colgrove, M. Ringle. R. Reynolds uncovered mishaps in the manuscript. I also wish to thank my scientific computation students of 1991 and 1993, for they bore the main weight of the whole experiment. In a sense P. Bragdon did, as President of Reed College, start this book ten years ago, with the foresight to take computation seriously as an academic imperative. In recent times S. Koblik, W. Haden, and E. McFarlane have found amicable, enlightened ways to support industrial algorithm development at Reed College.

For inspirational contributions to this work I thank L. Powell, C. Novak, L. Jobs, N. Bragdon, K. Leach, E. Leeb, H. Casabona, N. Stucki, C. Jacobs, J. Welch, C. Larson, S. Love, P. Lupino. Special thanks are due A. & A. & L. Barisich for providing, over the years, space and materials in support of this work; and to B. Barisich for solving transportation problems connected with book development. I am indebted to L. Buhler for the many logistical problems he solved during this work.

Lastly, I acknowledge incisive manuscript review and criticism from I. Vardi, P. Kuhn, G. Corliss, W. Titus, D. Kuethe, and an anonymous reviewer. What gets past such distinguished readers is about $\varepsilon > 0$. I must also acknowledge the asymptotically large patience of my editors/publishers P. Green, J. Benes, S. Purdy Pelosi, and A. Wylde. To say that the book would have been impossible without Mr. Wylde's strong dedication–to the world of publishing and to my own microscopic sector of it–would be a radical understatement, something like saying: "Numbers are infinitely dependable . . .for upon them, count you may."

R. E. Crandall
Portland, Oregon
1993

Contents

Projects in Scientific Computation

Preface vii

Subject-Project book structure viii

Project design viii

Project difficulty ix

Support software x

How to use this book for teaching xi

How to use this book for research xiii

Golf xiii

Acknowledgements xiv

1 Numbers everywhere 1

Selected topics in numerical analysis

1.1 Numerical evaluation 2

1.1.1 Evaluation of famous constants 6

1.1.2 Evaluation of elementary functions 16

1.1.3 Special functions 30

1.2 Equation solving 32

1.2.1 Matrix algebra 33

1.2.2 Non-linear equation systems 38

1.2.3 Differential equations 40

1.3 Random numbers and Monte Carlo 42

1.3.1 Generating random numbers 43

1.3.2 Numerical integration and Monte Carlo 47

2 Exploratory computation 51

Collected intra- and interdisciplinary projects

2.1 Mathematical problems 52

2.1.1 Planar geometry problems 52

2.1.2 Symbolic manipulation 55

2.1.3 Real and complex analysis 62

2.2 Nature-motivated models 68

2.2.1 Neural network experiments 70

2.2.2 Genetic algorithms and artificial life 77

2.3 Projects from biology 79

2.3.1 Population models 79

2.3.2 Physiology, neurobiology, and medicine 84

2.3.3 Molecular biology 89

2.4 Projects from physics and chemistry 92

2.4.1 Classical physics 93

2.4.2 Quantum theory 102

2.4.3 Molecules and structure 107

2.4.4 Relativity 110

3 The lure of large numbers 113

Projects in number theory

3.1 Large-integer arithmetic 114

3.1.1 Testing the operations 115

3.2 Prime numbers 117

3.2.1 Mersenne primes 117

3.2.2 Primes in general 120

3.3 Fast algorithms 129

3.3.1 Fast multiplication 129

3.3.2 Fast mod, division, inversion 133

3.3.3 Other fast algorithms 135

3.4 Factoring 136

3.4.1 Factoring algorithms 138

3.4.2 Status of Fermat numbers 145

4 **The FFT forest** 151

The ubiquitous FFT and its relatives

4.1 Discrete Fourier transform 152

4.1.1 Fundamental DFT manipulations 155

4.1.2 Algebraic aspects of the DFT 156

4.1.3 DFT test signals 160

4.1.4 Direct DFT software 162

4.2 FFT algorithms 165

4.2.1 Recursive FFTs 166

4.2.2 FFT indexing and butterflies 167

4.2.3 Complex FFTs, N a power of 2 174

4.2.4 Real-signal FFTs 177

4.2.5 FFTs for other radices 178

4.2.6 FFTs in higher dimensions 179

4.2.7 Applications of the FFT 181

4.3 Real-valued transforms 182

4.3.1 Hartley transform 183

4.3.2 Discrete cosine transform 185

4.3.3 Walsh-Hadamard transform 188

4.3.4 Square-wave transform 191

4.4 Number-theoretic transforms 193

4.4.1 Exploring number-theoretic transforms 194

5 Wavelets 197

Young arrivals in the transform family

5.1 Chords, notes, and little waves 199

5.1.1 Windowed Fourier transform 200

5.1.2 Continuous wavelet transform 203

5.2 Discrete wavelet bases 205

5.2.1 Example wavelet expansions 206

5.2.2 Mother function and its wavelet 208

5.3.2 Wavelets of compact support 212

5.3 Discrete wavelet transform 216

5.3.1 Fast wavelet transform algorithms 222

5.3.2 Applications of fast wavelet transforms 224

6 Complexity reigns 229

Chaos & fractals & such

6.1 Chaos 230

6.1.1 Quadratic map algebra 232

6.1.2 Bifurcation and chaos 236

6.1.3 Chaos models 240

6.1.4 Chaos, stability, and Lyapunov exponents 251

6.1.5 Applications of chaos theory 253

6.2 Fractals 254

6.2.1 Theory of fractals 258

6.6.2 Visualization of fractals 262

6.6.3 Fractal Brownian noise 273

6.2.4 Measurement of fractal dimension 284

7 Signals from the real world 293

Projects in signal processing

7.1 Data compression 294

7.1.1 Tour of lossless data compressors 297

7.2 Sound 307

7.2.1 Examples of sound processing 307

7.2.2 Examples of sound compression 314

7.3 Images 317

7.3.1 Examples of image processing 318

7.3.2 Image compression 326

Appendix 331

Support code for the book Projects

References 447

Index 457

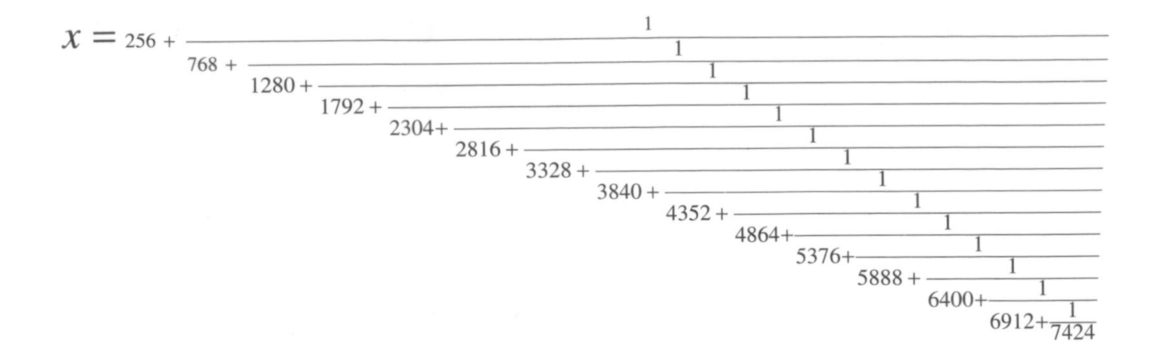

$$x = 256 + \cfrac{1}{768 + \cfrac{1}{1280 + \cfrac{1}{1792 + \cfrac{1}{2304 + \cfrac{1}{2816 + \cfrac{1}{3328 + \cfrac{1}{3840 + \cfrac{1}{4352 + \cfrac{1}{4864 + \cfrac{1}{5376 + \cfrac{1}{5888 + \cfrac{1}{6400 + \cfrac{1}{6912 + \cfrac{1}{7424}}}}}}}}}}}}}}$$

$$= \frac{822835854775861965165551196274654469634036521241600\,0}{321418620953838137515363682604278457765686476800\,01}$$

$$e \sim \left(\frac{x+1}{x-1}\right)^{128}$$

$$= \mathbf{2.71828182845904523536028747135266249775724709369995957}$$
$$\mathbf{49696762772407663035354759457138217852516642742746}76...$$

Evaluation of the number e via continued fraction optimization. The simple continued fraction x has relatively large elements lying in arithmetic progression. In turn, e is related to x via fundamental algebraic operations. The expansion shown for x yields more than 100 *good (boldface) digits of e, with only about* 60 *total multiplications involved. In fact, if the optimally efficient fraction is chosen, the number of good digits of e grows asymptotically faster than the number of operations.*

1 Numbers everywhere

Selected topics in numerical analysis

For obvious reasons, our knowledge of all phases in the development of number representations in different cultures is rather incomplete... The number 10 *represented a very natural choice, founded on the simple fact that we have* 10 *fingers. As a matter of fact,* 10 *strikes a good balance between the number of symbols in the system and the number of digits necessary to represent a given number... A decisive step forward was taken by the Hindus about A.D.* 500*; they used the ordinary digits* 0, 1,2,...,9 *with the important convention that the position of a digit in a number carries information about its value... This scheme turned out to be extremely practical...*

[*Froberg* 1985]

What is numerical analysis? ... A certain wrong answer has taken hold ... :"Numerical analysis is the study of rounding errors"... I propose the following alternative definition with which to enter the new century: "Numerical analysis is the study of algorithms for the problems of continuous mathematics"... The pivotal word is "algorithms."

[*Trefethen* 1992]

By analogy with Trefethen's notion we begin on the idea that numerical analysis is the study of algorithms also for the problems of *science*. We presently tour a selection of numerical analysis topics that to the author's mind require the type of thinking that scientific problems require.

We dwell momentarily on numerical analysis, moving to field-specific problems in Chapter 2. The reader interested in more thorugh treatments may wish to peruse established references, such as [Skeel and Keiper 1993] [Froberg 1985], or move on to intermediate level texts such as [Buchanan and Turner 1992][Greenspan and Casulli], or delve into more advanced treatments [Stoer and Bulisch 1993] or [Golub and Van Loan 1989]. Algorithm-specific texts such as the classic [Knuth 1981] and newer classic [Press et al. 1988] are also helpful in building a more thorough background.

Numerical evaluation

In our first Project we shall look at famous constants and elementary functions. Such study may not be explicitly scientific, but the exercise of evaluating these constants and functions is good preparation for field-specific investigations.

Generally speaking, what one might call a "moderately good" numerical evaluation algorithm will exhibit linear convergence; that is, the number of correct digits increases linearly in the number of accumulated operations. Another way to say this is that the remaining error decreases exponentially in the operation count. But there are faster algorithms that give rise to *quadratic* convergence, meaning that the remaining error is roughly squared at each iteration. This means that the number of correct digits essentially doubles with every iteration and so grows exponentially with operation count. In the following Project we look at methods of computing constants and evaluations of elementary functions; these methods range from "poorly convergent," having less than linear convergence, to the rapidly convergent quadratic and even higher-order-than-quadratic algorithms. The wide spectrum of algorithmic efficiencies mirrors well what happens in scientific fields, where computation speed can sometimes benefit immensely from theoretical preparation and forethought.

It should be mentioned that the time required to attain a desired accuracy is not the same as the required operation count. In assessments of algorithmic efficiency one can summarize linear convergence in a statement such as: "To get D accurate digits, one requires $O(D)$ arithmetic operations in D-digit precision arithmetic." Some authors will cast this as $O(D\ M(D))$ bit operations, where $M(D)$ denotes the number of bit operations required for a D-by-D-digit multiply. Think of one "operation" as a multiply of two high-precision D-digit numbers. Then this O notation makes sense: the $O(D)$ effectively might count the number of loop passes to complete a converging algorithm to some accuracy, each pass having a fixed number of high-precision multiplies; while $O(D\,M(D))$ is a better measure of *how long a time* the computation will take, because of course extra time is needed for higher-precision multiplies. In this notation, quadratic convergence means D digits of precision require only $O(\log D)$ high-precision operations.

The collection of known numerical evaluation algorithms is a healthy mix of very old, old, new, and very new ideas. A very old idea that still lives on at the heart of many numerical packages is the Newton method. The summary statement is so very simple: to solve the equation $f(x) = 0$, guess (adroitly!) an initial x and iterate:

$$x := x - \frac{f(x)}{f'(x)} \tag{1.1.1}$$

with the hope of rapidly approaching the desired solution. Note that if we actually had a solution at some iteration, then the quotient in (1.1.1) would vanish and the iteration would be perfectly stable. Simple as it appears, the Newton iteration can exhibit frightfully complex behavior, as we shall see in Chapter 6. But when initial guesses are sufficiently sane ones, solutions are approached rapidly. The method is still used today for such tasks as square root extraction, especially when high precision is desired. For one thing, if one has written a high-precision multiply routine, one can often take roots and even *divide* using the multiply within the Newton iteration.

Another old idea is that of Euler-Maclaurin summation, a means by which a sum can be approximated by an integral plus error terms [Froberg 1985][Stoer and Bulirsch 1993][Vardi 1991][Graham et al. 1989]. One version of the Euler-Maclaurin formula expresses an M-th order approximation of a sum by an integral:

(1.1.2)

$$\int_0^N f(x)dx - \sum_{k=0}^{N-1} f(k) = \frac{1}{2} f \,\Big|_0^N - \sum_{m=1}^{M} \frac{B_{2m}}{(2m)!} f^{(2m-1)} \,\Big|_0^N - \frac{B_{2M+2}}{(2M+2)!} N f^{(2M+2)}(y)$$

Here, f is assumed to be $(2M+2)$-times differentiable, y is some real value in $[0,N]$, and the B_{2m} are the Bernoulli numbers. It is known that if $f^{(2M+2)}$ does not change sign on $(0,N)$ then the error term has magnitude less than twice the magnitude of the first neglected term of the m-summation, and also has the same sign as the neglected term [Froberg 1985].

An example of a not-so-old evaluation method is the application of the elegant arithmetic-geometric mean (AGM) to computational problems. Lagrange, Gauss, and Legendre were familiar with the AGM in the 18th century, and in the 1970s Salamin and Brent resurrected these ideas (together with some new computer-scientific ideas) to yield the fastest known numerical evaluation methods for certain constants such as π, and certain transcendental functions [Borwein and Borwein 1987]. The AGM is defined as follows. For a pair of numbers a, b in $(0,\infty)$, obtain a new pair as the arithmetic and geometric means; i.e., perform the iteration:

$$\binom{a}{b} := \begin{pmatrix} \frac{a+b}{2} \\ \sqrt{ab} \end{pmatrix} \qquad (1.1.3)$$

Then the iterated pair will converge to a common limit which we denote AG(a,b). Furthermore, this convergence is generally quadratic, meaning the distance between the two members of the pair decreases faster than exponentially in the iteration count; more precisely, the number of digits required to specify the difference $|a-b|$ essentially doubles every iteration.

In the mid-twentieth century, modern numerical algorithms such as the Romberg, Aitken, and Richardson extrapolation methods began to appear. In these methods one guesses the final value of a limiting process according to its current behavior. The idea is appealing because there is minimal "waste"–even one's first few limit evaluations count for something. An idealized but illustrative example would be the following. Assume that the numerical limit of a sequence $\{y_n\}$ is desired, but that unbeknownst to the programmer, the sequence is secretly given by $y_n = a + b^n$, where $0 < b < 1$ so that the infinite-n limit of the sequence is just a. Then it is a matter of trivial agebra that:

$$b = \frac{y_{n+2} - y_{n+1}}{y_{n+1} - y_n}$$

$$a = \frac{b y_n - y_{n+1}}{b - 1}$$

which means that one can obtain the infinite-n limit just from knowledge of any three consecutive members of the sequence, even the first three members! Of course this trick works because the sequence has a very special form. Romberg and Aitken extrapolations, which use this trick in somewhat different ways to estimate limits on the basis of the geometric decay assumption, apply in practice to a vast generality of limit sequences. In fact, these methods provide efficiency gain in principle when terms beyond the unknown final limit comprise a power series in some parameter b. In this more general setting one naturally needs more algebra and more consecutive terms to arrive at the desired limit a. Theoretically, the schemes tend to enjoy close to linear convergence: the number of digits of accuracy goes roughly linearly with operation count. However, construction of the Romberg or Aitken tableaux involves very quick and easy operations, so for practical purposes such schemes can be superior to other linear convergence schemes.

Project 1.1.1 Evaluation of famous constants

Computation, a little theory, difficulty level 1-3.
Support code: Appendix "RombergLimits.ma," "AitkenLimits.ma," "Pi.ma," "ContFract.ma."

1) An interesting constant is the Euler (or Euler-Mascheroni) constant γ, of which one definition is:

$$\gamma = \lim_{n \to \infty}\left(1 + \frac{1}{2} + \frac{1}{3} + \ldots + \frac{1}{n} - \log n\right) \tag{1.1.4}$$

There is some mystery associated with this constant. It is not known for example whether γ is irrational. Calculate γ by performing the literal operations in the definition, until the first four digits read 0.5772. How many terms were required for this precision (i.e. what was n)? Next, perform the same calculation task with alternative definitions:

$$\gamma = \lim_{n \to \infty}\left(1 + \frac{1}{2} + \frac{1}{3} + \ldots + \frac{1}{n} - \log\left(n + \frac{1}{2}\right)\right) \tag{1.1.5}$$

$$\gamma = \lim_{n \to \infty}\left(1 + \frac{1}{2} + \frac{1}{3} + \ldots + \frac{1}{n-1} + \frac{1}{2n} - \log n\right)$$

and compare the apparent rate of convergence (number of accurate digits as a function of n) with respect to the experiment for the first form (1.1.4). Is there a better choice of logarithmic argument in the first of the two alternatives (1.1.5)? For example should one use $\log(n+a)$ for some $a \neq 1/2$?

A noticeably more efficient approach is to use the Romberg extrapolation method for limit evaluation. As intimated in the introduction to this Project, we attempt to project the ultimate value of the limit on the basis of the behavior at discrete indices. Denote by γ_m the m-th approximation to γ, say from the second formula of (1.1.5). It turns out that there is an asymptotic expansion of the form [Henrici 1977] :

$$\gamma_m \sim \gamma + a_1 m^{-2} + a_2 m^{-4} + \ldots$$

for large m. Simply knowing the fact of such an expansion enables accelerated convergence of numerical estimates. We choose a parameter $x = 1/4$ and imagine a column of values of γ_m for m running through powers of 2:

$$\begin{aligned}
\gamma_1 &\sim \gamma + a_1 + a_2 + \ldots \\
\gamma_2 &\sim \gamma + a_1 x + a_2 x^2 + \ldots \\
\gamma_4 &\sim \gamma + a_1 x^2 + a_2 x^4 + \ldots \\
&\vdots
\end{aligned}$$

Evidently the first two terms can be used to eliminate the a_1 term; in fact

$$\frac{\gamma_2 - x\gamma_1}{1-x} = \gamma - a_2 x + \ldots$$

and a similar removal of a_1 terms can be effected with any two consecutive column entries. Thus, a second column can be constructed, and from this a third and so on, with every entry of the j-th column missing asymptotic coefficients through a_j. By these means we create a tableau:

$$\begin{array}{llll}
\gamma_1\ (= q_{11}) & & & \\
\gamma_2\ (= q_{21}) & q_{22} & & \\
\gamma_4\ (= q_{31}) & q_{32} & q_{33} & \\
\gamma_8\ (= q_{41}) & q_{42} & q_{43} & q_{44} \\
\quad\vdots & & &
\end{array}$$

such that the diagonal entries will converge rapidly to the ultimate limit. Show that, if the left-hand column is established, the rest of the tableau should obey the recursion:

$$q_{ij} = \frac{q_{i,j-1} - x^{j-1} q_{i,j-1}}{1 - x^{j-1}} \qquad ; j = 2, \ldots, i$$

in order that the j-th column be missing the coefficients as claimed.

The Appendix code "RombergLimits.ma" computes the numerical tableau:

```
0.5
0.5568528194 0.57580376
0.5720389722 0.57710102   0.57718751
0.575915602  0.57720781   0.57721493   0.57721537
0.576890271  0.57721516   0.57721565   0.57721566   0.57721566
```

and, when taken out to higher precision can easily give approximations as good as:

$$\gamma = 0.5772156649015328606065120900824024310422\ldots$$

in a few workstation-seconds.

Here is a difficult but workable problem: If N is the total count of reciprocals and multiplications required to set up the Romberg tableau (we assume here the log() computations are negligible), what is an upper bound on the error of the lower-right diagonal entry, in terms of N? Results of this type should be roughly consistent with claims such as this: that 187 rows of the tableau do yield 10,000 correct decimal places for γ [Henrici 1977]. As to whether γ is irrational, this known approximation means that if γ is to be rational its denominator must exceed 10^{10000}. Take care to note the difference between the error as a function of operations and the error as a function of the number of rows of the tableau. Note that, by comparison, [Borwein and Borwein 1987] suggest a method for obtaining γ to D digits of precision in $O(\log^2 D)$ high-precision operations. This would imply a "sub-quadratic" convergence rate. Still, as mentioned in the introduction to this Project, the Romberg method may be the fastest available for some range of digit counts–perhaps a few dozen digits–even though the scheme can be asymptotically beaten.

2) Estimate the Euler constant of the last step, this time creating an Aitken extrapolation table. First show that if $y_n = a + b^n$ converges to a, then the modified sequence:

$$y'_n = y_n - \frac{(y_{n+1} - y_n)^2}{y_{n+2} - 2y_{n+1} + y_n}$$

converges more rapidly to a. This Aitken "Δ^2 formula" can then be applied to the new sequence to get terms y''_n and so on. Thus one builds up a tableau as with the Romberg scheme, although for the elementary Aitken method the number of columns is bounded. It is customary to enter the iterated sequence as a first column, save first- and second-differences as columns, whence the fourth column becomes the (usually) improved sequence. As soon as a new Aitken value is obtained, it is often advantageous to insert this as the next "original" data and start generating the sequence from this new estimate. An implementation of the Aitken tableau is found in Appendix code "AitkenLimits.ma."

3) Here we investigate the Euler-Maclaurin expansion (1.1.2). Once again we shall estimate the Euler constant (1.1.4). First show that

$$\gamma = 1 + \sum_{k=2}^{\infty} \left(\frac{1}{k} + \log\left(1 - \frac{1}{k}\right)\right)$$

Now perform numerical evaluation of γ by doing a direct sum from $k = 2$ up to some cutoff integer of choice, then approximating the sum from the cutoff to $k = \infty$ on the basis of (1.1.2). For given error bound on the final evaluation of γ, attempt to optimize with respect to the cutoff integer.

4) Compute in several ways the number log 2, using such representations as:

$$\log(1+x) = x - \frac{x^2}{2} + \frac{x^3}{3} - \frac{x^4}{4} + \ldots \qquad (1.1.6)$$

$$\log x = \lim_{q \to \infty} \frac{x^{2^{-q}} - x^{-2^{-q}}}{2^{1-q}}$$

Note that the latter formula can be evaluated in principle with successive square roots. As with the Euler constant of step (1) previous, the infinite sum formula is very inefficient. A more efficient method is again Romberg extrapolation, where we posit an initial column consisting as usual of our limit formula evaluated at exponentially increasing resolution:

$$q_{n1} = 2^{n-1} \sinh \frac{\log x}{2^{n-1}}$$

This relation appears at first glance to involve recourse to the value of the unknown log x itself, but these q values can be obtained from a convenient recursion:

$$q_{11} = \frac{x^2 - 1}{2x} \; ; \; q_{21} = \frac{x - 1}{\sqrt{x}} \; ; \; q_{n+1,1} = q_{n1}\sqrt{\frac{2q_{n1}}{q_{n1} + q_{n-1,1}}} \tag{1.1.7}$$

As exemplified in "RombergLimits.ma," log 2 can easily be obtained in this way to the precision:

log 2 = `0.693147180559945`...

in a fraction of a workstation-second. Again, a good exercise is to study the convergence rate. If this problem is approached theoretically, one may assume that a square root time is a constant multiple of multiply time at given precision. It will turn out that this evaluation of log 2, like the evaluation of any elementary function, can be done by other means to D digits in at most $O(\log^2 D)$ high-precision operations, as we discuss later in this Project.

5) Here we investigate methods for computing π. This task is probably the most celebrated evaluation problem of history; certainly it is about as old as such a problem could be, going back as far as the Babylonians and then to Archimedes who was able to provide important insight and at least three significant digits. On the basis of the previous two steps of this Project, one can correctly surmise that classical expansions, such as:

$$\frac{\pi}{4} = \tan^{-1} 1 = 1 - \frac{1}{3} + \frac{1}{5} - \frac{1}{7} + \frac{1}{9} - \ldots$$

or more computationally efficient forms such as:

$$\frac{\pi}{4} = 7\tan^{-1}\frac{1}{10} + \tan^{-1}\frac{1282831}{14587029}$$

will not be optimal, and indeed that is so. (Note also Brouncker's formula in step (7) below, which formula converges poorly.) For comparison purposes, the investigator might still want to try such classical expansions. Before going on to other schemes, it is of interest that the second $\tan^{-1}$ formula above can be applied with linear convergence in a certain modified sense: The number of significant (accurate) decimals is linear in the number of $\tan^{-1}$ expansion terms used. But what is the actual number of high-precision multiplies required to effect D digits of π in this way?

It is certainly possible to use the extrapolation methods of the previous steps (1) and (2). In fact, the recursion (1.1.7), if we start instead with $q_{11} = 1$, $q_{21} = \sqrt{2}$; gives the Romberg diagonal converging to $\pi/2$ [Henrici 1977], as exhibited in Appendix code "RombergLimits.ma." Again a 15-decimal value obtains in a fraction of a workstation-second. But this is the proverbial tip-of-the-iceberg. Again before going on, it is instructive to answer: What is the convergence rate for this method?

Modern schemes include AGM (arithmetic-geometric mean) methods, originally due to Gauss and Legendre, and the fantastic schemes of Ramanujan [Borwein and Borwein 1987][Chudnovsky and Chudnovsky 1987]. Here is just one AGM scheme which has truly fast convergence: Initialize three numbers

$$x = \sqrt{2}\ ; \qquad p = 2 + \sqrt{2}\ ; \qquad y = \sqrt[4]{2}$$

then iterate in the following order:

$$x := \frac{1}{2}\left(\sqrt{x} + \frac{1}{\sqrt{x}}\right)$$

$$p := p * \frac{x+1}{y+1}$$

$$y := \frac{y\sqrt{x} + \frac{1}{\sqrt{x}}}{y + 1}$$

Then the number p converges to π. Implement this algorithm and verify the claim that the absolute error $|\pi - \pi_n|$ is less than $10^{-2^{n+1}}$ [Borwein and Borwein 1987]. It follows that D digits of π can be obtained via $O(M(D) \log D)$ bit operations. The Appendix code example "Pi.ma" gives, after just four loop passes, the correct-as-it-stands value:

π = 3.14159265358979323846264338327950288 4197...

which is certainly faster convergence than we have seen so far in the current Project. The Appendix code when run will show that the number of good digits indeed doubles per every loop pass. There exist even faster methods, again using AGM computations. Such methods can exhibit even quartic convergence, for which perhaps a few dozen iterations (at huge pecision) will give π to tens of millions of digits.

But there are schemes for π that involve only fundamental operations and only a fixed number of square roots. One such is the Ramanujan-inspired formula [Chudnovsky and Chudnovsky 1987]:

$$\frac{1}{\pi} = 12 \sum_{n=0}^{\infty} (545140134n + 13591409) \frac{(-1)^n}{640320^{3n+1/2}} \frac{(6n)!}{(3n)!\,(n!)^3}$$

What is the convergence rate of this formula? Theoretically speaking, it does not converge quite as rapidly as does the AGM scheme; but the overhead of continual root-taking is happily absent, and so the above formula is a popular one, used in several commercial packages [Vardi 1993].

6) We have seen that square roots commonly appear in some of the fast algorithms, and one might wonder how to do square roots quickly. Consider the following generalized Newton iteration for evaluating $\sqrt{2}$:

$$x := ax + \frac{2-2a}{x} \tag{1.1.8}$$

Here, one fixes a constant parameter a in (0,1), then (adroitly!) chooses an initial x. The iteration will often converge on $\sqrt{2}$. Show first that if $x = \sqrt{2}$ then the iteration is stable: x does not change. Then study the rate of convergence as a function of the choice a. Does it ever make sense to change a dynamically, that is, at various times throughout an iteration? Note that when root-taking, including the problem of higher roots, is considered in the complex plane, the situation is frightfully complicated, as is seen with the Newton fractals of Chapter 6..

7) To "round out" our brief tour of fundamental constants, we turn to the problem of evaluating e. One may compare, say, three methods. First, there is the classical form:

$$e = \sum_{n=0}^{\infty} \frac{1}{n!}$$

Then there is a Newton iteration based on $f(x) = \log x - 1$:

$$x := x(2 - \log x) \tag{1.1.9}$$

where, as usual, an adroit first choice for x is in order. Of course, a fast logarithm, perhaps of the type discussed later in this Project, is essential to the success of this iterative scheme. Luckily logarithms can be evaluated without recourse to the value of e. A third option is to use the beautiful continued fraction expansion:

$$e = 2 + \cfrac{1}{1 + \cfrac{1}{2 + \cfrac{1}{1 + \cfrac{1}{1 + \cfrac{1}{4 + \cfrac{1}{1 + \cfrac{1}{1 + \cfrac{1}{6 + \ldots}}}}}}}}$$

An elegant way to evaluate any continued fraction of the general form:

$$x = a_0 + \cfrac{b_1}{a_1 + \cfrac{b_2}{a_2 + \dots}} \tag{1.1.10}$$

is to set four initial values:

$$p_0 = a_0 \ ; \quad p_{-1} = q_0 = 1; \quad q_{-1} = 0 \tag{1.1.11}$$

and to iterate:

$$\begin{aligned} p_n &= a_n\, p_{n-1} + b_n\, p_{n-2} \\ q_n &= a_n\, q_{n-1} + b_n\, q_{n-2} \end{aligned} \tag{1.1.12}$$

for $n = 1,2,3,\dots$ Then, when the limit is valid,

$$x = \lim_{n\to\infty} \frac{p_n}{q_n}$$

The Appendix code "ContFract.ma" carries out such computation for e and also for Brouncker's formula for π:

$$\frac{4}{\pi} = 1 + \cfrac{1}{2 + \cfrac{9}{2 + \cfrac{25}{2 + \cfrac{49}{2 + \dots}}}}$$

Besides the comparisons of the various methods for calculating e, there are interesting questions such as this: Is it better to compute e from its continued fraction, or from the continued fraction for $(e^2+1)/(e^2-1)$, or some other fraction involving e? This topic is discussed in step (8) next.

Note that continued fractions are involved somewhat in Project 2.1.3, and again touched upon in connection with chaos theory in Project 6.1.5.

8) Investigate the convergence rate of fractions $(e^{2/n}+1)/(e^{2/n}-1)$ for integers n. The simple continued fraction a sequence can easily be

guessed using the scf[ ] function in "ContFract.ma," but a more proper approach is to manipulate the continued fraction for tanh($1/n$), implicit in (1.1.14) below, until the numerator b sequence consists entirely of 1's. The point now is that one could compute, say, $e^{2/2^k}$ with a simple continued fraction and then square this number (k–1) times to get e. Thus the number of operations is $O(k+M)$, where M convergent levels of (1.1.12) need be calculated to get within some stated error bound. Using the fact that the absolute error for a simple continued fraction evaluated out to p_n/q_n is bounded by $1/(q_n q_{n+1})$, show that e can be computed in this way to D digits accuracy in at worst $O(\sqrt{D})$ operations. Certainly this is faster than linear convergence, but is not theoretically the best possible, as we see (for any elementary function evaluation) in the next Project. Still, note that the present ultra-linear method does not involve anything but the fundamental operations, e.g., no roots are involved. The frontispiece to this chapter sketches this story in a pictorial manner.

9) A fascinating number is the Khinchin constant:

$$K = \prod_{n=1}^{\infty}\left(1+\frac{1}{k(k+1)}\right)^{\frac{\log k}{\log 2}} \qquad (1.1.13)$$

It turns out that if one considers, for irrational x in (0,1), the simple continued fraction of x (the expansion (1.1.10) but with all numerator terms b_i equal to 1), then amazingly enough *almost all* x have the geometric mean:

$$\lim_{n\to\infty}(a_1 a_2 \ldots a_n)^{1/n} = K$$

Attempt to compute K by expanding random x into simple continued fractions (this is a difficult approach of dubious convergence). Consider instead the notion of expanding log K, using for example the formula of [Vardi 1991] which casts log K in terms of ζ ', the derivative of the Riemann Zeta function, which function is discussed in the next Project. Aside from this calculation of K, many simple sums can be cast in more convergent form by use of the Riemann Zeta [Froberg 1985].

10) Find continued fraction elements of e^γ, possibly rediscovering a very large a_i element as did Brent [Graham et al. 1989].

Project 1.1.2 Evaluation of elementary functions

Computation, a little theory, difficulty level 1-3.
Support code: Appendix "RombergLimits.ma," "AitkenLimits.ma," "ContFract.ma."

1) Investigate theoretically or numerically or both the convergence properties of certain continued fraction expansions for elementary functions:

$$\frac{x}{\tan^{-1} x} = 1 + \cfrac{x^2}{3 + \cfrac{4x^2}{5 + \cfrac{9x^2}{7 + \cfrac{16x^2}{9 + \dots}}}} \tag{1.1.14}$$

$$x \tan x = \cfrac{x^2}{1 - \cfrac{x^2}{3 - \cfrac{x^2}{5 - \cfrac{x^2}{7 - \dots}}}}$$

Note that just about everything elementary follows in principle from these two canonical cases. For example, the latter fraction can be used to get $\tanh x$, and thus e^x; likewise the first formula yields the log function via $\tanh^{-1}$. Recall, however, the remarks of step (8), Project 1.1.1, in regard to special exponential numbers $e^{2/n}$ which all can be expressed as *simple* continued fractions (i.e., having all b_j numerators equal to 1). Does it make computational sense to evaluate e^x, for rational $x = a/b$, by raising $e^{2/2b}$ to power a? In this regard note the frontispiece to the chapter.

2) Here we investigate a little further the Newton method upon which we have already touched. For the fundamental iteration (1.1.1) we have seen the square-root example (1.1.8) and the exponentiation example (1.1.9).

There is a classic reciprocation example, which shows that a reciprocal can be taken without the need for any explicit division. In fact, for given $a > 0$, we take $f(x) = 1/x-a$, whence the iteration

$$x := 2x - ax^2 \tag{1.1.15}$$

will, for the obligatory adroit choice of initial x, converge to $1/a$. Let $x = 1$ be the initial choice, and prove that, symbolically speaking, every iterate is a Taylor expansion polynomial of the function $1/a$ expanded around the point $a = 1$. Thus, the Newton method can sometimes "speed up" the process of taking Taylor expansions.

Now show that for real $a > 0$, an initial guess x_0 gives rise to eventual convergence if $(1-ax_0)^2 < 1$. Is the converse true? At any rate, such observations give rise to an airtight scheme for global Newton reciprocation, as follows:

- Given real a, start the procedure to find $1/|a|$ and adjust sign of a later.
- Now that $a > 0$, multiply or divide by powers of 2 until a is in $[1/2,1]$.
- Set $x_0 = a/2$ and iterate (1.1.15) until x is stable to within desired error.
- Recover possible power of 2 and possible sign factor to report $1/a$.

Note that "divide by 2" operations are not really using recourse to division because one may simply multiply by 0.5. Write a global reciprocation package, *sans* multiplication, that works for any non-zero input real number.

Next, consider the higher-order iteration:

$$x := x\left(1 + (1 - ax) + (1 - ax)^2\right) \tag{1.1.16}$$

and explain why this should generally converge more rapidly as a method for reciprocation. Test these various ideas numerically.

3) Write software to compute in the asymptotically fastest ways known, general elementary functions, meaning logarithmic and exponential functions. From these functions one may, in turn, use complex numbers to effect trigonometric functions and so on. It is known that any elementary function may thus be evaluated to D digits in $O(M(D) \log^s D)$

bit operations, or $O(\log^s D)$ high-precision operations for some s between 1 and 2 inclusive [Borwein and Borwein 1987].

One general software approach is to write a fast square root, then settle the logarithm problem, use this in a Newton iteration for the exp() function, finally to obtain the trigonometrics from the latter evaluated at certain complex arguments. For logarithms there is the following remarkable AGM formula:

$$\log x \sim \frac{\pi}{2}\left(\frac{1}{AG(1,\varepsilon)} - \frac{1}{AG(1,\varepsilon x)}\right) \tag{1.1.17}$$

which approximates log nearly to precision ε^2 [Borwein and Borwein 1987]. A *Mathematica* function that tests this idea might appear:

```
mylog[x_,n_] :=
    Pi/2 * (ArithmeticGeometricMean[1.0,10.0^(-n)]) -
        ArithmeticGeometricMean[1.0,10.0^(-n) x])
```

One would pass the precision exponent n and expect a little less than about $2n$ correct digits.

4) Explain in theoretical terms why the fundamental Newton recursion (1.1.1) tends to give rise to quadratic convergence, i.e., the remaining error is essentially squared on each iteration. Such a treatment must assume of course that the iteration is converging in the first place.

Next, investigate the interesting variants of the Newton iteration. One variant is:

$$x := x - A\frac{f(x)}{f'(x)} \tag{1.1.18}$$

where A is an acceleration constant; while another variant is Halley's iteration which involves n-th derivatives:

$$x := x + (n+1)\frac{\left(\frac{1}{f}\right)^{(n)}}{\left(\frac{1}{f}\right)^{(n+1)}} \tag{1.1.19}$$

which for $n = 0$ gives the standard Newton iteration, but for $n > 0$ gives higher-order iterations such as (1.1.12). In particular, it is sometimes claimed that for some problems, various $A \neq 1$ will actually accelarate convergence in (1.1.18).

Yet another variant which can sometimes show *cubic* convergence is the nested scheme:

$$x := x - \frac{f(x)}{f'(x)} - \frac{1}{f'(x)} f\left(x - \frac{f(x)}{f'(x)}\right) \tag{1.1.20}$$

Experiments with this scheme should involve a second variable, call it $y = x - f/f'$, so that x becomes $y - f(y)/f'(x)$.

5) Investigate Pade approximants as a computational means of numerical evaluation of elementary functions [Press and Teukolsky 1992][Borwein and Borwein 1987]. In Pade schemes one looks for an approximation to a function f:

$$f(x) = \sum_{k-0}^{\infty} f_k x^k \sim \frac{\sum_{n=0}^{N} a_n x^n}{1 + \sum_{n=1}^{N} b_n x^n} \tag{1.1.21}$$

by demanding that the Pade quotient agree with f at $x = 0$, and that the first $2N$ derivatives of the Pade quotient also agree with the respective derivatives of f at $x = 0$. Show that these conditions are equivalent to:

$$a_0 = f_0 \tag{1.1.22}$$

$$\sum_{k=1}^{N} b_k f_{N-k+m} = -f_{N+m} \quad ; m = 1, 2, \ldots, N$$

$$\sum_{k=0}^{m} b_k f_{m-k} = a_m \qquad ; m = 1, 2, \ldots, N$$

with b_0 defined to be zero. Clearly we have $2N+1$ equations in the same number of unknowns $\{a_0,\ldots,a_N\}$ and $\{b_1,\ldots,b_N\}$. Then implement in software a Pade scheme which knows formal derivatives of an elementary function, then proceeds to develop solutions to the above relations in order to build a Pade quotient which may then be used for fast approximation. Press and Teukolsky give the algebraic function:

$$f(x) = \left(7 + (1+x)^{4/3}\right)^{1/3}$$

as an example of how the Pade system can sometimes magically give an approximation, in their case even for $N = 4$, that has wider range of good behavior than does the original power series. These authors also point out that, even though the system (1.1.22) involves a Toeplitz (near-diagonal) matrix, this matrix tends to be near-singular, so that full LU-decomposition or other numerically stable solvers may be called for.

The following step, and also potentially step (4), Project 3.2.2 involve the Pade fraction concept.

6) Investigate the efficiency of Lagrange's continued fraction:

$$\frac{kx}{1-(1+x)^{-k}} = 1 + \cfrac{\frac{x(1+k)}{1\cdot 2}}{1 + \cfrac{\frac{x(1-k)}{2\cdot 3}}{1 + \cfrac{\frac{2x(2+k)}{3\cdot 4}}{1 + \cfrac{\frac{2x(2-k)}{4\cdot 5}}{1 + \ldots}}}}$$

for approximate evaluation of the algebraic form on the left. In particular, take $k = 11/2$ and show via plots or tables that:

- Involvement of the first six numerators (b terms) of the fraction yield a rational polynomial expression as a cubic divided by a quadratic;

• Over the range $-0.75 < x < 0$ this rational approximation is superior to the binomial expansion of the left-hand side through term x^5.

Investigate the similarity (or difference) between Lagrange's fraction and Pade approximants of step (5) previous.

7) Here we work with yet another attractive and potentially useful formula for π:

$$\pi^2 = 72 \sum_{n=1}^{\infty} \frac{\left(2-\sqrt{3}\right)^n}{n^2\binom{2n}{n}}$$

It is instructive first to try to guess a more general identity: Replace the number $(2-\sqrt{3})$ with a variable x^2, and think of the left-hand side as proportional to the square of an arcsine. Describe the rate of convergence (i.e., how many summands are required to obtain D good digits of π^2?). Describe a method for computing π^2 in this way, but only doing rational arithmetic, evaluating $\sqrt{3}$ just once, at the end of the computation. Assuming the correct arcsine identity, describe means for bettering the "14 good digits per summand" convergence of the Ramanujan formula of step (5), Project 1.1.1. This can be done, but at the expense of some fixed number of square root operations; the Ramanujan formula requires only one root be taken.

This class of identity is important in Apery's proof that $\zeta(3)$ is irrational [Guy 1980]. There is an identity for $\zeta(3)$ as a multiple of the sum:

$$\sum_{n=1}^{\infty} \frac{(-1)^{n-1}}{n^3\binom{2n}{n}}$$

and it is a worthwhile test of one's numerical processor to find the implied constant of proportionality. Such an exercise–to guess an identity on the basis of numerical output–is neither facetious nor ignoble. In fact it was discovered purely numerically that the following identity is likely to be true:

$$\zeta(4) = \frac{\pi^4}{90} = C \sum_{n=1}^{\infty} \frac{1}{n^4 \binom{2n}{n}}$$

where C is evidently a *rational* constant. One good way to find this conjectured constant is to expand a numerical version of it as a continued fraction. The author does not know whether this obvious rational guess for C has yet been proven rigorously correct.

Now we turn to numerical evaluation of special functions, by which we mean scientifically important functions beyond the elementary class (of log, exp, sin, cos, $\tan^{-1}$, etc.). Special functions are commonly used in both pure and applied science. It will turn out that most special functions, even of complex argument, can be evaluated with linear convergence; i.e., the number of good digits grows roughly linearly in the computation effort. Furthermore this convergence can be effected via a bounded operation count irrespective of the complex argument. Thus, it is possible with a total of N high-precision multiplies to provide at least CN good digits of the Bessel function $J_0(z)$, where the constant C is independent of complex z. The basic method for converging in this uniform fashion is "asymptotic breakover," in which we switch between ascending series and asymptotic series representations (or perhaps amongst additional representations) depending on z. The author conveyed results for the breakover method to J. Keiper for some of the original special function calls in the first editions of *Mathematica.* It should be mentioned that even for these special functions the ideas of accelerated convergence and extrapolation of the previous Project are sometimes applicable to the evaluation series chosen.

Asymptotic breakover technique is well exemplified in the case of the error function, defined:

$$erfc(z) = \frac{2}{\sqrt{\pi}} \int_z^{\infty} e^{-t^2}\, dt \tag{1.1.23}$$

The function admits of an ascending series expansion:

$$erfc(z) = 1 - \frac{2z}{\sqrt{\pi}} \left(1 - \frac{z^2}{3 \cdot 1!} + \frac{z^4}{5 \cdot 2!} - \frac{z^6}{7 \cdot 3!} + ...\right) \tag{1.1.24}$$

and an asymptotic expansion:

$$erfc(z) = \frac{e^{-z^2}}{z\sqrt{\pi}} \left(1 - \frac{1}{2z^2} + \frac{1 \cdot 3}{2^2 z^4} - \frac{1 \cdot 3 \cdot 5}{2^3 z^6} + ...\right) \tag{1.1.25}$$

and a continued fraction representation:

$$erfc(z) = \frac{e^{-z^2}}{\sqrt{\pi}} \cfrac{1}{z + \cfrac{1/2}{z + \cfrac{1}{z + \cfrac{3/2}{z + \cfrac{2}{z + ...}}}}} \tag{1.1.26}$$

where the pattern in the numerators after the first "1" runs {...1/2,1,3/2,2,5/2,3,7/2,4,...}, consisting of consecutive pairs $(n-1/2,n)$. It is often the case that the error accrued by stopping a series is bounded by the magnitude of the last term used; at least this is true for real arguments z for the error function. Assume non-negative real argument z. It can be shown that for the error function's ascending series (1.1.24) the remaining error H_N (overall error in computed *erfc*() after summing through the term z^{2N-2} in the parentheses of (1.1.24)) is bounded by the magnitude of the first neglected term; i.e. the error satisfies:

$$|H_N| < \frac{z}{N^{3/2}} \left(\frac{z^2 e}{N}\right)^N \tag{1.1.27}$$

By the same token, for real positive z the corresponding error for the asymptotic series (1.1.25) is bounded by:

$$|A_N| \; < \; \frac{1}{z}\left(\frac{z^2 e}{N}\right)^{-N} \tag{1.1.28}$$

The "breakover" idea is to choose one series or the other depending on which error bound is smaller. It can be shown in this way that for D good digits, the total number of terms required is proportional to D.

The linear convergence of the breakover method applies equally well to a wide variety of special functions. The author believes that the proper starting point of the general theory is the Mellin-Barnes representation for the Whittaker functions of the second kind:

$$W_{k,m}(z) \; = \; \frac{z^k e^{-z/2}}{\Gamma(-L)\Gamma(-L')} \frac{1}{2\pi i} \oint \Gamma(s)\Gamma(-s-L')\Gamma(-s-L)\, z^s\, ds \tag{1.1.29}$$

where $L = k + m - 1/2$, $L' = k - m - 1/2$, and the contour runs from $-i\infty$ to $+i\infty$ and separates the poles of $\Gamma(s)$ from the poles of the other two Γ's. This representation appears formidable for good reason: so many special functions, not to mention their asymptotic properties, follow from it. In modern times it has been possible to apply (1.1.29) to the computational problem for special functions along the following lines. The W function can be cast in terms of the Whittaker function of the first kind:

$$W_{k,m}(z) \; = \; \frac{\Gamma(-2m)}{\Gamma(-L)} M_{k,m}(z) \; + \; \frac{\Gamma(2m)}{\Gamma(-L')} M_{k,-m}(z) \tag{1.1.30}$$

with M in turn defined by:

$$M_{k,m}(z) \; = \; z^{m+1/2} e^{-z/2} \, {}_1F_1\left(m - k + \frac{1}{2};\; 2m+1;\; z\right) \tag{1.1.31}$$

where ${}_1F_1$ is one of the standard hypergeometric functions:

$${}_1F_1(a;\, b;\, z) \; = \; 1 \; + \; \frac{az}{b} \; + \; \frac{a(a+1)z^2}{b(b+1)} \; + \; \ldots \tag{1.1.32}$$

$$ {}_2F_0\left(a;\ b;\ \frac{1}{z}\right) = 1 + \frac{ab}{z} + \frac{a(a+1)b(b+1)}{z^2} + \ldots $$

Evidently (1.1.30) can be written in terms of ascending series for most values of the parameters k, m. On the other hand the asymptotic expansion of W can be carried out on the basis of (1.1.29), to yield, at least formally:

$$ W_{k,m}(z) \sim z^k e^{-z/2}\ {}_2F_0\left(-L;\ -L';\ -\frac{1}{z}\right) \tag{1.1.33} $$

where it is understood that this asymptotic series generally does not converge.

The breakover conditions for these special functions can be handled in a uniform–if complicated–manner, as follows. The author has derived from the contour representation (1.1.29) the following result, of which the error function case (1.1.28) is special. If we take $N-1$ terms of the asymptotic series (1.1.33), if $N > Re(L) + 1/2$, $Re(L') + 1/2$, and if $|arg(z)| < 3\pi/2$; then the total error in the truncated asymptotic series is bounded as:

$$ |A_N| < |C_N T_N| \tag{1.1.34} $$

where T_N is the first neglected series term and the multiplier C_N is:

$$ C_N = \frac{N}{\Gamma(N-L)} \int_0^1 (1-y)^{N-1}\, dy \oint_0^{\infty e^{ia}} \frac{s^{N-L} e^{-s}\, ds}{s\left(1+\frac{sy}{z}\right)^{N-L'}} \tag{1.1.35} $$

where a is any angle such that $|arg(z) - a| < \pi/2$. The contour integral runs outward along the infinite ray from the origin, at inclination angle a. Again a formidable expression, but the benefits to be reaped are substantial; as many functions may now be evaluated to rigorous precision by adroitly switching between ascending series (1.1.30) and asymptotic series (1.1.33).

Special functions that can be evaluated numerically using these techniques include:

Error function: (1.1.36)

$$erfc(z) = 1 - erf(z) = \frac{e^{-z^2/2}}{\sqrt{\pi z}} W_{-\frac{1}{4},\frac{1}{4}}(z^2) \qquad ; |arg(z)| < \frac{\pi}{2}$$

$$erfc(z) = 2 - erfc(-z) = \frac{1}{\sqrt{\pi}} \Gamma\left(\frac{1}{2}, z^2\right)$$

The latter formula can be used always to bring $arg(z)$ within proper range for convenient asymptotic error bounds.

Incomplete Gamma function: (1.1.37)

$$\Gamma(a,z) = \int_z^\infty t^a e^t \frac{dt}{t} \qquad ; Re(a) > 0$$

$$\Gamma(a,z) = z^{\frac{a-1}{2}} e^{-\frac{z}{2}} W_{\frac{a-1}{2},\frac{a}{2}}(z)$$

$$\Gamma(a,z) = \Gamma(a) - \frac{z^a}{a}\, {}_1F_1(a; 1+a; -z) \qquad ; a \neq 0, -1, -2, \ldots$$

$$\Gamma(-n,z) = \frac{(-1)^n}{n}\left(\Gamma(0,z) - \frac{e^{-z}}{z}\, {}_2F_0\left(1;\ 1;\ -\frac{1}{z}\right)_{n-1}\right)$$

$$\Gamma(0,z) = -Ei(-z) = -\gamma - \log(\pm z) - \sum_{k=1}^{\infty} \frac{(-z)^k}{k!\ k}$$

$$\Gamma(a,z) = \cfrac{z^a e^{-z}}{z + \cfrac{1-a}{1+\cfrac{1}{z+\cfrac{2-a}{1+\cfrac{2}{z+\ldots}}}}}$$

In the $\Gamma(-n,z)$ formula, the hypergeometric series is truncated at the $z^{-(n-1)}$ term inclusive. In the exponential-integral (Ei) formula, the sign of the log argument is minus if and only if $arg(z) = \pi$. The final continued fraction is correct when it converges.

Bessel functions: (1.1.38)

$$H_\nu{}^{(1)}(z) = J_\nu(z) + iY_\nu(z) = \sqrt{\frac{2}{\pi z}}\, e^{-\pi i\left(\frac{\nu}{2}+\frac{1}{4}\right)} W_{0,\nu}(-2iz)$$

$$J_\nu(z) = \left(\frac{z}{2}\right)^\nu \sum_{k=0}^{\infty} \frac{\left(-z^2/4\right)^k}{k!\,\Gamma(k+\nu+1)}$$

The first formula for the Hankel function yields an asymptotic series for J_ν or Y_ν, on the basis of (1.1.33). As is well known, such asymptotic series truncate automatically when ν is an integer plus 1/2; in which case the resulting finite set of terms comprises an exact expression for the relevant Bessel function.

Another class of special functions comes from analytic number theory. There is an interesting connection between these and other special functions, notably the incomplete gamma function. Define the sums:

(1.1.39)

$$F(s,a,y) = \sum\nolimits'_{n \varepsilon Z} \frac{e^{2\pi i y n}}{|n+a|^s} \quad ; \; G(s,a,y) = \sum\nolimits'_{n \geq -\lfloor a \rfloor} \frac{e^{2\pi i y n}}{|n+a|^s}$$

where a is presumed real and the prime (') notation on the summations means one ignores any singular summands. (It turns out to be convenient for asymptotic analyses with a and s complex to define the denominators as $((n+a)^2)^{s/2}$.) Now in terms of these sums we have:

Riemann Zeta function: (1.1.40)

$$\zeta(s) = \sum_{n=1}^{\infty} \frac{1}{n^s} = \frac{1}{2} F(s,0,0)$$

Polylog function: (1.1.41)

$$P(z,s,v) = G\left(s, v, \frac{\log z}{2\pi i}\right)$$

Lerch-Hurwitz Zeta function: (1.1.42)

$$\zeta(s,a) = G(s,a,0)$$

Beta function: (1.1.43)

$$\beta(s) = \sum_{n=0}^{\infty} \frac{(-1)^n}{(2n+1)^s} = 2^{-s} G\left(s, \frac{1}{2}, \frac{1}{2}\right) = Im\left(G\left(s, 0, \frac{1}{4}\right)\right)$$

The Beta function is related to a special case of Dirichlet L-functions for which the general summand is $a_n n^{-s}$, where a_n often is given number-theoretic importance. Analysis in the original style of Riemann can be applied (for example, start by writing F as an integral of a theta function) to obtain functional equations involving two new sums:

(1.1.44)

$$Q(s,a,y) = {\sum_{n\varepsilon Z}}' \frac{e^{2\pi i y n}}{|n+a|^s} \Gamma\left(\frac{s}{2}, \pi(n+a)^2\right)$$

$$R(s,a,y) = {\sum_{n\varepsilon Z}}' \frac{e^{2\pi i y n}(n+a)}{|n+a|^{s+1}} \Gamma\left(\frac{s+1}{2}, \pi(n+a)^2\right)$$

in the following form. For convenience denote a function value $D(s,a,y)$ by just the symbol D, whence we obtain two functional relations:

(1.1.45)

$$\Gamma\left(\frac{s}{2}\right)F = 2\pi^{s/2}\left(\frac{-\delta_a}{s} + \frac{\delta_y}{s-1}\right) + Q +$$

$$+ \pi^{s-1/2} e^{-2\pi i a y} Q(1-s, y, -a)$$

$$2\Gamma\left(\frac{s+1}{2}\right)(G-F/2) \;=\; R + i\pi^{s-1/2}e^{-2\pi i a y}R(1-s,y,-a)$$

where δ_x means 1 if x is an integer, 0 otherwise. These relations allow some interesting identities with which one may test special function software. Some such identities are, first, for the Riemann Zeta function

$$\Gamma\left(\frac{s}{2}\right)\zeta(s) \;=\; \frac{\pi^{s/2}}{s(s-1)} + \sum_{n=1}^{\infty}\left(\frac{\Gamma\left(\frac{s}{2},\pi n^2\right)}{n^s} + \pi^{s-1/2}\,\frac{\Gamma\left(\frac{1-s}{2},\pi n^2\right)}{n^{1-s}}\right) \tag{1.1.46}$$

then for the Euler constant:

$$\gamma \;=\; -2 + \log 4\pi + 2\sum_{n=1}^{\infty}\left(\frac{\Gamma\left(\frac{1}{2},\pi n^2\right)}{n\sqrt{\pi}} + \Gamma\left(0,\pi n^2\right)\right) \tag{1.1.47}$$

Special evaluations that yield interesting closed forms include:

$$\begin{aligned}
G(s,0,1/4) \;&=\; -\pi^2/48 + iC \qquad ;\ C \;=\; \textit{Catalan constant}\\
G(1,1/2,1/2) \;&=\; \pi/2\\
F(1,0,1/2) \;&=\; -\log 4\\
F(2,0,0) \;&=\; \pi^2/3\\
P(z,1,0) \;&=\; -\log(1-z)
\end{aligned} \tag{1.1.48}$$

although it is possible to come up with many more forms. Such formulae tend to allow fairly rapid convergence. Often one can obtain D good digits for a number-theoretic function in $O(D^a)$ operations, for some positive real number a.

Project 1.1.3 Special functions

Theory, possible computation, difficulty level 3-4.
Support code: Appendix "ContFract.ma," "ZetaGamma.ma," "Erfc.ma."

1) Use the error bounds (1.1.27), (1.1.28) and the reflection formula for *erfc*() in (1.1.36) to show that for any real z, $erfc(z)$ can be calculated to D digits in cD high-precision multiplies, where c is an absolute constant.

2) Here we consider means to evaluate the error function over the entire complex plane. First, show that in the error bound (1.1.34) one has the inequality $|C_N| < 1$ when $|arg(z)|$ is in $[0,\pi/4]$, and $|C_N| < 2\sqrt{(N+1)}$ when $|arg(z)|$ is in $[\pi/4,\pi/2]$. What can be said now about the number of operations required to compute $erfc(z)$ for arbitrary complex z? Write software that accordingly evaluates $erfc(z)$ to D good digits for *any* complex z. Note that the Fresnel integrals from wave optics can be expressed in terms of error functions of complex argument.

Note that step (6) below is a more general task, with more complicated error analysis required.

3) Work out software that computes Bessel functions by breakover switching. Bounds are not easily expressed, and are best derived first for the Hankel function $H_\nu^{(1)}$, then carried over to J_ν or Y_ν. It helps to assume for convenience in analysis that $Re(z) > 0$ and that the index N of a first-neglected term in either the ascending or asymptotic series satisfies $N(N + Re(\nu)) > |z|^2/2$, and also $N > 1 + Re(\nu)$. Though the asymptotic breakover criteria are cumbersome for this class of functions, the reward is a good one: efficient, worry-free means of evaluating any of the Bessel functions to rigorous precision, for scientific applications.

4) Derive, or guess by running routines from "ContFract.ma," the simple continued fraction expansion of certain ratios of modified Bessel functions; viz. $I_0(1/k)/I_1(1/k)$ for positive integers k. Infer then a continued fraction (where the fraction elements are not necessarily

integers) for more general ratios. Finally, using handbook relations between I and J functions, infer some sort of continued fraction for the general ratio of two Bessel functions: $J_\nu(z)/J_{\nu+1}(z)$. Does this fraction have any practical use for numerical evaluation? If the answer is yes, then the proper thing to do is to determine exactly when to switch to the fraction and ignore both ascending and asymptotic series.

5) Work out means to evaluate the standard gamma function $\Gamma(z)$ to D specified digits. One approach is first to use the handbook gamma reflection formula to ensure $Re(z) \geq 1/2$, then to engage in the following computation. Write the classical Stirling-Binet series for the gamma function as:

$$\Gamma(z) = \sqrt{\frac{2\pi}{z}}\, z^z e^{-z} e^{\sum_{j=0}^{N-1} T_j} e^{A_N} \tag{1.1.49}$$

where the Binet series terms are defined in terms of Bernoulli numbers as [Henrici 1977]:

$$T_j = \frac{B_{2j+2}\ z^{-1-2j}}{(2j+1)(2j+2)}$$

and the error exponent A_N is whatever complex number renders (1.1.49) exact. Then prove the following result. For $Re(z) \geq 1/2$, assume that for given D we have $\pi|z| > 2 + (D/2) \log 10$. Then for $N = [2 + (D/2) \log 10]$ the error term satisfies $|A_N| < 10^{-D}$. Use this theorem to evaluate gamma by successive use of $\Gamma(z+1) = z\Gamma(z)$ to "build up" the magnitude of z until the error theorem can be applied, then use the Stirling-Binet series, then normalize the final value according to the inverse "build-up" factor. Once again the question should be asked: "How many complex arithmetic operations are required to yield D good digits for the gamma function?"

6) Work out software that numerically evaluates the incomplete gamma function (a special case of which is the error function). It is perhaps a good idea to work out software first for real arguments. A good question is, should the continued fraction representation for the incomplete gamma be invoked somewhere "in between" the application of ascending and asymptotic series; and if so, what are breakover criteria? One approach is

to cast the continued fraction into "simple" form (all b numerators equal 1) and assume that the error after computing n convergents is of order $1/(q_n q_{n+1})$.

7) Work out software to evaluate the Riemann Zeta function using the incomplete gamma representation (1.1.46). Alternatively, use the Euler-Maclaurin notion starting from (1.1.2) and the explicit sum for the Zeta function, establishing a cutoff summation index in the style of step (3), Project 1.1.1. Which of these schemes is more efficient for evaluating the Riemann Zeta function? In particular, what are the operation counts for given D-digit precision? It should be possible for example to evaluate the representation (1.1.46) to D good digits in time $O(D^2)$.

These two approaches should be compared with the power series method of [Keiper 1992], who shows how to expand Zeta around $s = 1$.

8) Evaluate the Catalan constant on the basis of (1.1.48). The remarks and questions of step (7) previous apply again.

9) Evaluate the Euler constant on the basis of (1.1.47). The remarks and questions of the previous two steps are relevant. There may be an even faster way to get this constant using special functions; viz., the *Ei* relation of (1.1.37) which involves γ. Can this approach be any better than that of step (3), Project 1.1.1?

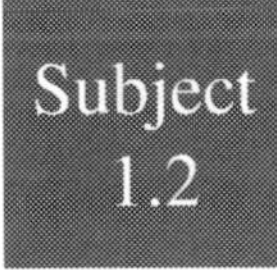

Equation solving

The idea of N simultaneous linear equations in N unknowns is certainly an old one, but is nevertheless a good testing ground for ideas and software optimization studies. We envision a set of N linear equations in the form:

$$Av = w \qquad (1.2.1)$$

where A is an N-by-N matrix, v is a vector of N unknowns, and w is a vector of N constants. Then the symbolic observation that $v = A^{-1}w$ suggests that matrix inversion is the key. Indeed, multiplication and inversion of matrices are fundamental to large (linear) system computation.

Project 1.2.1 Matrix algebra

Theory, possible computation, difficulty level 1-3.
Support code: Appendix "Strassen.ma," "strassentest.c," "strassen.[ch]," "NewtonMatrixInverse.ma," "inverse.c," "Eigenvalue.ma."

1) In order to investigate problems posed in this and the next Project, it is essential to have some kind of matrix algebra software. Consider writing (especially if a commercial package is not available!) all or parts of a matrix package that supports algebra for arbitrary N-dimensional vectors and N-by-N (square) matrices. A typical checklist of operations one might support runs:

- multiply matrix by vector
- multiply matrix by matrix
- add/subtract matrices
- multiply constant times matrix
- dot-product vector by vector
- return the norm (length in N-dimensions) of a vector
- transpose a matrix
- invert a matrix
- return the determinant of a matrix

A good symbolic processor can be used to test such routines on random matrices and vectors. One very good test is to set up rotation matrices (generally matrices whose inverse and transpose are equivalent), and test that, when such a matrix is applied to a vector, that vector's length is invariant.

2) The following task may be a "tired and homely" one, but some investigators learn much by doing it: write software to solve N equations in N unknowns. Clearly, if one has implemented one's own matrix

inverse as in the general step (1) previous, one has essentially finished this solver task already.

3) Analyze the complexity of matrix multiplication and conclude that direct multiplication should require $O(N^3)$ scalar multiplies. Investigate Strassen's algorithm, which uses the following clever identity:

$$\begin{bmatrix} a & b \\ c & d \end{bmatrix}\begin{bmatrix} e & f \\ g & h \end{bmatrix} = \begin{bmatrix} p+s-t+v & r+t \\ q+s & p+r-q+u \end{bmatrix}$$

where seven parameters are defined:

$$\begin{aligned} p &= (a+d)(e+h) \\ q &= (c+d)e \\ r &= a(f-h) \\ s &= d(g-e) \\ t &= (a+b)h \\ u &= (c-a)(e+f) \\ v &= (b-d)(a+h) \end{aligned}$$

It is of interest that this identity also holds *when a-h are themselves matrices*. Show that recursive application of the Strassen identity, to successively quadrant-divided matrices, results in a matrix multiplication algorithm requiring $O(N^{\log 7/\log 2})$ scalar multiplies. How many adds are required?

Implement a Strassen algorithm as exemplified in the Appendix code "Strassen.ma." A good exercise is to insert a global counter to count precisely how many internal multiplies (or adds) result. Study the effect of stopping the recursion, not at the lowest, scalar-multiplication level but at some level at which direct matrix multiplication is called.

Now consider the inverse problem, that is to find the matrix quadrants $\{e,f,g,h\}$ such that

$$\begin{bmatrix} a & b \\ c & d \end{bmatrix}\begin{bmatrix} e & f \\ g & h \end{bmatrix} = \mathbf{1}$$

What will work is the sequence of assigments:

$$
\begin{aligned}
P &= 1/a \\
Q &= cP \\
R &= bP \\
S &= cR \\
T &= S - d \\
U &= 1/T \\
f &= RU \\
g &= QU \\
V &= Rg \\
e &= P - V \\
h &= -U
\end{aligned}
$$

Implement a recursive Strassen inverse, in the style of the Appendix code, with the recursive multiplication just explained to be used to some depth. Show that the inverse algorithm is again of complexity $O(N^{\log 7/\log 2})$.

4) Work out matrix inversion software that uses the Newton method in the form of Schulz' iteration [Stoer and Bulirsch 1993] reminiscent of the scalar reciprocal problem (1.1.11):

$$X := 2X - XAX$$

The hope is that an adroit initial choice of matrix X will converge via this iteration to the matrix A^{-1}. It turns out that convergence is guaranteed for initial guess matrix X_0 if

$$lub(\mathbf{1} - AX_0) < 1$$

where generally the *lub* norm is defined as the maximum possible dilation ratio for vectors v:

$$lub(M) = max_{v \neq 0} \frac{|Mv|}{|v|}$$

Furthermore, the convergence is quadratic in this situation. Note the "one-dimensional" case of this scenario, step (2), Project 1.1.2. One approach to rapid convergence is to prenormalize A to satisfy the *lub* condition on $\mathbf{1}$–$A^2/2$, then adopt initial guess $X_0 = A/2$.

5) For continued fraction representations of the form (1.1.10) for numbers x, find a way to express the n-th convergent p_n/q_n in terms of the determinant of a matrix having only diagonal and once-off-diagonal terms (i.e., three diagonal strips of terms).

6) Investigate iterative schemes for eigenvalue-eigenvector solution. The goal, given a matrix A, is to solve:

$$Av = \lambda v$$

for both an eigenvector v and its associated eigenvalue λ. One scheme is to guess an initial λ_0 and a vector v_0, then to perform an iteration loop in this order:

- Solve the linear system $(A - \lambda_i \mathbf{1})\, u = v_i$ for the vector u.
- Set $v_{i+1} = u/|u|$ and set $\lambda_{i+1} = \lambda_i + |v_i \cdot v_i| \,/\, |\, u \cdot v_i|$.

Explain why this scheme sometimes converges (v_i -> v, λ_i -> λ), and beware of pitfalls in software implementation [Press et al. 1988].

7) Investigate the "power method" for isolating the leading (largest in absolute value) eigenvalue of a matrix A; we assume all eigenvalues are real. One guesses an initial vector v_0 and proceeds according to the following loop steps:

- Set vector $u = (1/m)\, v_i$ where m is the largest (in magnitude) component of v_i.
- Set $v_{i+1} = Au$.

In this scenario, under reasonable assumptions [Greenspan and Casulli 1988], the intermediate vector u should approach the eigenvector of the leading eigenvalue, the latter approached by the m values. Explain why this works (an illustrative exercise in linear algebra) and study numerically how effective this scheme is. An elementary starting version is Appendix code "Eigenvalue.ma." Does this scheme always converge, and is the convergence rapid, as one expects from one-dimensional Newton iterations?

8) Write (or use pre-existing) matrix manipulation software to effect a package that performs, on arbitrary input data, statistical least-squared-error regression of the linear (best-fit straight line) variety, or even higher-order (best-fit polynomials), and reports numerical curve parameters.

Now we turn to non-linear equation systems such as (one variable) polynomial equations or non-linear coupled systems. Consider the multi-dimensional Newton method, with which we endeavor to solve N equations in N unknowns:

$$f_i(x_1, x_2, \ldots x_N) = 0; \qquad i = 1, 2, \ldots, N \tag{1.2.2}$$

One generalization of the one-variable iteration (1.1.1) involves the Jacobian matrix:

$$J = \begin{bmatrix} \frac{\partial f_1}{\partial x_1} & \cdots & \frac{\partial f_1}{\partial x_N} \\ \vdots & & \vdots \\ \frac{\partial f_N}{\partial x_1} & \cdots & \frac{\partial f_N}{\partial x_N} \end{bmatrix} \tag{1.2.3}$$

within the iteration:

$$x := x - J^{-1}(x) f(x) \tag{1.2.4}$$

where (one has to be careful here) x denotes the *vector* $\{x_i\}$ and $f(x)$ denotes the *vector* of functions (1.2.2) evaluated at vector x; while J^{-1} is the matrix inverse of J evaluated at vector x.

Good references exist stating known theorems [Stoer and Bulirsch 1993] on convergence of the multidimensional Newton iteration. In particular, the method, when it is converging properly, can be expected to enjoy at least quadratic convergence, with the absolute error between the solution vector and the current iterate x decreasing roughly as the square of the previous such error.

Project 1.2.2 Non-linear equation systems

Theory, possible computation, difficulty level 2-3.
Support code: Appendix "PolynomialSolve.ma," "MultiNewton.ma."

1) Investigate what is essentially a complete polynomial solver for the situation that all roots of $p(x)$ are distinct and real. The method of Maehly [Stoer and Bulirsch 1993] can find, using a clever, dynamically modified Newton method, all the (presumed real) roots. A sample implementation is found as Appendix code "PolySolver.ma."

2) Implement software to solve N simultaneous and (possibly) non-linear equations in N unknowns, using the multidimensional Newton method with Jacobian matrix. In particular, obtain convergence to one of the solutions of:

$$a^3 - b^2 = 1$$
$$a^2 + c^4 = 4$$
$$bc = -2$$

for example:

$\{a,b,c\} =$ `{1.663307249659,-1.897812129375,1.053845093011}`

A sample implementation is found as Appendix code "MultiNewton.ma."

3) Find the exact analytic "dimension" D at which V_D, the volume of the unit D-ball is maximum. The formula is given in step (4), Project 1.3.2.

The typical educational progression, in regard to numerical differential equation solvers, runs something like this:

- Euler's method;
- modified Euler and midpoint methods;
- higher-order Runge-Kutta methods;

• variable-step methods.

As with other predominant themes in numerical analysis, we cannot do justice here to the breadth of the field, referring the reader instead to the references mentioned in the introductory remarks to this chapter. What we can say is that methods vary essentially in two ways: first, the order to which they approximate the true solution to a differential equation; and second, their computational complexities. For approximation order, assume that time t evolves on the unit interval, and that such time is discretized into time steps $h = 1/N$. One says that the *approximation order* of a method is the largest power n such that the numerical solution $x(t)$ for a differential equation differs from the true analytic solution by the n-th power of h:

$$|x_{true}(t) - x(t)| \sim O(h^n) \tag{1.2.5}$$

This is sometimes called global truncation error. Of course, variable-time-step methods require modification of this definition. But the main point is that methods abound for various approximation orders. As expected, the higher-order solvers tend to be more computationally complex. In a sense, variable-time-step is the obvious way to reduce complexity, by allowing less iteration in regions where the differentiable functions or trajectories change relatively smoothly.

Euler's method solves:

$$\frac{dx}{dt} = f(x, t) \tag{1.2.6}$$

subject to initial condition that $x(0)$ be given, by the most obvious initial choice and iteration:

$$x_0 = x(0) \tag{1.2.7}$$
$$x_{i+1} = x_i + h\, f(x_i, t)$$

This is an order-1 method in the sense that the solving error (1.2.5) can be expected to be of basic magnitude h. On the other hand, there is a higher-order method called the midpoint method, which goes:

$$
\begin{aligned}
x_0 &= x(0) \\
a &= f(x_i, t) \\
b &= f(x_i + ha/2, t+h/2) \\
x_{i+1} &= x_i + hb
\end{aligned}
\tag{1.2.8}
$$

This method turns out to have order 2. Then we have the "workhorse of the industry," the classical Runge-Kutta. The iteration runs:

$$
\begin{aligned}
x_0 &= x(0) \\
a &= f(x_i, t) \\
b &= f(x_i + ha/2, t+h/2) \\
c &= f(x_i + hb/2, t+h/2) \\
d &= f(x_i + hc, t+h) \\
x_{i+1} &= x_i + (h/6)(a + 2b + 2c + d)
\end{aligned}
\tag{1.2.9}
$$

This method is an order-4 method. The tradeoff of moderate complexity with healthy order-4 approximation power has rendered this system quite common, although for exacting professional work (imagine for instance the importance of precise spacecraft trajectories!) there exist significant further enhancements of this rudimentary Runge-Kutta.

Project 1.2.3 Differential equations

Theory, possible computation, difficulty level 2-3.

1) Work out an argument that explains why the Euler method is is a first-order, i.e. $O(h)$ method.

2) Write an Euler method solver for first-order-in-time systems and show how the order-1 approximation property can be detected numerically. One differential equation valuable for testing (with this Euler method or for that matter any method) is:

$$
\frac{dx}{dt} = -cx
$$

where c is a constant. Since exact solution is $x(0)e^{-ct}$, one can easily check the solver output with the truth. It might be a good idea to graph the *difference* between the exact and numerical solutions.

3) Work out the approximation order for the standard Runge-Kutta solvers (1.2.9).

4) Implement a classic Runge-Kutta solver (1.2.9), and compare numerical results with lower-order methods. One efficient and useful programming style is to call a Runge-Kutta update function, to which one passes initial time t_0 , initial coordinate $x(t_0)$, step size h, and some *finite* elapsed time interval t_1-t_0. The function is then expected to return a good approximation to $x(t_1)$. As long as the calling and loop overhead in this function remain negligible, the interval t_1-t_0 can be a short one. The advantage to this design is that future experimental enhancements such as variable step size h can be attempted without writing a new function.

5) Explore the modern world of sophisticated solvers. For example, the rudimentary "workhorse" Runge-Kutta of step (4) previous can be markedly enhanced. The Runge-Kutta-Felberg (RKF) schemes with variable step size and sharp error estimates are of special interest, but there are many approaches that attempt to "tune" the solver dynamically on the basis of error criteria and/or functional complexity. One may consult the more advanced references mentioned just prior to Subject 1.1.

6) Here is how handle some higher-order differential systems: just force the system (when possible) to be a list of first-order differential equations, then run through the iteration sequence using vector notation. For example, in (1.2.9) the classical Runge-Kutta scheme can be made multidimensional by making the initial choice, the intermediates a,b,c,d, and the iterated x_i all be vectors. Put another way, only the time t and time-step h remain scalar. Use this principle to write a solver for Newton's Law for a unit mass:

$$\frac{d^2x}{dt^2} = F(x,t)$$

where F is a force function. The main idea is to create a second variable, call it $p(t)$, which is the derivative dx/dt, giving two coupled differential equations:

$$\frac{dp}{dt} = F(x,t) \ ; \qquad \frac{dx}{dt} = p$$

and may think of a pair of right-hand functions $\{F, p\}$ to which a two-dimensional Runge-Kutta method can be applied.

One may recognize the coupled pair of equations here as Hamilton's equation of motion. It is surely a wonderful thing that the Hamiltonian formalism, which is old and monumental, lends itself so naturally to modern numerical solvers [Stump 1986]. It is even possible in some cases to solve or at least approximate the Hamiltonian solution symbolically [Crandall 1991].

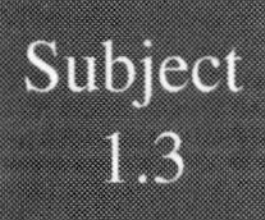

Random numbers and Monte Carlo

The basic random number from which almost all computer-based random generation springs is the uniform, or equidistributed random variable. Once this random variable is in hand, more general probability distributions can be obtained numerically. In what follows we shall assume a probability density $p(x)$ and a probability distribution function $P(x)$ such that:

$$p(x) \geq 0 \ ; \qquad \int_{-\infty}^{\infty} p(x)\, dx = 1 \tag{1.3.1}$$

$$P(x) = \int_{-\infty}^{x} p(u)\, du$$

Thus, P is monotonic non-decreasing on $(0,\infty)$. For example, an equidistributed random variable is taken to have density function as the characteristic constant function on (0,1), so that in this case P is a linear ramp.

The most direct theoretical way to create a random variable with a prescribed distribution is the inverse method. One creates numbers:

$$y = P^{-1}(x) \tag{1.3.2}$$

where x is taken to be equidistributed on (0,1). It is not hard to see that the y have probability distribution P.

A second method, which is actually quite practical, is the rejection method. Here, one simply finds a density q such that $p(x) \le cq(x)$ for some constant $c \ne 1$. Now take a random x equidistributed on (0,1), and take a z from the distribution pertaining to q. If $cx > p(z)/q(z)$ then reject z, and repeat the random choice of a $\{x,z\}$ pair. But if the inequality is "$\le$" instead, accept the number z as the random variable. It turns out that z has both the density p and the distribution P, as desired.

There are excellent texts on random number generation, from the classic [Knuth 1981] to modern random number and Monte Carlo methods [Niedereiter 1992].

Project 1.3.1 Generating random numbers

Theory, computation, difficulty level 1-3.

1) Study means for generating equidistributed random numbers on (0,1). For most C compilers a floating-point equidistributed random variable x in (0,1) can be obtained via:

```
#define DEN ((double)((1<<31)-1))
double x = random()/DEN;
```

although one should always check documentation for the range of any random() call. Above, it is assumed that random() returns an unsigned integer that stays within the inclusive range 0 to 2^{31}.

As far as generating algorithms go, the most popular is the linear congruential method [Knuth 1981][Niederreiter 1992], but there are other interesting methods. Just one such that is entertaining to investigate is the iteration:

$$x := 1/x + a \pmod p$$

where p is prime, a is constant, and the inverse is interpreted as 0 if $x = 0$. It turns out that this is fairly effective for certain pairs (a,p). Another interesting scheme is the "stripped down" linear congruential method, in which we simply do:

$$x := ax \pmod m$$

but the parameters must be very carefully chosen [Press and Teukolsky 1992]. Implement and test this generator with $a = 16807 = 7^5$, and $m = 2^{31}-1$.

2) Using the inverse method, show that a Poisson probability density $p(t) = e^{-t}$, for t on $(0,\infty)$, can be effected by creating numbers $\{-\log x\}$ where x is taken from an equidistribution on (0,1).

If proper audio equipment is available, use this technique to simulate a geiger counter. One assumes that radioactive material produces clicks in such a way that:

- The mean rate in clicks/sec. is a constant r (as measured over very long times;
- If a click has just occurred, the probability that the next click has *not* happened in elapsed time t is e^{-rt}.

3) Use the rejection method to create software that produces a random variable with Gaussian distribution of given mean and variance.

4) Investigate the "polar method" for generating Gaussian random variables. First, obtain two independent random numbers u, v equidistributed in (0,1). Then form:

$$x = \sqrt{-2\log u} \;\cos 2\pi v$$

$$y = \sqrt{-2\log u} \;\sin 2\pi v$$

Show that both x and y are Gaussian distributed with mean 0, variance 1, and, furthermore, that they are statistically independent. It can help to think of (x, y) as the coordinates on the complex plane, in which case $2\pi v$ is simply a random angle. Implement the scheme and run experiments that reveal Gaussian-like histograms. This and related schemes are discussed in [Black and Kennedy 1989]. Of particular interest is the "ziggurat" method, which divides regions of the Gaussian curve into "slow" and "fast," rendering the method one of the fastest known.

5) Though at first the following technique appears inefficient, it has actually been implemented with success. The idea is to obtain a normal random variable with Gaussian density p by computing the inverse of the error function in (1.3.2). Regardless of how well one has implemented error functions from Project 1.1.3, say, such a generator is expected to be slow, right? Not necessarily so. A method intended for supercomputers has been worked out, in which the error function's domain is split into 32 regions, over each of which a kind of splining takes place [Marsaglia 1991]. The method is claimed to be quite practical.

6) Work out theory and numerical tests of the following interesting problem in statistical extrema. Let x on $(-\infty,\infty)$ have Laplacian probability density:

$$p(x) = \frac{1}{2a} e^{-|x|/a}$$

and consider, after N samples of x are drawn, the random variable:

$$M = max\{|x_0|, \ldots, |x_{N-1}|\}$$

then the expectation of M should be asymptotic for large N to $(a \log N)$. This observation has application in signal processing, where one might like to estimate the maximum deviation between signal elements, given a stream of N elements. If one does have a good idea of the extremal value of x, the problem of quantization (how many binary bits to allocate for x) becomes easier.

Armed with random numbers, one may perform Monte Carlo integration along the following lines. For multidimensional integrals of the form:

$$\int \dots \int f(x_1, \dots, x_D)\, dx_1 \dots dx_D$$

one may drop a cloud of N random D-dimensional vectors $\{x_i\}$ into the domain of integration, and simply add up the values of f at those points. The integral will be approximated then by that sum divided by N. It is not hard to see that the average error in such a procedure should decrease statistically with N; in fact it should decrease as $O(1/\sqrt{N})$. The inviting and originally exciting thing about Monte Carlo is that the O behavior is independent of the dimension D (although the implied big-O constant may depend radically on D). This is to be contrasted with straightforward D-dimensional integration on a regular grid. The error in this procedure depends on both D and N, in what could be thought of as the worst possible form: $O(1/N^{1/D})$.

One of the great modern computational discoveries (which the author believes has not yet found its full brace of applications) is *discrepancy integration* [Niederreiter 1992][Wozniakowski 1991][Traub and Wozniakowski 1992]. This scheme comes under the heading "quasi-Monte Carlo integration," which refers to schemes in which random numbers are replaced with specially generated point clouds. In this technique, one avoids a random sequence of integration points and chooses instead a "hard-coded" sequence. Before we go any further, it is to be stressed that discrepancy integration is a multidimensional approach, and does not compete in any significant way with classical methods over one-dimensional domains (except that it can happen that one's discrepancy generator in software is faster than built-in random() functions!). Let us say, then, that we intend to perform three-dimensional integration. One may set up a discrepancy sequence as follows. For integer $n = 1,2,\dots,N$, determine the n-th random point in 3-space by its three coordinates:

$$x_n = 0.p_0p_1p_2\ldots$$
$$y_n = 0.q_0q_1q_2\ldots$$
$$z_n = 0.r_0r_1r_2\ldots$$

where the p's q's and r's are the reversed base expansions of the counting integer n in three different prime bases, call them P, Q, R. For example, this notation means that

$$n = p_0 + p_1P + p_2P^2 + \ldots$$

with a similar expansion implied for bases Q and R; all three base expansions are of the same counting integer n that indexes the chosen 3-space point. This beautiful theory allows more general bases, but choosing three small primes such as 2,3,5 works nicely. Above all, the most impressive result is that discrepancy integration results in errors of the relatively miniscule magnitude, typically $O(1/(N \log^{D-1} N)$. This is virtually $1/N$ decay for small to moderate dimension size D, so that in a real sense multiple dimensions are no longer so stultifying as they used to be.

Project 1.3.2 Numerical integration and Monte Carlo

Theory, possible computation, difficulty level 1-3.
Support code: Appendix "quasi.c," "sphere.c."

1) Implement, for purposes of understanding and comparison, one of the many possible one-dimensional classical integration schemes. Examples are:

- Simpson's rule integration
- Gauss quadrature
- Romberg integration

All of these fundamental methods have found their descriptive way into virtually all numerical analysis texts. But the enterprising reader could in principle start immediately to re-invent (for example) Romberg integration by applying the Euler-Maclaurin summation rule (1.1.2) with $M = 0$ (i.e., no correction terms, which means just Simpson's approximation) with N running through powers of 2; thus constructing the left-hand column of a Romberg tableau as in step (1), Project 1.1.1.

A satisfying one-dimensional integration exercise is use one or more of these classical integration schemes to find the arc length of the ellipse whose polar representation is:

$$r = \frac{1 - e^2}{1 - e\cos\theta}$$

where the eccentricity $e = \sin(\pi/12) = \sqrt{(1-\sqrt{3}/2)/2}$. The exercise is appealing in that, first, it serves as a good test of integration software; and second, it involves one of the relatively few historical resolutions of an integral of the elliptic class. For this particular eccentricity the exact arc was given by Ramanujan as [Whittaker and Watson 1972]:

$$s = \sqrt{\frac{\pi}{\sqrt{3}}}\left\{\left(1 + \frac{1}{\sqrt{3}}\right)\frac{\Gamma\left(\frac{1}{3}\right)}{\Gamma\left(\frac{5}{6}\right)} + 2\frac{\Gamma\left(\frac{5}{6}\right)}{\Gamma\left(\frac{1}{3}\right)}\right\}$$

which has numerical value $s \sim 6.17660198765869346 5...$, with which one can compare integration runs.

2) Provide convincing arguments as to the order of integration error, as a function of N sampling points, via the regular grid, and for Monte Carlo in D dimensions.

3) Work out direct Monte Carlo integration software that finds D-dimensional integrals by the simple expedient of dropping random points–say equidistributed over the relevant domain–and simply adding these up.

One class of integrals worthy of numerical test is:

$$\int_0^1 \cdots \int_0^1 \sin^2\left(\frac{\pi}{4}(x_1 + \ldots + x_D)\right) dx_1 \ldots dx_D = \frac{1}{2} - \frac{2^{3D/2}}{2\pi^D} \cos\frac{D\pi}{4}$$

Another good test is to estimate the volume of the unit D-dimensional ball (the integrand function is just then the characteristic constant function over the interior ball). For a more efficient approach to these particular integrals, see step (4) next.

4) Implement D-dimensional discrepancy integration, as exemplified in Appendix code "quasi.c," which is a multidimensional integration library, and "sphere.c," which calculates the volume of the unit D-ball. Time this against direct Monte Carlo in terms of the number of dropped points. On the basis of tests such as those suggested for step (2) previous, show explicit numerical data that reveals the stupendous improvement the discrepancy method allows over direct Monte Carlo. The exact formula for the volume of the unit D-ball is:

$$V_D = \frac{\pi^{D/2}}{\Gamma\left(\frac{D}{2} + 1\right)}$$

which is quite useful for test purposes. It is a tribute to quasi-Monte Carlo discrepancy integration and its fascinating history [Niederreiter 1992] that one may, as an exercise, approximate π by finding a numerical volume for the 3-ball. The present author finds that, using p-adic expansion primes $\{2,3,5\}$, dropping 10 million points into the unit cube resulted in $\pi \sim 3.141593\ldots$ The direct Monte Carlo approach will typically be much less accurate than this.

$$\frac{\frac{\partial^2 f}{\partial r^2} + \frac{1}{r}\left(1+\left(\frac{\partial f}{\partial r}\right)^2\right)\frac{\partial f}{\partial r}}{\left(1+\left(\frac{\partial f}{\partial r}\right)^2\right)^{3/2}} = K(1-f)$$

$$f(r) = -\frac{Kr^2}{4} - \frac{K^2(K+1)r^4}{64} - \frac{K^3\left(3K^2+10K+1\right)}{2304}r^6 - \ldots$$

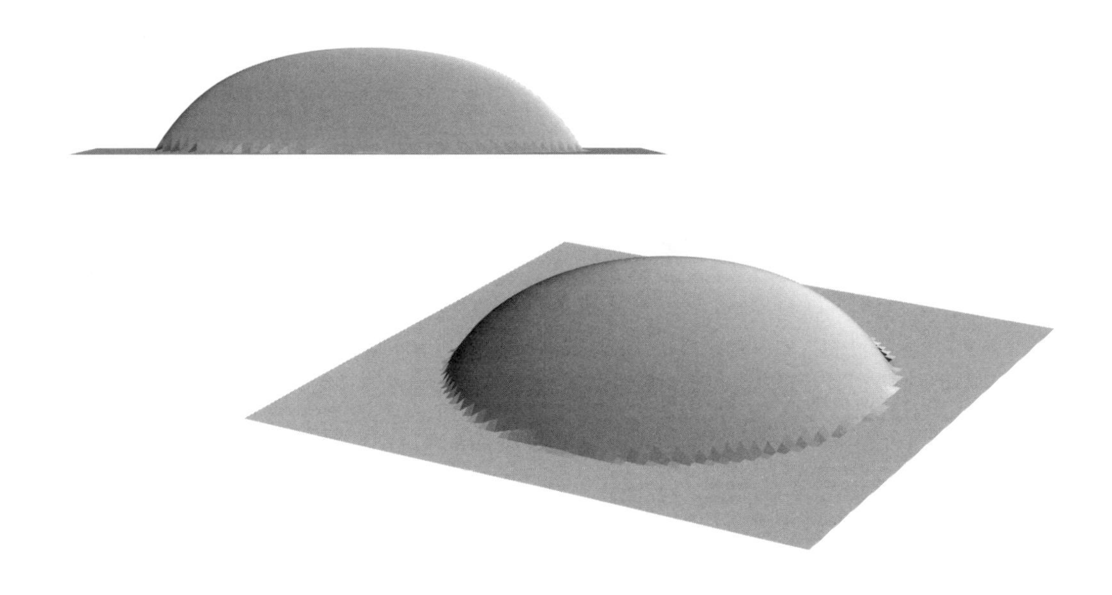

Symbolic power series solution of the classical differential equation for the surface of a liquid sessile drop. By equating coefficients for curvature (left-hand side of first equation) and for pressure-tension (right-hand side) one obtains a plottable series, as well as a practical relation between drop mass and wetting angle.

2 Exploratory computation

Collected intra- and interdisciplinary projects

> *It* [*the word* software] *started as a mildly witty mirror image of hardware. For non-specialists (and, unfortunately, for many computer people as well) the term has come to imply that software is somehow less important, perhaps less noble, than hardware. Get the hardware in good shape, it is thought, and the software can be taken care of somehow or another. This is a naive, mischievous assumption.*
>
> [*Karin and Smith* 1987]

Herein we encounter some intra- and interdisciplinary projects that were chosen to be representative of what one encounters in actual research. Perhaps an example of each of the two problem classes would serve to set the stage. One example of an interdisciplinary project would be the collective study of quantum Hamiltonians, Gaussian statistical matrix ensembles, and the Riemann Zeta function. These areas are connected in a fascinating way, a way still not entirely understood. One might call such a field "theory of chaotic eigenvalues," but that title is risky since the precise role of chaos in the overall study is yet unclear. In such investigation one may expect to peruse texts and papers pertinent across the disciplines of quantum theory, matrix algebra, and complex analysis. An example of an *intra*disciplinary project is the one we encounter first in this chapter: a collection of plane geometry problems. In that project, not much is required beyond knowledge of certain features and facts of plane geometry, so except for some elementary forays into supporting fields, the plane geometry project stays within one field.

Subject 2.1 Mathematical problems

The author finds elementary two-dimensional geometry problems especially instructive, particularly for students or investigators who retain some lingering fear of computation. One advantage of planar geometry exercises is that they obviously and readily transfer to a blackboard, paper, or any of those time-honored two-dimensional recording media with which we live. Beyond these visual advantages, geometry problems can feel (psychologically, to the problem solver) like intradisciplinary problems for which one stays within one field to work out the dilemma.

Project 2.1.1 Planar geometry problems

Computation, a little theory, graphics, difficulty level 1.

1) Buffon's needle problem is to drop a needle, infinitely thin and of length L, onto a plane surface that has been ruled with parallel lines separated by the same L. The Buffon needle is to fall randomly, with random orientation. Thus, the needle may or may not cross a ruled line.

Argue first that the probability the needle crosses a line is $2/\pi$. Thus, the Buffon needle is a humble apparatus for estimating π. Next, model this experiment in software, but *without recourse to the value of* π. At first this last constraint seems stultifying, because one naturally wants to choose a random angle of rotation for the final resting state of the needle, and so is tempted to use a floating-point number $2\pi x$ where x is equidistributed in (0,1). But that is forbidden, so it requires some extra thought to report an honest value of π from this experiment.

An interesting question is: is there a ruled "floor" that will allow yield π with more rapid convergence? Work out the probability of crossing a line in cases where the Buffon needle's length is not the same as the line spacing. What about the probability of crossing a line if the ruled lines run both vertical and horizontal, rendering a graph-paper effect? Of course, there is always the generalization that the Buffon needle device attains width or curvature.

2) The "grapestake" problem runs as follows. A field of stakes, one at each integer lattice pair, is viewed from the origin. A stake is called "visible" if and only if no closer stake blocks it. For example, in Figure 2.1.1, stake (2,4) is blocked by (1,2) and therefore is colored dark.

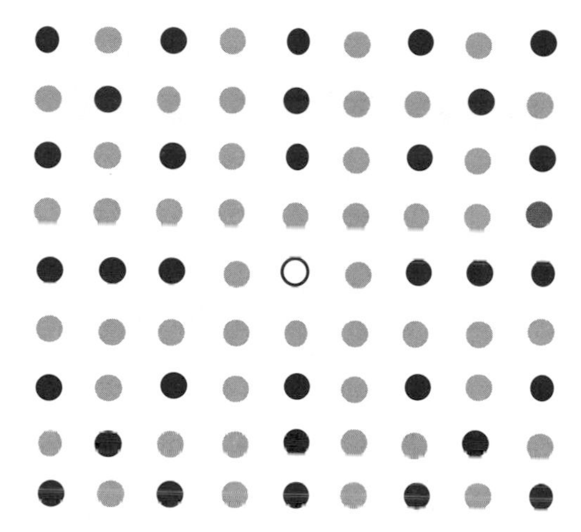

Figure 2.1.1: Setting for "grapestake" problem. Gazing from the origin, stakes on the integer lattice are "invisible" (dark) if blocked by some other stake, otherwise visible (light). The question is, on very large scales what is the probability that a random stake is visible?

Using software, estimate the probability that a random stake is visible. Note the symmetries in the problem, in fact one can bring the problem down to a calculation over one octant (half a quadrant) of the stake field. Also, knowledge of the theoretical answer can be exploited in order to group integer pairs for blocked stakes in computationally efficient ways.

The exact theoretical answer is that the probability a stake is visible, when considered, say, over ever-increasing centered square areas, is $6/\pi^2$. One arrives at this answer by interpreting the probability that two random integers are relatively prime as the infinite product:

$$\left(1-\frac{1}{2^2}\right)\left(1-\frac{1}{3^2}\right)\left(1-\frac{1}{5^2}\right)\left(1-\frac{1}{7^2}\right)\cdots$$

It is illuminating to gaze longingly, not at stakes but at this formula, and see just how its construction tells the probability story. This infinite product is $1/\zeta(2)$, which is how we get $6/\pi^2$.

3) Work out software that reports the area of a polygonal figure whose vertices are given as in Figure 2.1.2:

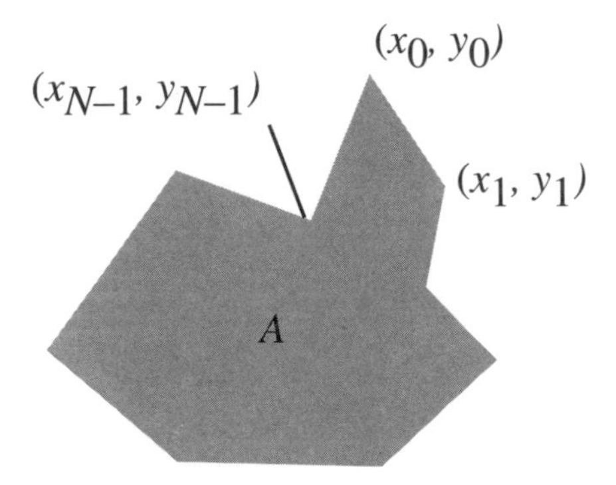

Figure 2.1.2: Setting for elementary planimetry problem. The polygonal curve has an area A which is directly and simply estimable from knowledge of the vertex coordinates.

The idea is to use the planimetry equation:

$$A = \frac{1}{2}\sum_{k=0}^{N-1}(x_k y_{k+1} - x_{k+1} y_k)$$

where periodic indexing is used, e.g., x_N is the same as x_0. Prove this relation and implement it in software. If a graphics mouse is available, one can immediately report areas of drawn figures. One approach to a proof is to consider the cross product of every consecutive pair of vectors that run from origin to respective vertices.

4) Work out software that finds polygonal area in a manner somewhat different from the approach of step (3) previous; namely, use the lattice-area formula that pertains to polygons *all of whose vertices lie on integer coordinates.* The elegant and useful formula is:

$$A = I + \frac{B}{2} - 1$$

where I is the number of strictly interior lattice points, B is the number of boundary points (vertices and any other lattice points precisely skewered by the sides of the polygon). Prove the formula by induction (it helps to consider simple general assemblies of squares or triangles to get the drift of the theory).

For the next Project, a good symbolic processor should be available if the reader wishes to implement the ideas. The steps of the Project were chosen to convey the flavor of symbolic problems that have been brought to the author's attention by interested investigators.

Project 2.1.2 Symbolic manipulation

Theory, possible computation, difficulty level 2-4.

1) Recall, as in Subject 1.1, that the Newton iteration:

$$x_{n+1} = x_n/2 + a/(2x_n)$$

will, for appropriate choice of initial value x_0, converge to the square root of a. Using some form of symbolic processing, rediscover the closed form solution of [Gray and Knill 1992], which states that for the case when the initial guess is just $x_0 = a$, the exact value of a general iterate is:

$$x_n = \frac{(1+\sqrt{a})^{2^n} + (1-\sqrt{a})^{2^n}}{(1+\sqrt{a})^{2^n} - (1-\sqrt{a})^{2^n}}\sqrt{a}$$

It may look as if recourse to the (unknown!) root of a is being used, but in fact the formula, when properly reduced, never has any remaining root symbols. Thus, in principle the general iterate x_n can be written out as a series in terms of combinatorial brackets.

One interesting study is to use this symbolic result to determine the precise convergence properties of the Newton iteration. For example, just how close is the square of the closed form expression to the goal, namely a?

Another interesting analysis is to think of the Newton iteration as generating a continued fraction (continue symbolically updating the denominator $2x_n$), in which case one should be able to arrive at the closed form using the theory of convergents as described in Project 1.1.1.

Yet another way to arrive at closed form iterates is to observe the elegant rule (communicated to the author by [Buhler and Gleason 1993]) that if

$$x = \sqrt{a}\,\frac{1+d}{1-d}$$

then

$$\left(x + \frac{a}{x}\right)\Big/2 = \sqrt{a}\,\frac{1+d^2}{1-d^2}$$

i.e., the Newton iteration merely increments the power of d appearing. Use this to find a closed form for the n-th iterate for *any* initial choice x_0. After that, what can be said about the overall stability of the Newton method?

2) Here is a symbolic problem from the engineering field of color reproduction [Graffagnino 1993]. Given three colored inks whose concentrations are $\{c,m,y\}$ (standing for cyan, magenta, yellow) which are to be mixed together on a physical medium, one might like to correct for vision, mosaic effects, so on. One ends up with a non-linear simultaneous equation system as follows. Denote:

$$\begin{aligned}
f_1 &= (1-c)(1-m)(1-y)\\
f_2 &= c(1-m)(1-y)\\
f_3 &= (1-c)m(1-y)\\
f_4 &= (1-c)(1-m)y\\
f_5 &= (1-c)my\\
f_6 &= c(1-m)y\\
f_7 &= cm(1-y)\\
f_8 &= cmy
\end{aligned}$$

Then the idea is to think of f as a vector having these eight components, and solve any three equations:

$$\begin{aligned}
A &= X \cdot f\\
B &= Y \cdot f\\
C &= Z \cdot f
\end{aligned}$$

where $\{X,Y,Z\}$ are each 8-component vectors of coefficients and $\{A,B,C\}$ are three constants. Implement a multidimensional Newton method as in Project 1.2.2 in order to solve this system on demand, given the 27 constants. The solution $\{c,m,y\}$ would be, for example, the analog amounts of the colored inks that would actually be placed on the medium. It is amusing and indeed satisfying to imagine the Newton iteration taking place within a color printer mechanism!

An open question, as far as the author knows, is: Can one find a symbolic exact solution to the three equations? It might be advantageous to exploit the obvious symmetries in the assignments of the f_i.

3) The kind of problem the previous color space example represents can be frustrating, especially if one does not know whether to "keep waiting" for one's symbolic processor to "cough up." Here is a system that does have an answer; not too simple and not too complex, but perhaps beyond the normal power of pencil and paper. Solve symbolically the following system of three equations:

$$\begin{aligned}
j &= (a-X)(b-Y)(c-Z)\\
k &= (d-X)(e-Y)(f-Z)\\
l &= (g-X)(h-Y)(i-Z)
\end{aligned}$$

where lower-case symbols $a,\ldots,l$ are constants and X,Y,Z are the unknowns. Consider preliminary theoretical simplifications such as elimination of some of the constants by scaling and translating in the variable space. This is in fact a good example of how human forethought can take some computational load off one's symbolic computing machinery.

4) Here is a chromatographic model from the field of equilibrium chemistry, proposed to the author by [Leddy 1990]. Solve the set of ten simultaneous non-linear equations:

$$\begin{aligned}
aV &= YW \\
bT &= UX \\
cR &= SX \\
dQ &= YX \\
e &= YZ \\
Y+S+U &= W+X+Z \\
f &= X+T+R+Q \\
g &= T+U \\
h &= S+R \\
i &= V+W
\end{aligned}$$

where lower-case variables $a,\ldots,i$ are constants and the ten upper-case variables $Q,\ldots,Z$ are the unknowns. In particular, it was desired to know the concentration X. The author was stunned to find that this system can be solved exactly, in the sense that one obtains *a tenth-degree polynomial that X must satisfy*. One open question is, which root of this horrendous polynomial is the correct one?

5) Here are some symbolic integration problems that were originally solved by the author and collaborators along the same phenomenological path as indicated here, ending eventually in rigorous proofs [Mayer and Buhler 1993] . Define what we call a "cascade integral" by:

$$I_k = \int_{-\infty}^{\infty} \frac{du}{1+u^2} \frac{\partial}{\partial u} \frac{1}{1+u^2} \frac{\partial}{\partial u} \frac{1}{1+u^2} \cdots \frac{\partial}{\partial u} \frac{1}{1+u^2}$$

where k occurrences of the derivative operator are understood, and each derivative operates in the usual way on everything to its right. The simplest case is $I_0 = \pi$. Using symbolic integration, work out a list of exact ratios I_{k+1}/I_k and conjecture a general ratio. Finally, prove the conjecture (unless by some miracle you have actually gotten your symbolic processor to so prove!). This cascade integral turns out to figure centrally in a solution to the Schroedinger equation for a linear potential. In fact, for potential $V(x) = x^n$, with n a positive integer, one can define the cascade integral as involving k occurrences of $(\partial/\partial u)^n$ and in this way address the problem of such a power potential [Crandall 1993]. Thus, a spectacular open problem–kind of a "holy grail" of quantum theory–namely, the resolution of the quartic potential $V = x^4$ (e.g. finding exact eigenvalues) can be brought down to the problem of evaluating:

$$J_k = \int_{-\infty}^{\infty} \frac{du}{1+u^2} \left(\frac{\partial}{\partial u}\right)^4 \frac{1}{1+u^2} \left(\frac{\partial}{\partial u}\right)^4 \frac{1}{1+u^2} \cdots \left(\frac{\partial}{\partial u}\right)^4 \frac{1}{1+u^2}$$

where k fourth-derivative operators are understood. This integral has unknown closed form. The finding of such would open a kind of treasure chest.

Another integral that was conjectured/proven along the same basic symbolic processing path is:

$$L_k(a) = \int_{-\infty}^{\infty} \frac{du}{\prod_{j=0}^{k} \left(a^2 + (u+j)^2\right)}$$

This too can be given a closed form in terms of k and a, via symbolic ratio listing and a little ingenuity.

6) Here we arrive at a closed form expression for a certain matrix inverse. Define T_m as the m-by-m matrix:

$$T_m = \begin{bmatrix} 1 & t & t^2 & t^3 & \cdots & t^{m-1} \\ t & 1 & t & t^2 & \cdots & \\ t^2 & t & 1 & t & \cdots & \\ t^3 & t^2 & t & \ddots & & \\ \vdots & \vdots & \vdots & & & \\ t^{m-1} & & & & & 1 \end{bmatrix}$$

i.e., the pq-th element is $t^{|p-q|}$. The problem, which comes from quantum path integral theory, is to find the exact (symbolic) inverse of the matrix $(\mathbf{1}-zT_m)$. Now on the assumption that $D_m = det(\mathbf{1}-zT_m)$ satisfies a recurrence relation (with changing index m), use symbolic processing to deduce that relation, finally obtaining the closed form:

$$D_m = t^m(U_m(w) - (1+z)tU_{m-1}(w))$$

where U is the Chebyshev polynomial of the first kind, and $w = (1-z+(1+z)t^2)/2t$. A harder step now is to analyze matrix minors to deduce a general expression for the inverse's diagonal elements $(\mathbf{1}-zT_m)^{-1}{}_{aa}$ and the off-diagonal elements $(\mathbf{1}-zT_m)^{-1}{}_{ab}$ with $a > b$. The final result is an exact expression for the inverse $(\mathbf{1}-zT_m)^{-1}$ from which quantum theoretic conclusions can be drawn by assigning meaningful values to the parameters t and z.

7) Find a set of *distinct* positive odd integers R such that

$$\sum_{r \varepsilon R} \frac{1}{r} = 1$$

You will need R to contain at least nine members [Guy 1980]. Numerical cases of this problem may possibly be assailable via symbolic manipulation together with a genetic algorithm as in Project 2.2.2.

8) Here is a fascinating problem of long history and great importance to modern algebra [Koblitz 1984]: find congruent numbers. These are integers n that happen to be possible areas of right triangles whose sides are rational. For example $n = 5$ is congruent because the right triangle with sides $a = 3/2$, $b = 20/3$, $c = 41/6$ has area 5.

The author has found that one fascinating approach is to combine notions of non-linear equation systems, genetic algorithms, and Diophantine relations. Choose some reasonably composite integer D and ask for solutions to the *analog* system (x,y,z are now continuous variables):

$$x^2 + y^2 - z^2 = 0$$

$$xy/2 - n = 0$$

$$\sin^2 \pi D x + \sin^2 \pi D y + \sin^2 \pi D z = 0$$

The trigonometric sum vanishes when each of $\{x,y,z\}$ is a rational whose denominator divides D. A solution to this coupled system can in principle be found approximately with real arithmetic. The author tried $n = 5$, $D = 30$, and applied a multi-dimensional Newton iteration (as in Appendix code "MultiNewton.ma,") on genetic trials (as discussed in Subject 2.2) until one of the offspring converged to the set $\{3/2, 20/3, 41/6\}$ to within 20 decimal places. Obviously, at the end of such a scheme one needs to check symbolically the Diophantine relations for exactitude. But the point is, apparently one can find congruent numbers with this multifaceted approach.

For a hard problem, show that $n = 157$ is congruent by exhibiting a right triangle having rational sides and area 157. The fanciful "D denominator" approach above is not recommended for this case, unless an extremely fast, high-precision package is available! The reference [Koblitz 1984] explains some not necessarily terminating algorithms for finding congruent numbers, and also explains Tunnell's 1983 theorem that almost completely solves this ancient problem.

The possibility of Diophantine numerics using such continuum methods is intriguing. One general approach would be to give floating-point solutions "scores" depending on "how rational" they are. For example, during one of the author's runs for $n = 5$ the respective simple continued fractions for $\{x,y,z\}$ were:

$$x = 1 + \cfrac{1}{1 + \cfrac{1}{1 + \cfrac{1}{25842359227 + \ldots}}}$$

$$y = 6 + \cfrac{1}{1 + \cfrac{1}{1 + \cfrac{1}{1 + \cfrac{1}{2584194402 + \ldots}}}}$$

$$z = 6 + \cfrac{1}{1 + \cfrac{1}{4 + \cfrac{1}{1 + \cfrac{1}{697489808 + \ldots}}}}$$

which is how the exact solution {3/2, 20/3, 41/6} was guessed.

Next we turn to a difficult field: that of modeling analytic problems on computers. It is here that, in modern times, rigorous results have sometimes been obtained via numerical runs.

Project 2.1.3 Real and complex analysis

Theory, possible computation, difficulty level 3-4.

1) Investigate computational aspects of the Riemann Zeta function. The fundamental open problem in the field is to resolve the Riemann hypothesis, which states that all zeros of $\zeta(s)$ in the strip $0 < Re(s) < 1$ have real part 1/2. It is now known that the first 1,500,000,001 zeros with positive imaginary part lie exactly on that half line, that they are all simple zeros, and furthermore that they lie on the line segment $0 < Im(s) \leq 545439823.215$ [van de Lune et al. 1986]. An interesting account of Riemann's original computational efforts, which were followed by those

of Siegel, is found in [Hajhal 1987]. A readable account of properties of the Zeta function is [Edwards 1974].

Though we have encountered at least two ways of calculating values of $\zeta(s)$ in the previous chapter, there are special considerations for s in the critical strip. [Wagon 1991] describes tools for such computations. In particular, an interesting exercise is to ascertain the "Lehmer phenomenon," whereby two very close zeros occur in the region of $s = 1/2 + 7005\, i$.

A second fascinating study concerns the distribution of zeros and the connection with Hamiltonian physics and chaos theory. One pretends that the imaginary coordinates of the Riemann zeros are eigenvalues of a Hamiltonian. Then there are asymptotic arguments that involve Gaussian unitary matrix ensembles and result in good predictions as to the statistics of the Riemann zeros [Berry 1988].

It is also computationally interesting to find zeros of tremendous imaginary part. [Odlyzko 1993b] has for example found about one hundred million critical zeros near the 10^{20}-th critical zero, which itself lives at 1/2 + 15202440115920747268.6290299... Such calculations are relevant for statistical tests of aforementioned ensemble theories.

2) Some interesting problems in real analysis, all quite worthy of further investigation, have been presented by [Varga 1990]. One is to *disprove* the Bernstein conjecture, which dates from 1913, and states the following. For the function $f(x) = |x|$ on x in $[-1,1]$, denote by E_n the infimum of the maximum deviation $|f(x)-p(x)|$ over $[-1,1]$ and over all polynomials p of degree n. For given n there is a "best approximating" polynomial and E_n is its worst-case deviation from f on the interval. Now Bernstein proved, with some labor, that:

$$\lim_{n\to\infty} 2nE_{2n} = \beta$$

where this "Bernstein constant" β is between 0.278 and 0.286. Bernstein then went on to notice that perhaps $\beta = 1/(2\sqrt{\pi}) = 0.28209...$ As fate would have it, the implicit conjecture is false. What is impressive is that the modern negative proof involved actual numerical computation. Varga and collaborators have been able to obtain high-precision values for β, running into the dozens of digits, by the expedient of Richardson extrapolation (a generalization of the Aitken method of Subject 1.1). Thus, good computational projects include independent calculation of β, and attacks on the many open problems associated with Bernstein approximation.

3) Provide a rapid and elegant proof that $e^{\sqrt{2}}$ is irrational, in the following way. First, find the exact simple continued fraction (all b numerator terms are 1 in (1.1.10)) for the real number:

$$\omega = \sqrt{2}\,\frac{e^{\sqrt{2}}+1}{e^{\sqrt{2}}-1}$$

One way to find the precise continued fraction is to obtain numerically some number of fraction elements, guess the pattern, and then prove the result. Observe that the a sequence for this fraction is not eventually periodic, so by the Lagrange theorem [Corless 1992], ω cannot be a "quadratic surd" of the form $q+r\sqrt{s}$, where q,r,s are all rational. Conclude that $e^{\sqrt{2}}$ is irrational.

4) To the author's knowledge, not a single simple continued fraction is known (i.e., has a known pattern of a_i) for an algebraic number of degree greater than two. [Guy 1980] asks whether *every* such $\geq$ 3-rd degree number has unbounded $\{a_i\}$. In particular, study the continued fraction for one of Ulam's numbers, i.e., roots z for the coupled quartic system:

$$z = \frac{1}{z+y}$$

$$y = \frac{1}{1+y}$$

For example, does it appear that the geometric mean of the list $\{a_1, a_2, \ldots\}$ is approaching the Khinchin constant K of step (9), Project 1.1.1?

5) Here we investigate the powerful technique of Poisson summation. When one wishes to sum a series of values $f(n)$ over D-dimensional integer vectors n it is often advantageous first to transform the sum via the

Poisson identity:

$$\sum_{n \varepsilon Z^D} f(n) = \sum_{m \varepsilon Z^D} \int e^{2\pi i m \cdot n} f(n)\, d^D n$$

Here each integral is D-dimensional and during integration the variable n is treated as a D-dimensional vector of reals. This formula is valid for a wide class of functions f. One might ask, what is the benefit of replacing the left-hand sum over D-tuples of integers with yet another sum likewise over D-tuples? The answer is that often the right-hand side has superior convergence properties. As an example, show first that

$$\sum_{n=-\infty}^{\infty} \frac{1}{n^2 + a^2} = \frac{\pi}{a} \coth \pi a$$

and note that the coth() function can be evaluated, using for example methods of Chapter 1, much more rapidly than the sum. Another interesting exercise is to show from this identity, by symbolic expansion of coth(), that $\zeta(2) = \pi^2/6$. Similarly, one may find a closed form for the sum of $1/(n^4 + a^4)$, deduce from this $\zeta(4) = \pi^4/90$, and so on.

Another good exercise is to find the relevant Poisson identity for the theta function:

$$\theta(t) = \sum_{n \varepsilon Z} e^{-tn^2}$$

and then use asymptotic breakover methods from Chapter 1 to compute $\theta(t)$ in bounded time, for any complex t having positive real part.

Step (6) next is an example of the applicability of multi-dimensional Poisson summation.

6) Investigate the celebrated Gauss circle problem, which asks for a bound on the error term:

$$E(r) = \#(r) - \pi r^2$$

where $\#(r)$ is the number of lattice points lying properly inside a circle of radius r. This is one of the hard problems in number theory, and is connected with the Riemann Zeta function and related complex functions. Gauss knew that $E(r) = O(r)$, so that the error is certainly no worse asymptotically than the circumferences of the circles. But it can be shown that $E(r) = O(r^{2/3+\varepsilon})$ using analytic formulae for E. In fact sharper results are known [Guy 1980][Hardy and Wright 1978]. The pioneering work of Van der Corput in the 1930s on certain sums of oscillating terms has given us $E(r) = O(r^{66/100+\varepsilon})$, so we know that the 2/3 power is not best possible. In fact, modern bounds such as $O(r^{24/37+\varepsilon})$ can be found in the literature, with the rational power seemingly creeping slowly downward over the years. Conversely Hardy and Landau showed that $E(r)$ is *not* $o(r^{1/2}\log^{1/4} r)$, so the relevant power is somewhere between 1/2 and about 24/37 (or whatever is this year the current upper bound). In fact the error $E(r)$ is known both to exceed $r^{1/2}$ and to fall below $-r^{1/2}$ infinitely often.

What can computation do to lend insight into this problem? We speak not simply of counting the lattice points inside circles, but to verify and analyze the behavior of oscillating sums with which some of the aforementioned proofs have been effected. For example, it can be shown via two-dimensional Poisson summation that for any positive λ, and r^2 not the sum of two integer squares (i.e. r is not the distance to any lattice point), the circle error is given exactly by:

$$E(r) = e^{-\frac{\pi^2 r}{\lambda}} + r \sum_{m \,\varepsilon\, Z^{2'}} \left\{ \frac{J_1(2\pi r|m|)}{|m|} e^{-\lambda m^2} + I(r, |m|, \lambda) \right\}$$

where

$$I(r, |m|, \lambda) = \frac{1}{\pi} \int_0^{\pi} dt \frac{r - m\cos t}{r^2 + m^2 - 2rm\cos t} e^{-\frac{\pi^2}{\lambda}\left(r^2 + m^2 - 2rm\cos t\right)}$$

The summation is over integer lattice vectors $m = (a,b)$ such that the magnitude $|m| \neq 0$. For a theoretical exercise, it is worthwhile to show that, for large r, we can break down the error as:

$$E = O(1) + O\left(\frac{r^{1/2}}{\lambda^{1/4}}\right) + O\left(r\lambda^{1/2}\right)$$

corresponding to the leading exponential, Bessel, and I-integral terms respectively. By optimizing λ we find $E = O(r^{2/3})$, one of the classical non-trivial results. Actually, by applying modern exponential sum methods to the Bessel term, one can rediscover the advent of powers < 2/3. It is also interesting to analyze just how the exact formula for E becomes the lattice point count minus an area, in the two limits $\lambda \to 0$ and $\lambda \to \infty$.

But there remains an interesting computational exercise, which is to compute the *integer* number of lattice points $\#(r)$ as $\pi r^2 + E$, with E computed using the Bessel formula and some choice of the λ parameter. What a nice idea–to check a formula from analytic number theory by trying to attain an exact integer answer. The summation can be made more efficient by noting that the number of representations, $r_2(n)$, of an integer n as the sum of two squares can be brought into play:

$$\sum_{m \in Z^2,} X(|m|) = \sum_{n=1}^{\infty} r_2(n) X(\sqrt{n}) = \sum_{n=1}^{\infty} 4X(\sqrt{n}) \sum_{d\, odd,\, d \mid n} (-1)^{(d-1)/2}$$

Here is an example check of the exact Bessel formula. The author found, using *Mathematica* calls for the Bessel functions and numerical integration for I, that for radius $r = 7.1$, parameter $l = 1/2$, and taking n in the r_2 formula above from 1 through 100, that:

$$E + \pi(7.1)^2 \sim 161.0000167343428...$$

Indeed there are precisely 161 lattice points inside the circle of radius 7.1. Both the exact E expansion and the software are being sharply tested by this quest for integer lattice point counts.

An interesting problem is to count in this fashion the points inside very large radii. It may be possible to use asymptotic breakover criteria, as discussed in Chapter 1, to speed up such computations. An interesting question is, how great an area in the "reciprocal lattice" (the lattice over which the m vector index runs) is necessary to resolve, for given r, this integer lattice point count? Note, for example, that a good portion of the I integral contribution arises from a region of the m-lattice in which $|m| \sim r$.

Yet another computational exercise is instructive, albeit highly inefficient. Note that $n = 1 \pmod 4$ is prime if and only if $r_2(n) = 8$. Thus, such an n is signified by a discontinuity of 8 in E as one moves between closely spaced radii $\sqrt{n} \pm \varepsilon$. So let us prove with the analytic formula for E that $n = 13$ is prime. We compute, for parameter $\lambda = 1$:

$$
\begin{aligned}
E(\sqrt{13} + 0.001) - E(\sqrt{13} - 0.001) &= \\
&45.000000046... - 37.000000040... \\
&\sim 8
\end{aligned}
$$

verifying that 13 is indeed prime.

Regardless of method, the problem of counting lattice points inside given boundaries is challenging. As early as the work of [Keller and Swanson 1963] it was known how many lattice points lie within $r \leq 259750$. It would be interesting to see just how far this radius can be taken in the 1990s. A related problem, which is to count all positive integers n less than x that have $r_2(n) \geq 1$, has been addressed by [Shiu 1986].

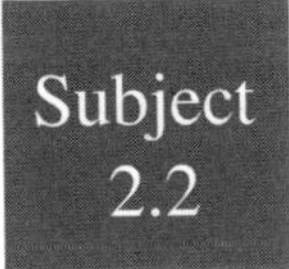

Subject 2.2 Nature-motivated models

On the basis of the title to this Subject, certainly two computational fields must come to mind: neural networks and artifical life studies. The manner in which living neurons respond to pulse trains, evidently according to some sort of learned weighting scheme, is the original motivation for neural networks. On the other hand, this author would claim that artificial life studies comprise a means to determine just how much of evolution, or the complexity of living systems, can be said to arise out of pure combinatorics.

Neural networks go back roughly to mid-century; naturally this was also the approximate era of neurological breakthroughs such as discovery of ion channel action potentials and so on. One way to build a modern idealized neural network suitable for computation is to start with a "perceptron:"

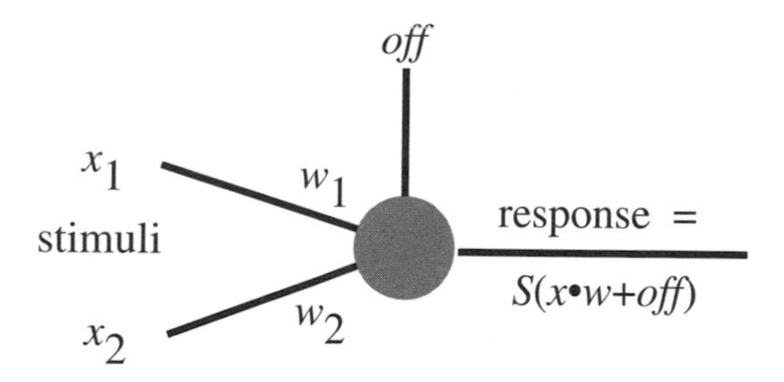

In the model, the "stimuli" x_1 and x_2 are real numbers (usually in $(-1,1)$), the "weights" w_i and "offset" *off* are real numbers; and the perceptron (gray dot) puts out the S function applied to the sum of *off* and the dot-product $x_1w_1 + x_2w_2$. This S function is some kind of monotonic non-linear saturating function. It is usually chosen so that for large arguments it is +1 and for small arguments it is –1. A common choice is the sigmoid function [Hush and Horne 1993]:

$$S(z) = \frac{1}{1 + e^{-\beta z}}$$

which indeed runs from 0 (very negative z) to the value 1/2 at $z = 0$, to the value 1 at very large positive z. The "gain" β is often taken to be 1, but can be considered to be yet another tuning parameter.

To understand the point underlying computational neural networks, consider the problem of deciding whether the pair (x,y) lie in a certain region of the real plane. The question could concern a curvy footprint-shape, say, and we wish the network to respond with "yes" (i.e. +1 output) if and only if the input (x,y) lies in the footprint; otherwise we want the response "no" (–1). Such a neural network can be built, but it needs to be a multi-layer perceptron (MLP); that is, a network-connected set of individual perceptrons. Figure 2.2.1 shows a two-layer MLP, where we have changed notation to be more in line with the Appendix code example "Neural.ma."

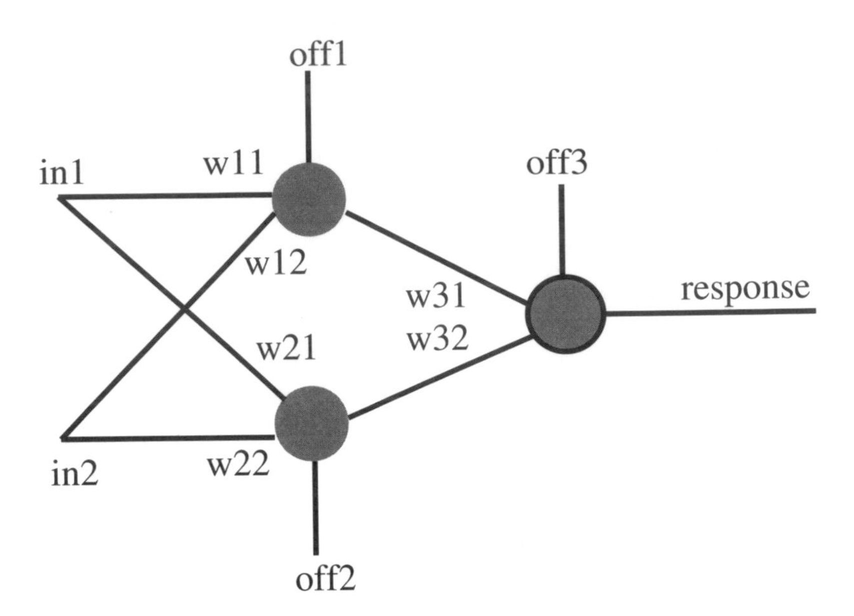

Figure 2.2.1: Neural network built as a two-layer perceptron. Each perceptron response unit (gray dot) outputs a response which is a sigmoid function of a fixed offset plus the vector dot-product of the weights with the inputs.

Project 2.2.1 Neural network experiments

Computation, difficulty level 2-4.
Support code: Appendix "Neural.ma."

1) Find weights and offsets required to force a *single* perceptron to behave like one of these logic gates: and, or. For example, in the "and" case, one wants the response output to be nearly +1 only when both inputs are reasonably positive; otherwise the response is to be –1.

Then generate a convincing argument that an exclusive-or gate cannot be done with just one perceptron. A two-layer ex-or solution is discussed in the next step.

2) Implement, along the lines of example Appendix code "Neural.ma," a

two-layer MLP network. Sample weights for the MLP to mimic an "exclusive-or" gate are given by [Hush and Horne 1993] as:

```
w1 = {{-2.69, -2.80},-2.21};
w2 = {{-3.39,-4.56},4.76};
w3 = {{-4.91,4.95},-2.28};
```

where, in reference to Figure 2.2.1, the weight-offset parameters for each perceptron are stored in one nested list. Clearly there is the programming option to have all perceptrons in a master list. The referenced authors state that some 1000 training passes using back-propagation (see step (3) next) were required to settle on these parameters. Figure 2.2.2 shows the response resulting from their two-layer perceptron.

Figure 2.2.2: Neural network response, designed to mimic an exclusive-or logic gate, with weights given in step (2). When both x and y are small (bottom corner of plot) the response is negative (dark), as it is when both x and y are large; in keeping with the exclusive-or operation.

3) In step (2) previous we analyzed a simple, two-layer hard-coded MLP network but we did not actually solve *a priori* for the system weights. Study back-propagation, which is one technique that often achieves results. Methods for resolving networks on the basis of training inputs are discussed in the excellent review articles [Lippman 1987] [Hush and Horne 1993].

The author enjoys thinking of back-propagation as a kind of statistical version of multidimensional Newton iteration. In fact, there are various steps in back-propagation algorithms where the concepts of simultaneous non-linear system theory, Jacobian matrices, and so on are involved. For sigmoid transfer functions (as opposed to digital transfer) this is as expected, for, like the Newton methods, neural network behavior depends–in any region of parameter or solution space–at least locally on the ensemble of system derivatives.

Figure 2.2.3 shows (this time a contour plot of the response surface is used) the result of the weight system:

```
w1 = {{-0.603417, -0.708207}, -1.27779};
w2 = {{1.25621, 2.89778}, -2.3439};
w3 = {{-2.65504, -2.88652}, 2.74878};
```

which provides an approximate answer to the question: Is $x + 4y < 2$? The contour plot shows the final response of the network with these weights. In the large, lower-left area under an imagined curve $x + 4y = 2$, the plateau is positive and almost flat: The network is responding "yes" in that region.

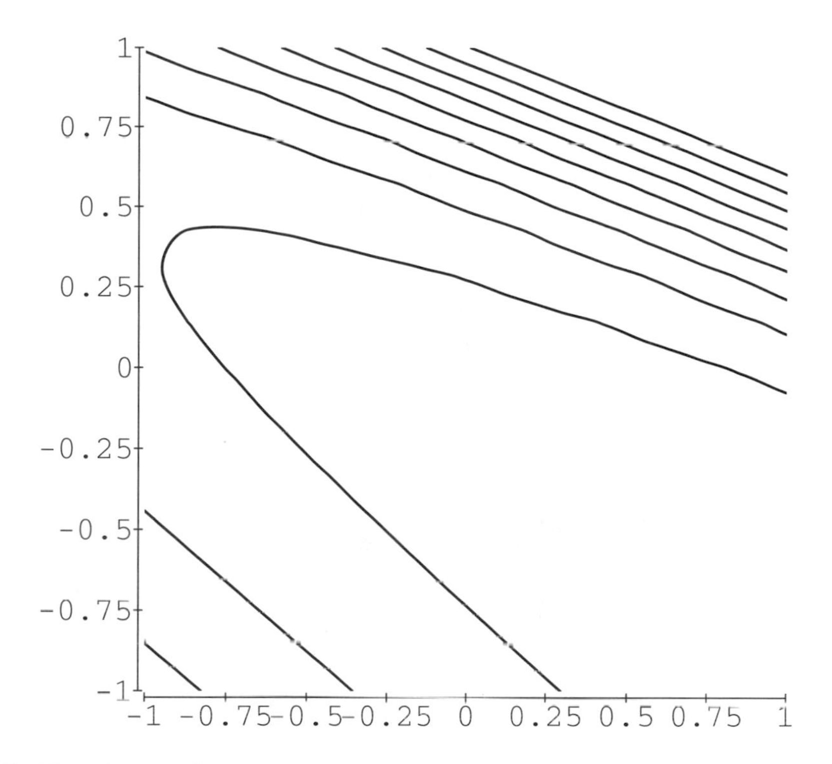

Figure 2.2.3: Neural network response, designed to answer the question: Is $x + 4y < 2$? The contour plot shows equi-heights of the final response for the network weights given in the text. Whenever $x+4y$ is clearly greater than 2, the response has correctly fallen off precipitously to the negative (upper right tight contours).

4) Study the connection between adaptive neural networks and pattern recognition. A good reference is [Pao 1989].

5) Study the field of "fuzzy logic." Adaptive fuzzy systems are a potential competitor with neural networks. There are certainly many similarities, but adaptive fuzzy systems re-arrange themselves internally. A good reference is [Kosko 1992].

6) As an alternative to back-propagation of step (3), consider genetic algorithm methods for solving for optimal weights and offsets of a multiplayer network. The idea would be to start with many random $3N$-dimensional vectors as parents, and let the network "evolve," with these vectors "mutating," according to fitness criteria connected with total error

in the desired output function. Genetic algorithms are discussed in Project 2.2.2 next.

7) Investigate some of the many interdisciplinary applications of neural networks: signal processing (adaptive filters), adaptive resonance systems, simulated annealing, and so on. An excellent source of such projects is [Freeman and Skapura 1992].

Next we turn to genetic algorithms and artificial life. A genetic algorithm is one in which we instill at least these two notions from evolutionary theory:

- Random mutations;
- Selection according to fitness.

There are many applications of such algorithms [Forrest 1993]. We study in the following Project the problem of function optimization. Figure 2.2.4 shows the basic scenario.

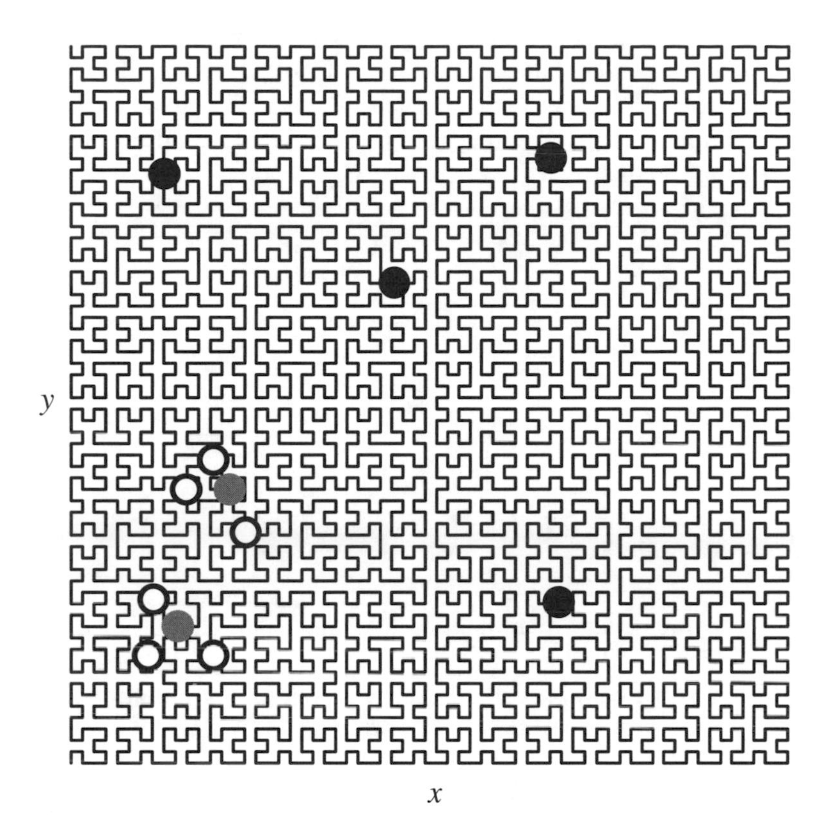

Figure 2.2.4: One possible genetic algorithm scenario. The goal is to optimize a certain function $f(x,y)$. The variable pair x,y is mapped onto a one-dimensional genetic mutation space, in this case a Hilbert space-filling curve. Initially, six positions (gray and black dots) are randomly chosen. Then the four "losers"–i.e., those with lowest values of f, are extinguished (black). The two winners (gray) are allowed to reproduce offspring (white dots), whose positions along the genetic space are randomly distributed with mean zero about the respective parent's coordinate. The whole selection procedure now continues for the six new (white) dots.

The Hilbert curve chosen in the Figure has the feature–as do many space-filling curves–that a differential change in the arc coordinate amounts to a lower-order change in the x,y coordinates. Another way to say this is, as one counts off arc length, the curve tends to stay for a long time in its current spatial region. The Hilbert expedient, which is a form of Gray coding (see frontispiece to Chapter 6), is intended to model the changing of individual bits in a genetic code. Another way to model bit mutations is to write any coordinate pair x,y as a binary union; e.g., $x = 7$, $y = 3$, in a 3-bit-by-3-bit plane, would be 111011. Then one applies a one-dimensional Gray code to each of the three-bit halves. Bit mutations (simple bit flips) then occur in Gray code space, after which a new xy bit string is constructed. A three-bit Gray code is:

Binary	Gray
000	000
001	001
010	011
011	010
100	110
101	111
110	101
111	100

for which every increment or decrement is a change of one and only one Gray bit. Whether this scheme or a Hilbert, or other curve be used, the idea is to avoid bit-carry effects when invoking mutations.

Artificial life models include those models in which *implicit* evolution occurs. That is, there is no *a priori* fitness function or assumed death rate. As explained in [Bedau and Packard 1992], one can construct computational models of complex adaptive systems, and measure in various ways what one can call evolution.

Figure 2.2.5 shows two successive frames from an artificial life movie. The figure is displayed to lend some clarity to the following Project of constructing an artificial life model. Clarity is precious in regard to such models, because there are just so many degrees of freedom, so many parameters, as in life.

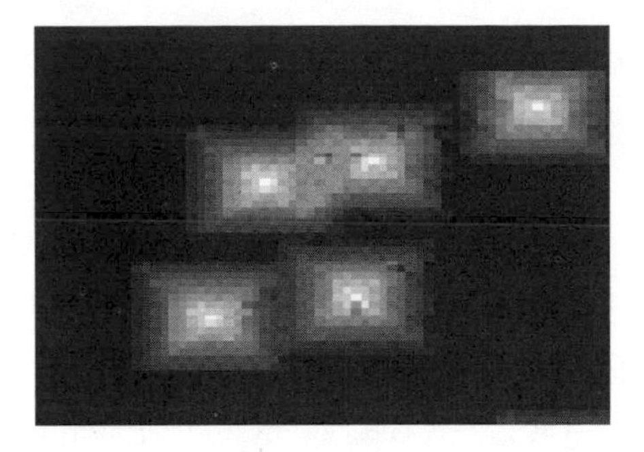

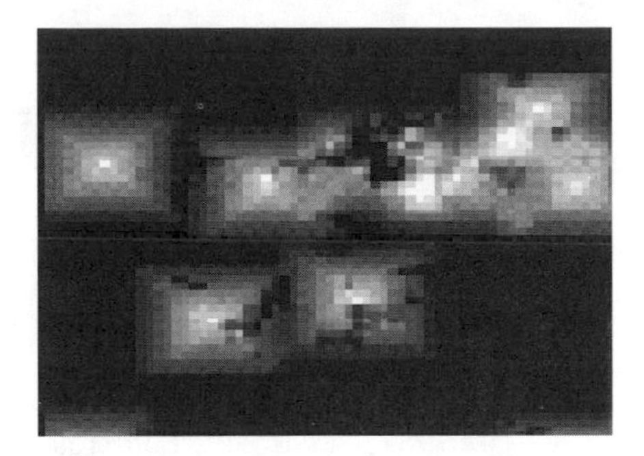

Figure 2.2.5: Scenes from an artificial life movie [Bedau and Seymour 1993]. Five food stores (rectangular dollops at left) show initial evidence of consumption by foraging organisms. Later, new food has appeared and some older dollops are well eaten.

Project 2.2.2

Genetic algorithms and artificial life

Computation, some graphics, difficulty level 3-4.
Support code: Appendix "Genetic.ma."

1) Implement a genetic algorithm with which to maximize functions in N variables. Appendix code "Genetic.ma" finds the maximum of a function in three variables, using the three-dimensional Hilbert curve for the one-dimensional mutation space. The first six generations for one particular run had, for the top eight fitness parents, the fitness scores (1 – maximum):

```
{0.856, 0.856, 0.864, 0.879, 0.907, 0.911, 0.957, 0.978}
{0.952, 0.961, 0.961, 0.965, 0.974, 0.978, 0.978, 0.996}
{0.978, 0.982, 0.982, 0.987, 0.987, 0.991, 0.991, 0.996}
{0.987, 0.987, 0.991, 0.991, 0.991, 0.996, 0.996, 0.996}
{0.991, 0.991, 0.991, 0.991, 0.991, 0.991, 0.996, 1.}
{0.996, 0.996, 0.996, 0.996, 0.996, 0.996, 1.   , 1.}
```

Thus by the sixth generation there have appeared two perfectly fit parents and six nearly perfectly fit ones.

Some interesting options for further study beckon:

- Attempt the modification of mutation rate and study how convergence to optimal solutions is affected;
- Work out an algorithm for slowing down mutation rate when one is near optimal;
- Use, instead of a space-filling mutation map, the binary-union method with Gray codes as described in the introduction to this Project.

2) Construct an artificial life model along the following lines [Bedau 1993]. Assume organisms will live on a 128-by-128 lattice. Establish a real-valued function on the lattice, which will be the food function. As in Figure 2.2.5, this function can be changed dynamically by dropping parcels of additional food. One can drop food in this way at random. The organisms are the food sink for the system, constantly decreasing the food function value at their current eating locations.

Now say that each organism can sense five immediate food locations (at the adjacent compass positions and where the organism sits), each with two bits of resolution (four levels of food concentration), for a total of 1024 possible food numbers. The key to this model is that an organism's "genes" amount to a lookup table, with 1024 food-driven pointer entries. Each entry dictates such information as, for example, how far to jump. So organisms, after jumping, consume the food present, then jump according to the table.

Of course there is a tax on jumping (it takes energy). One can track an organism's internal food supply, and modify jumping or eating behavior according to that supply.

Whenever an organism's internal food supply crosses some threshold, it reproduces, and *offspring inherit their parent's lookup tables*, except, of course, for possible mutations,which can be noisy or somehow ordered.

Analyses to consider after setting up a computation model are:

- Sample global measures, such as $F(t)$, the total amount of uneaten food at time t; and $P(t)$, the organism population;
- Plot experimental relations between F and P;
- Experiment on what level/type of mutation is optimal, in terms of maximizing P in some appropriate sense.

If a more concrete problem is desired, one may endeavor to replicate experiments from the literature [Bedau and Packard 1992].

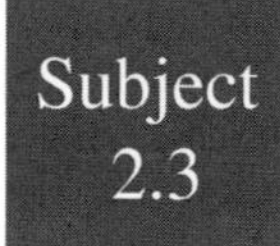

Subject 2.3 Projects from biology

We start this Subject with some biological phenomena the modeling of which involves symbolic or numerical methods in connection with matrices and differential equations. Two very natural–and might we say current–topics along these lines are population biology and epidemiology. The reason these subjects come up immediately as computationally interesting is undoubtedly that large numbers, continuous time, and complex dynamics are central to the phenomena.

Project 2.3.1 Population models

Computation, a little theory, difficulty level 2-4.
Support code: Appendix "PolynomialSolve.ma."

1) Elementary annals of population biology are replete with variants on the celebrated "rabbit problem." If the reader has not been through this, the combinatorics are instructive. Rabbits reproduce only on *odd numbered* birthdays 1,3,5,7,9,... Starting with one rabbit, show that on successive birthdays of this original rabbit we find respectively (starting with the zero-th birthday of the original):

1,2,3,5,8,13,...

rabbits. This is part of the celebrated Fibonacci sequence which is conveniently defined $F_0 = 0$, $F_1 = 1$, and for $n > 1$:

$$F_n = F_{n-1} + F_{n-2} \tag{2.3.1}$$

There are so many fascinating paths to take from here. We shall take just a few of these.

Show first from the recurrence that the rabbit population grows in a Malthusian manner; i.e., that for large n:

$$F_n \sim e^{rn}$$

for an appropriate constant $r > 0$. One way to show this is first to prove (perhaps using a symbolic programming language!) that:

$$F_n = \frac{\left(\frac{1+\sqrt{5}}{2}\right)^n - \left(\frac{1-\sqrt{5}}{2}\right)^n}{\sqrt{5}} \qquad (2.3.2)$$

Another channel of proof is to insert the exponential *ansatz* into the recursion relation and solve for r. In any case, one obtains everything in terms of the Golden Ratio $(1+\sqrt{5})/2$ which dates back to antiquity, and is known to be involved not just with idealized rabbits but many living systems, such as sunflowers and nautilus shells.

It is worthwhile to study the more general recurrence:

$$G_n = AG_{n-1} + BG_{n-2}$$

First, give a "rabbit-like" birthday-reproduction rule that gives this relation. Second, determine the asymptotic Malthusian growth parameter r. Third, express r in terms of a continued fraction, noting that division of the G recurrence relation through by G_{n-1} starts the generation of such a fraction. Note also the recurrence relations for the fraction convergents, as in step (7), Project 1.1.1.

Here is an intriguing computation question: If, instead of reproducing on every odd birthday, rabbits reproduce *statistically*, with a probability p on odd birthdays and q on even birthdays, how do simulated populations grow? Do they always appear to admit of an asymptotic growth rate? If two independent rabbit populations are started, what is the statistical behavior of the *difference* between the two total populations with time?

2) Investigate the classical theory of Erlang-Steffenson extinction. The basic question asked is as follows. If an individual and its descendants each beget n offspring with respective probability p_n, then what is the probability that the lineage will eventually become extinct?

The precise theoretical answer is intriguing [Jagers 1975]. It turns out that the extinction probability is the smallest non-negative root of the equation:

$$-x + \sum_{k=1}^{\infty} p_k x^k = 0$$

If $p_1 = 1$ then the probability of extinction is taken to be $x = 0$, and each descendant always has a descendant. But if $p_1 < 1$, it is a theorem that probability of extinction is less than 1 if and only if the derivative of the summation, evaluated at $x = 1$, is greater than 1.

One good exercise is to prove the theorem. Another is to plot extinction probability for many pairs (p_1, p_2); or even triples. Note that Appendix code "PolynomialSolve.ma" can be of some use here. Yet another exercise is to model the descendant scenario stochastically, throwing the proverbial random die for each prospective parent to determine the number of offspring (including the possibility of none; hence extinction). It would be a considerable achievement to obtain, via such a Monte Carlo approach, general numerical agreement with the theoretical prediction.

3) Investigate the possibility of numerical "chaos control." As we shall see eventually in Project 6.1.1, the Verhulst (sometimes called the Ricatti) logistic equation:

$$\frac{dp}{dt} = rp(1-p) \qquad (2.3.3)$$

is, when discretized as on a computer, subject to chaos for certain pairs of values of the r parameter and the time step dt. Here, the nomenclature is population p (normalized to be 1 at carrying capacity) at time t, with constant asymptotic Malthusian growth rate r. On the other hand, the continuous form here admits of exact solution. An interesting question then is, how do we solve the continuous equation numerically, with confidence, when we *know* that intrinsic chaos may develop in any discretized solution?

Investigate ways of solving such a continuous differential equation without harboring chaos in the process. There are some interesting ideas along these lines, such as reinterpreting the discretized right-hand-side of the differential equation. For example, the natural discretization for chosen time step dt is:

$$p_{n+1} = p_n + rp_n(1 - p_n)dt$$

which is already known to give rise to chaos. So one good research question is, does either of the alternative discretizations:

$$p_{n+1} = p_n + rp_n(1 - p_{n+1})dt$$
$$p_{n+1} = p_n + rp_{n+1}(1 - p_n)dt$$

avoid chaos? Such questions are discussed in [Wang et al. 1992].

4) Describe in general the population dynamics for the discrete iteration:

$$p_{n+1} = \frac{a{p_n}^M}{1 + {p_n}^N} \tag{2.3.4}$$

where a is constant and M, N are non-negative integers. A good starting example that is not completely trivial is $M = N = 2$ [Hoppensteadt and Peskin 1992].

5) Here we investigate some features of the fundamental Euler-Lotke equation of demographics:

$$1 = \int_0^\infty e^{-rt} L(t)B(t)dt \tag{2.3.5}$$

Here, $L(t)$ is the probability that a newborn has survived to age t, and $B(t)$ is fertility of a parent at age t. The solution r turns out to be the asymptotic Malthusian growth rate parameter for the population. Generally speaking, the whole population profile $p(x)$ for age x is assumed frozen, except that the total population (area under this frozen demographic shape p) grows roughly as e^{rt}.

Three interesting challenges are: First, determine the theoretical stable $p(x)$ profile in terms of L and B functions. Second, work out numerical (or much harder, symbolic!) means for solving the Euler-Lotke relation for r. Here is one illustrative, exactly solvable case. Assume probability of survival to age t and fertility at age t are respectively:

$$L(t) = e^{-at} \quad ; \quad B(t) = te^{-bt}$$

where a and b are constant parameters. Describe qualitatively how the shape of the L and B curves make biological sense. Give the exact solution r, give the stable population profile $p(x)$, and state conditions on a,b parameters such that a stable population is possible in the first place.

Third and finally, model the population dynamics numerically, starting with an initial population profile vs. age, and literally iterating according to the postulated statistics embodied in the L, B functions. One would start with an array $p(x)$ at time $t = 0$, and iterate by forcing statistical attrition according to L, and adding values to the birth position $p(0)$ according to B. What is picturesque about this particular numerical setting is that one gets to "advance" the age of members of the array by simply shifting each element along the positive x axis.

In the last two decades, a great deal of mathematics and computational science has been applied to physiological models. We investigate next some avenues that are particularly instructive and satisfying. The primary class of computational tools, at least for the following Project, is the numerical differential solver class.

Project 2.3.2

Physiology, neurobiology, and medicine

Computation, difficulty level 2-4.

1) The human heart always presents an excellent computational challenge, especially when data-gathering (recording of EKG signals) has occurred. But there are also interesting models and simulations possible. One interesting aspect of the heart is its evident dynamical complexity. It is a kind of non-linear dynamically driven dipole oscillator.

First, consider the problem of a time-dependent electric dipole buried in a volume conductor (that would be the heart buried in tissue). This dipole creates a surface potential, which is what is seen in the laboratory or operating room. It appears possible to gather high-fidelity information about the dipole, not by using traditional point electrodes, but by using a "body surface Laplacian" scheme [He and Cohen 1992]. These authors derive the following formula. Let a dipole of vector moment p be buried at depth $-z$ below the xy plane. Below this plane is a half-infinite medium of conductivity σ. Denote by r_p the vector position $(x,y,0)$; and denote by r_i the vector position $(0,0,-z)$. Then the surface (xy plane) potential is given by:

$$\phi(x,y) = \frac{1}{2\pi\sigma}\frac{(r_p - r_i)\cdot p}{|r_p - r_i|^3} \qquad (2.3.6)$$

and the Laplacian is computed as:

$$\nabla^2\phi = \frac{\partial^2\phi}{\partial x^2} + \frac{\partial^2\phi}{\partial y^2}$$

The clever idea for electrocardiography is now *to measure the Laplacian of* ϕ at the body surface. The authors attempt this with electrodes in circular arrangement and grid meshes of elctrodes, exploiting the approximation that the Laplacian at a point on the surface is about proportional to the potential at that point minus the average potential in the surrounding region. At any rate, one can apparently measure the Laplacian of ϕ in the laboratory or operating room. Here is a good computational exercise: work out plots of the Laplacian at various points (x, y), as a function of time, when the heart is a buried dipole that has moment p in a fixed direction, but strength which is some fabricated pulse. One goal of electrocardiography is to find that time-dependent dipole strength; after all, it is ultimately the heart property one wants, not the skin's potential. Essay on how similar the Laplacian plots are to the pulse function for the dipole. Work out computational means by which, given the time-dependent Laplacian data, one can backtrack to find the pulse function numerically.

2) Here we apply some numerical methods to a specific, highly abstracted neuron model. Solve numerically what could be called a phenomenological "Hodgkin-Huxley oscillator:"

$$\frac{d\theta}{dt} = \sum_{n=0}^{\infty} A_n \cos n\theta \qquad (2.3.7)$$

for various choices of (finite) coefficient sets of A_n's. Amazingly enough, phase-lagged dynamical solutions $\cos(\theta+\phi)$, for constant phase lag ϕ, looking very much like Hodgkin-Huxley action potentials occur even for such simple choices as $A_0 = A_1 = 1$ and all other coefficients vanishing [Hoppensteadt and Peskin 1992]. These authors also present a voltage-controlled oscillator (VCON) model which has a relation to the Van Der Pol oscillator of nonlinear dynamics, and exhibits chaotic phenomena such as devil's staircases; some of which phenomena are discussed in Chapter 6. Figure 2.3.1 shows a typical numerical plot of a phase-lagged solution $\cos(\theta+\pi/4)$ for the differential equation:

$$\frac{d\theta}{dt} = 1 + \cos\theta \qquad (2.3.8)$$

subject to initial condition $\theta(0) = 3.25$.

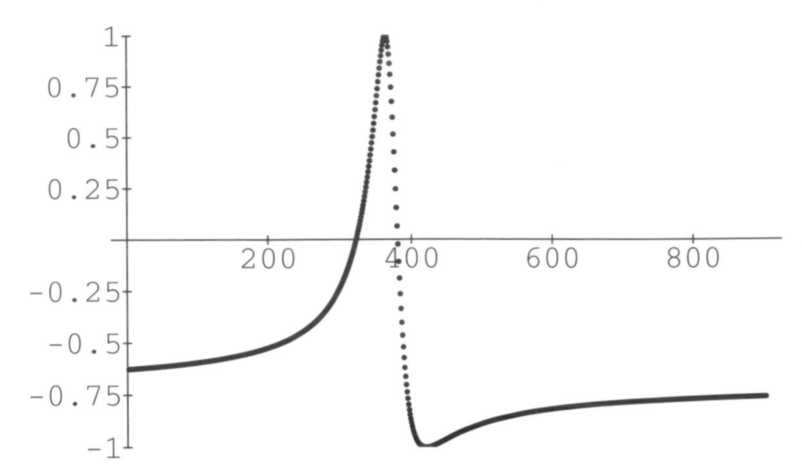

Figure 2.3.1: A synthetic action potential pulse arises from the phenomenological differential equation $d\theta/dt = 1 + \cos\theta$. The horizontal axis is time, while the vertical axis is not θ but the phase-lagged quantity $\cos(\theta+\pi/4)$. The initial condition was $\theta(0) = 3.25$.

These "Hodgkin-Huxley oscillator" models have interesting thresholding properties; e.g., experience reveals that a plot like Figure 2.3.1 requires some care in the choice of magnitude for initial value $\theta(0)$, *lest there be no pulse development at all.* But this threshold effect is what one would expect for any good neuron propagation model, phenomenological or otherwise.

A satisfying exercise is to show that (2.3.8) admits of exact solution [Griffiths 1993]:

$$\cos\left(\theta + \frac{\pi}{4}\right) = \frac{1}{\sqrt{2}} \frac{1 - (t - t_0)^2 - 2(t - t_0)}{1 + (t - t_0)^2} \tag{2.3.9}$$

where t_0 is a constant of integration, related in a complicated way to the choice of initial angle. Find exact expressions for the value of $t-t_0$ for pulse minimum and maximum. The existence of these formulae for the extrema shows the thresholding: If one's initial condition is such that the pulse extrema have *already occurred* (in the sense that t_0 happens to be sufficiently negative), the evolution of the differential equation (2.3.8) will show no pulse development.

Besides demonstrating a threshold effect one can also create periodic pulse trains by including more of the general harmonic terms in the differential equation. Figure 2.3.2 shows numerical results for a case where the next higher cosine harmonic is present.

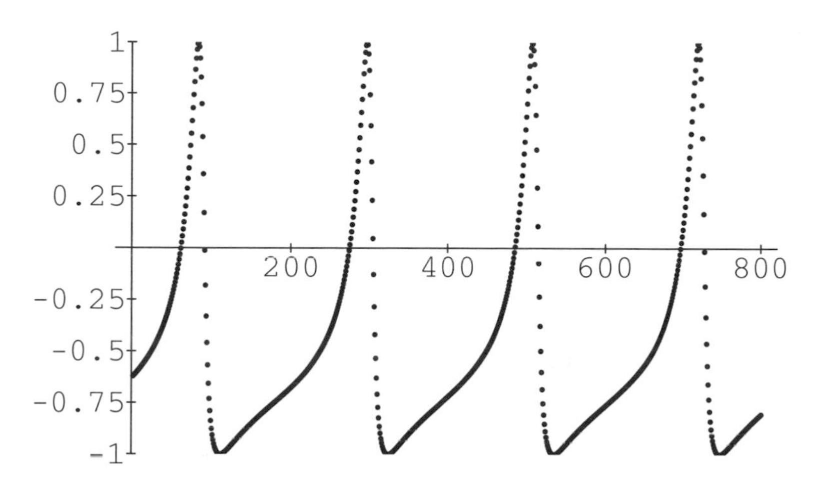

Figure 2.3.2: Action potential pulse train is obtained by including another phenomenological term. This time the equation $d\theta/dt = 1 + \cos\theta + 0.05\cos 2\theta$ is numerically solved. The vertical axis and initial condition are the same as in Figure 2.3.1.

It is interesting that such a small perturbation $(\cos 2\theta)/20$ can cause the advent of infinite pulse trains. The author is unaware of closed form solutions to such higher-order harmonic models.

3) Implement a less phenomenological Hodgkin-Huxley model, which–in the original style of the named researchers–uses explicit ion channel terms. One striking result of such numerical models is a prediction for the velocity of action potential propagation. A sample numerical study is given in [Crandall 1991].

4) It often happens that laboratory data, especially from gel gradient runs, appears to possess a few not-too-distinct peaks. In such cases it is common that precise locations of peaks is desired. Say that some laboratory data appears thus:

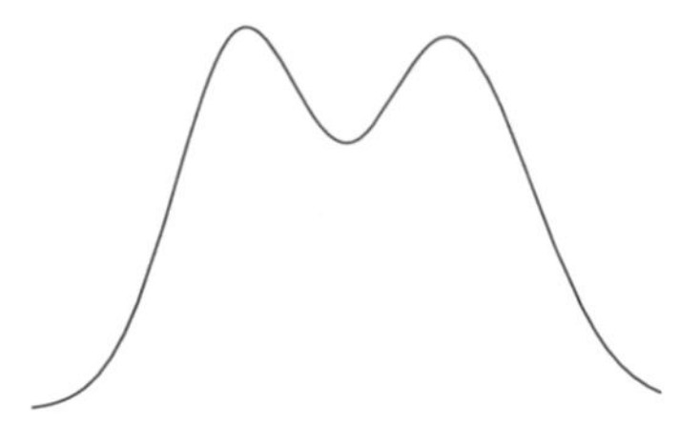

The problem is, if the two peaks shown are the result of some kind of diffusion, at some hidden level in the experiment, then a more precise peak assessment would involve what the above distribution *used to be* prior to its diffusion. In the analytical case above (a sum of two Gaussians) one could in principle "backward diffuse" the process to yield:

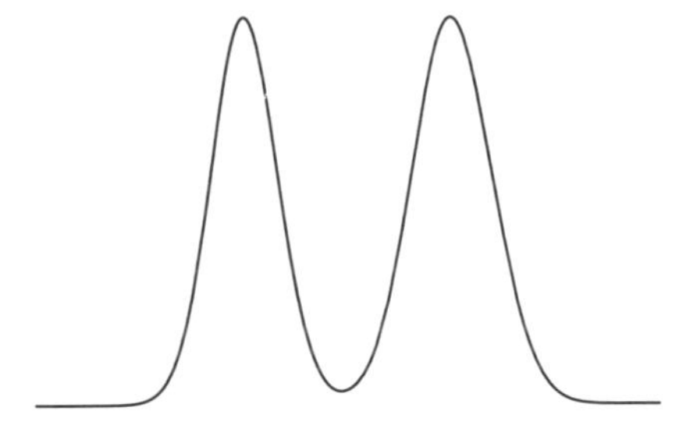

The "pre-diffusion" peaks are taller, but, more importantly, they are more horizontally separated than the data suggested. The simple fact is that two diffused Gaussians that overlap sufficiently pull each other's peaks inward.

Write software that takes initial data, assumes it is composed of sums of Gaussians, and "inverse diffuses" said data, to reveal a sharper peak structure. This is a hard computational problem, chiefly because typical data do not enjoy a perfect Gaussian superposition; so that pathologies develop quickly during the backward procedure. Still, theory and applications can be worked out for some situations [Crandall et al. 1987].

5) Investigate one of the many fluid-flow models, for example for circulation in the heart [Peskin 1977], or for the cochlea of the ear [Beyer 1992]. The latter authors solve incompressible Navier-Stokes equations in two dimensions, using modern differencing and matrix techniques. The results are in good agreement with experimental data on basilar

membrane phase and displacement.

6) Investigate neural network models of neurological response. Though we look at some elementary neural networks in Subject 2.2, some modifications must be made to model, for example, retinal response [Barlow et al. 1993]. These authors analyze a large neural network as retina model, using parallel processing on a large scale. The present author believes it likely that not too much biologically meaningful work can be done along these lines, without recourse to massive parallelism–at least not in this century.

Next we turn to a domain of obviously massive complexity: molecular biology. It is difficult to do too much without supercomputing or at least parallelism. However, as the first step of the Project shows, one can start with elementary visualization and in this way gain intuition in regard to the complexity of biomolecular structures.

Project 2.3.3 Molecular biology

Computation, graphics, difficulty level 3-4.

1) Elementary visualization of DNA amounts to a good introduction to scientific visualization in general. What is needed is:

- programming capability: Object-oriented programming is probably best for such visualization, because there is literally a hierarchy of molecular "objects;" viz., atoms, genes, sequences, molecules;
- means for visual display, preferably color display; the color not so much for aesthetic purposes, but in order to differentiate objects;
- DNA structure and coordinate data from a reputable publication.

Sample data that actually indicate spatial positions and bond angles can be found in such as [Arnott et al. 1969][Arnott and Hukins 1972]. The author used data, from the first reference, in a NEXTSTEP application that contains group objects, such as adenine, thymine, guanine, cytosine; pairing rules for base pairs A-T, G-C; and actual coordinate data for atomic sites. Thus, a sequence of coordinates and colored-sphere radii is built up, to be displayed back-to-front via an Euler-angle rotation method, a z-sorting method, and a display method. The result of a typical base-pair sequence is shown in Figure 2.3.3.

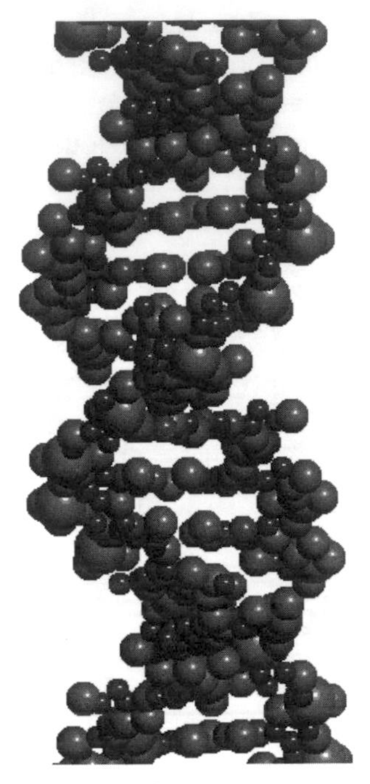

Figure 2.3.3: Elementary visualization of DNA using an object-oriented display program. No explicit molecular biology is involved, but the exercise is valuable in that many object-oriented computational methods are called for: data reading, list-building, 3-space rotation, back-to-front sorting, and display.

2) Investigate the various fascinating computational models that have taken hold in the field of molecular biophysics; for example, molecular interactions, diffusion, Monte Carlo techniques, and so on. [Briggs and McCammon 1992] discusses some of the issues, including an interesting supercomputer-level study involving actual physical coordinates for a rhinovirus (essentially, cold virus).

3) Study the computational aspects of molecular conformation. The central theme of this work is to assess the (usually vast) array of possible conformations of a given molecule, for example a polypeptide. A generalized potential energy function of the type commonly used in

computation is [Ripoll and Thomas 1992]:

$$E = \sum_{bonds} K_R (R - R_0)^2 + \sum_{angles} K_\theta (\theta - \theta_0)^2 +$$

$$+ \sum_{dihedrals} \frac{V_n}{2} (1 + \cos(n\phi - \gamma)) + \sum_{i<j} \left[\frac{A_{ij}}{R_{ij}^{12}} - \frac{B_{ij}}{R_{ij}^6} + \frac{q_i q_i}{\varepsilon R_{ij}} \right] +$$

$$+ \sum_{H-bonds} \left[\frac{C_{ij}}{R_{ij}^{12}} - \frac{D_{ij}}{R_{ij}^{10}} \right]$$

where the terms mean respectively:

E = (linear bonds) + (polar torsion) +
+ (azimuthal torsion) + (Van der Waals + electrostatic) +
+ (H-bond energy)

Thus, there is a conformal hypersurface with, commonly, a vast number of local minima into which the molecule could conceivably fall. The named reference discusses Monte Carlo methods for ferreting out these conformational extrema. They find in supercomputer-level runs that, for example, even a decapeptide GLU-VAL-VAL-PRO-HIS-LYS-LYS-MET-HIS-LYS has dozens of locally stable conformations of obvious character; many more that are more subtle but still representing local minimum values of E, and so on. Incidentally the study also reports an almost-linear speedup for certain conformation calculations that are parallelizable.

4) Explore the Human Genome Project, a worldwide interdisciplinary effort the goal of which is to provide a complete map of human genes. In the United States, the Human Genome Project began in the mid 1980s with the explicit goal of obtaining the genetic code for all genes in the human genome. In the words of [Pinkerton 1993]:

> The Project is funded from two sources, the Department of Energy, and the National Institues of Health. DOE's primary interest is in obtaining in the raw sequence information necessary to carry out its congressional mandate to understand environmentally induced genetic variation in the population of the United States. NIH, on the other hand, is primarily in understanding

the code, and applying such knowledge to further research on human diseases like Alzheimer's Disease, cancer, and more recently, AIDS.

It should be pointed out that controversy surrounding the efficacy of, and need for, the project rages on [National Forum, Spring 1993].

The human genome is comprised of some three billion nucleotides, arranged as chromosomes; in turn these sequences encode proteins. Beyond the immense cataloging, biological data-gathering, and other more classical activites, computation is hindered by the fact of proteins not having a straightforward bijective relation to the sequencing; in fact, a protein construction can depend on more than one sequence. Thus, the project involves several difficult computational problems, particularly in the areas of data storage, database searching, distributed database access, pattern recognition, and exploratory analysis of unstructured data.

Beyond the pure-computational aspects, some organizational and biologically oriented descriptions can be found in [Cantor 1990][Watson 1990].

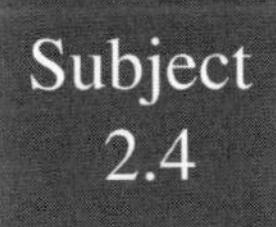

Projects from physics and chemistry

It may seem upside-down to say it this way, but in the fields of computational physics and chemistry we find more than anywhere else that tests of software exist all around us, as physical structure and dynamical effects. It is said this way to underscore the contrast between these computational fields and, say, the numerical studies of Chapter 1. For in the numerical work, especially that portion which models continuous mathematics, the physcial world does not routinely supply us with test data.

Because the physical world has so many degrees of freedom, so many particles in a typical ensemble, and so on, it is important in these computational sciences that we ponder, develop, and refine fast algorithms. The next set of Projects involve some relatively efficient (or even genuinely "fast" (meaning virtually best-possible in algorithmic terms) schemes for physical modeling.

Project 2.4.1

Classical physics

Computation, theory, graphics, difficulty level 1-4.
Support code: Appendix "threebody.c," "Equipotential.ma," "Sessile.ma."

1) For the "true" pendulum whose displacement angle $\theta(t)$ satisfies:

$$m \frac{d^2\theta}{dt^2} = -\frac{g}{L} \sin\theta$$

where m is bob mass, L is rod length, and g is Earth's gravitational acceleration, the exact period P can only be written as an elliptic integral. This stands in contrast to the "false" elementary classroom value $P = 2\pi\sqrt{(L/g)}$, on the assumption that $\sin\theta \sim \theta$. Instead, the period for the true pendulum is dependent on initial (from rest) release angle θ_0:

$$P = \sqrt{\frac{8L}{g}} \int_0^{\theta_0} \frac{d\theta}{\sqrt{\cos\theta - \cos\theta_0}}$$

Perform the following analyses and tasks, which to the author's mind comprise an excellent interdisciplinary tour of elementary computational mechanics:

• Derive the formula for P using conservation of energy;
• Show that:

$$P = \sqrt{\frac{16L}{g}} K\left(\frac{\sin\theta_0}{2}\right)$$

where K is a complete elliptic integral of the first kind;
• Show that there is an exact solution of the differential equation, of the form:

$$\theta(t) = a + b\tan^{-1}\left(e^{-ct}\right)$$

for some appropriate constants a,b,c; and state the period P;

- Numerically integrate to find P for $m = L/g = 1$ and release angle $\theta_0 = \pi/2$, to get $P \sim 7.416298...$;
- Obtain this last numerical value for P by using a known series representation for the elliptic integral K;
- Obtain a value for P (extremely rapidly!) using the identity:

$$P = \frac{2\pi\sqrt{L/g}}{AG\left(1, \cos\frac{\theta_0}{2}\right)}$$

where AG is the arithmetic-geometric mean of Subject 1.1 [Borwein and Borwein 1987].

- Numerically solve the differential equation using, say, classical Runge-Kutta (see step (5), Project 1.2.3), and in this way obtain yet another estimate for P.

For a more sophisticated computational exercise, consider obtaining the FFT (algorithms covered in Chapter 4) of evolved pendulum data $\theta(t)$, and compare this with theoretical deductions, which deductions among other things must include the fact of anharmonicity in the true pendulum (and therefore some FFT components not solely fundamental).

There we have it: the pendulum, perhaps the simplest and most elegant mechanical system on the planet; yet its proper study suggests projects that hop all over the computational spectrum!

2) On the basis of its historical grandeur and celebrated analytical intractibility (for $N>2$), the N-body problem seems a worthwhile exercise in computational physics. Let N masses have dynamical coordinates $\{r_i(t): i = 0,...,N-1\}$, where r_i is generally a vector, and for the physical world is a three-dimensional vector. One imagines a force law such that pairwise mutual vector forces $F_{ij} = -F_{ji}$ exist and drive the motion. For an astrophysical arrangement such as stars in a galaxy, one may posit the Newtonian gravitational force:

$$F_{ij} = -GM_iM_j \frac{r_i - r_j}{|r_i - r_j|^3} \tag{2.4.1}$$

where G is the universal constant of gravitation and the M_i are respective body masses. Thus, the N vector equations of motion are:

$$M_i \frac{d^2 r_i}{dt^2} = \sum_{j \neq i} F_{ij} \quad ; \ i = 0, \ldots, N-1 \tag{2.4.2}$$

The natural computational question concerns the algorithmic complexity of updating the equations of motion to within some specified accuracy. How many operations are required to advance (2.4.2) one time step, say one "Eulerian" time step using the scheme (1.2.6), (1.2.7)? The straightforward algorithm would use, say, time step $h = dt$, turn (2.4.2) into $2N$ vector differential equations involving acceleration and velocity, denoted say a_i and v_i respectively, and run:

$$\begin{aligned} a_i &= \sum_{j \neq i} F_{ij} \quad ; \ i = 0, \ldots, N-1 \\ v_i &= v_i + a_i dt \\ r_i &= r_i + v_i dt \end{aligned} \tag{2.4.3}$$

Of course, one should probably use a Runge-Kutta or other modern scheme, but the Euler method here illustrates the complexity problem, which is that there are $N(N-1)/2$ possible forces as summands in the acceleration term. This number of forces comes from the fact of $N-1$ summands in each of N cases of the sum (2.4.3), with division by two because of the antisymmetry relation $F_{ij} = -F_{ji}$. So, naively, one must compute (2.4.1), which takes a finite number of operations, roughly N^2 times. Thus, on the face of it, the N-body computational update problem has complexity $O(N^2)$.

Note the comments in Project 1.2.3, especially step (5). For an elementary example of a classical solver, the Appendix code "threebody.c" was written as "just about the shortest" readable program for the three-body problem.

But on to the complexity issue. Just as with the FFT we study in Chapter 4, and just as with certain sorting algorithms, the naive inventory of $O(N^2)$ operations does not take certain redundancies into account. In fact, *two closely spaced bodies have almost the same effect on a remote body.* This means that the forces are not completely independent.

First, consider a scheme by which the relevant galactic domain that contains N stars in 3-space is divided into M small boxes, with each box thought of, and approximated as, just one composite particle with mass equal to the sum of the masses inside it. Argue that, if the whole galaxy remains more or less roughly equidistributed, the update problem then has complexity $O(N^2/M)$. Of course, equidistribution is not essential because one could in principle partition the galaxy into non-uniformly sized boxes on *every* time step update via some fast procedure. Schemes of this sort, which for better accuracy can involve dipole-expansion of the gravitational potential due to a many-particle box, tend to be $O(N^s)$ algorithms, for some $s < 2$. This is not the best possible, but it can be more than an order of magnitude improvement over the naive approach [Bahm 1992].

As the astute connoiseur of algorithms might surmise, there exist schemes at least as good as $O(N \log N)$ for the update problem, with the implicit big-O constant depending on desired accuracy and perhaps on the pathology of the star distribution. Some of these schemes use multipole spherical harmonic expansions for enhanced accuracy. Implement one of these modern schemes, such as the hierarchical scheme of [Barnes and Hut 1986] or see [Katzenelson 1989]. The basic idea is to keep subdividing boxes recursively; in fact, it is effective to construct a hierarchy in the form of an octal tree, in which each body of the galaxy appears somewhere as a leaf.

One is doing quite well if one can show a "movie" in the form of several hundred time steps for say 1 to 10 million stars. The modern upper limit would seem to be about 100 million stars. It is staggering to think of how hard it would be to model a galaxy of that size if one had not access to an $O(N \log N)$ algorithms.

Parallelization of the N-body and other computational problems is the theme of an excellent expository article [Hillis and Boghosian 1993].

3) Next we move to classical electrostatics (or its analytical equivalent in two dimensions: fluid flow) and consider the problem of plotting equipotentials and field lines on the xy plane. The computational problem has a history going back at least to Whittaker's method of 1977 [Horowitz 1990], in which one uses the fact that, if vector v is tangent to a field line, then $v \times E = 0$. One can therefore assume a jump dx in the plotted field line and do the corresponding jump $dy = (E_y/E_x)\, dx$. To control jump magnitudes and so on, one common prescription is constantly to compute $|E|$ and iterate:

$$dx = -\delta \frac{E_x}{|E|}$$

$$dy = -\delta \frac{E_y}{|E|}$$

where δ is a prechosen constant jump length. This method is still in use today, but suffers from non-intuitive density of lines. The question arises: Where should a line's drawing start, if one demands the final overall impression that line *densities* correspond to field strength?

A second approach is to endeavor to perform contour plots of the potential function (whose orthogonal bundle will be the electric field lines). Contour plots of equipotentials can be effected with *Mathematica* lines:

```
ContourPlot[v[x,y],{x,-1,1},{y,-1,1}]
```

where $V(x,y)$ is the potential, resulting in plots such as Figure 2.4.1.

Figure 2.4.1: Equipotential lines for a wire quadrupole. The computational task of plotting the associated field lines is interesting and tractable [Horowitz 1990].

However, it is not so clear how to obtain the orthogonal field lines. One exercise, therefore, is to use existing software somehow to plot the orthogonal field.

But the new method of [Horowitz 1990] deserves attention. Given a fine grid of potential values, one establishes first a discrete broken line (a line staying on the grid points) of equipotential going through a given point (x,y). This is not hard to do, and for this equipotential line one obtains a series of discrete points $s_i = (x_i, y_i)$ spanning the line. The key is that the flux through the line is now just the integral, along the arc, of the strength of E, because E is always normal to the line of equipotential. Thus, we get from this flux the number of lines N we should plot in order to indicate average field strength. Each of these lines can be given a starting location via a simple formula that forces partial line integrals to agree with partial flux values taken over successive pieces of the arc, for a total of about N pieces. In this way Horowitz obtains a set of starting points, and *then* uses the Whittaker jump method described above.

Implement the Horowitz method or some other means of properly representing field-line density. Apply this to some potentials of interest.

There is the electrostatic dipole of the *Mathematica* example above, but here is a beautiful example in which the real part of a complex function is used to plot the celebrated Karman "vortex street", in which two infinite rows of vortices rotate oppositely. The potential will be:

$$V(x,y) = Re\left\{ i \log\left(\sin\pi \frac{z}{a} - i\frac{b}{a} \right) - i \log\left(\sin\pi \frac{z}{a} - \frac{a}{2} + i\frac{b}{a} \right) \right\}$$

where $z = x + iy$, a is the spacing between consecutive vortices, and b is the spacing between the vortex rows. The Horowitz method yields a wonderful fluid flow-line diagram showing all vortices and the overall flow trend.

4) Work out software to solve the Laplace equation

$$\nabla^2 V = 0$$

for V, subject to given boundary conditions. First, it is instructive to write a software solver for the xy plane, and furthermore to work on a square grid, the outer boundaries of which receive forced initial conditions. The method of "relaxation" consists of the following simple algorithm:

- Replace every interior grid value with the average of its neighbors;
- Repeat until no interior grid points are changing, i.e. they are all stable.

Argue that this provides a discrete solution to the Laplace equation. Compare such relaxation solutions with exact solutions to the Laplace equation. Such solutions are not hard to work out; one method is simply to write down a nice solution and see what it is at the boundaries, which can then become one's boundary conditions for a relaxation run.

Then, augment the method, for example using the refinements of [Gash 1991] that include rigorous statements about accuracy of the final solution.

Here is an application of three-dimensional Laplace solvers. Consider computing the Madelung constant of step (1), Project 2.4.3 by using the fact that it is the value, at the origin (0,0,0), of the solution to the Laplace equation over a unit cube (vertices {±1/2, ±1/2, ±1/2}) whose boundary values are as follows: At a position on the cube surface, the forced potential is $-1/r$ where r is the distance from the origin.

5) Find some number of terms of a symbolic power series solution $f(r)$ to the sessile drop equation:

$$\frac{\frac{\partial^2 f}{\partial r^2} + \frac{1}{r}\left(1 + \left(\frac{\partial f}{\partial r}\right)^2\right)\frac{\partial f}{\partial r}}{\left(1 + \left(\frac{\partial f}{\partial r}\right)^2\right)^{3/2}} = K(1 - f)$$

in which the left-hand side is mean curvature and the right-hand side is a combined pressure/surface tension term. The f function is to be the height of the surface of a cylindrically symmetric liquid drop. The frontispiece shows a typical solution for $K = 1/10$. One problem for which the author does not know an answer is how to continue the solution past the "caustic;" that is, when the drop is massive enough to be about to "fold under" itself (i.e. $f(r)$ has diverging slope), how does one continue the surface solution?

One interesting continuation is to symbolically integrate to find the mass underneath a given height, then to find the wetting angle as the arctangent of the surface slope, and thus uncover a relation between mass and wetting angle.

6) Consider the two-dimensional (sometimes called "conical") pendulum, which is a bob on a rod of length L, which rod is connected to a ceiling pivot so that the bob can move over a spherical surface. Write a differential solver for this syetem (which can be modeled with four first-order equations). Check the solver in the degenerate case that the bob is released from rest without any azimuthal velocity, in which case one is simply solving the one-dimensional pendulum of step (1) previous.

Next, however, is a serious computational challenge: verify numerically the beautiful relationship between orbital precession and orbital area. The basic principle is that, if the bob is allowed to swing in an approximately elliptical orbit (precisely elliptical in the harmonic approximation!), then *the rate of precession is essentially proportional to the area.*

7) Show using a differential solver that the effect of the moon on a tightly orbiting Earth satellite is to cause the latter's orbit to elongate in *a particular direction* with respect to the Earth-moon line.

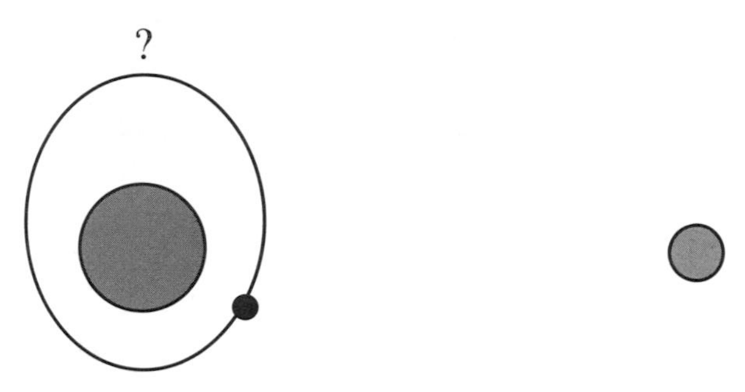

Assume that the moon is not orbiting significantly fast, so that the forces to which the satellite is subject are of the form (2.4.1) with Earth and moon positions stationary. Assume further that the initial satellite orbit is a circle, in whose plane lives the Earth-moon line. Thus the coordinates $\{x(t), y(t)\}$ of the satellite completely specify the motion.

It is also of interest to work out a perturbation theory for the moon's effect on the initially circular orbit, then to solve symbolically such a system and in this way see just how the satellite orbit elongates in its peculiar way.

One can take these ideas further, and model the Saturn system. One can take ring particles to be affected by orbiting moons. Theory says that when the ring particle's orbital angular velocity is a rational multiple of a moon's angular velocity, resonance effects may occur. Such a model has achieved some success if, for example, the Cassini Division–a black gap in the main ring–eventually appears. This gap is situated in a resonant manner with respect to the moon Mimas. The author has found that the Cassini Division will appear if one iterates as follows:

- Let the orbiting moon perturb the (the initially circular) orbits of all ring particles;
- After some amount of time, "reset" the orbits by starting each ring particle into a new circular orbit, but using the current radius;

The intent of the second step is twofold. First, it prevents computer runaway of perturbed particles (even if such "whiplash" runaway actually occurs at Saturn!); and second, it is an effective, albeit rudimentary model for particle collision and damping. What happens is that particles at the Cassini Division tend to be perturbed elsewhere by resonance, where they begin new circular orbits, and so on. As with the Earth-moon problem, the perturbation theory for such a system can be solved exactly; and with the rudimentary damping can predict a ring density having local minima at resonance gaps.

Quantum theory, at least from the computational standpoint, differs from classical theory in that one must forget about point masses and static fields, and concentrate on such notions as wave functions, certain discrete phenomena, and time evolution of ensembles.

For the next Project we confine ourselves to the nonrelativistic theory. The Schroedinger equation for numerical studies can be adopted in dimensionless form (atomic units):

$$i\frac{\partial \Psi}{\partial t} = -\frac{1}{2}\nabla^2 \Psi + V(x)\Psi \qquad (2.4.4)$$

in which $V(x)$ is the potential at vector position x, while the traditional physical entities $h/2\pi$ and m have been set equal to 1.

Project 2.4.2 Quantum theory

Computation, graphics, difficulty level 1-4.
Support code: Appendix "SloshingPacket.ma."

1) It may be useful to gain some intuition in regard to the time evolution of wave packets. For a one-dimensional harmonic oscillator potential $V(x) = \omega^2 x^2/2$ it is possible to derive an exact bound wave packet solution to the Schroedinger equation (actually, a packet with Gaussian initial data $\Psi(x,0)$) and obtain for that special packet a probability density:

$$|\Psi(x,t)|^2 = \frac{\exp\left(-\frac{1}{2}\,\frac{(x-a\cos\omega t)^2}{\cos^2\omega t+\frac{\sin^2\omega t}{\omega^2}}\right)}{\sqrt{\cos^2\omega t+\frac{\sin^2\omega t}{\omega^2}}}$$

Write software to animate this packet. An example is Appendix code "SloshingPacket.ma." Note, either from the exact formula or from animation experiments, that the packet sometimes "pinches down" at the origin, as in Figure 2.4.2; but–depending on the value of ω–sometimes "peaks up" instead. This is one of many examples of how quantum mechanics can be non-intuitive. An interesting question is: what is the meaning of the "classical limit" in this context? That is, assuming a macroscopic mass on a spring is an initially Gaussian packet, what is the classical meaning of the "pinch" (or is it a "peak") when the motion passes the origin?

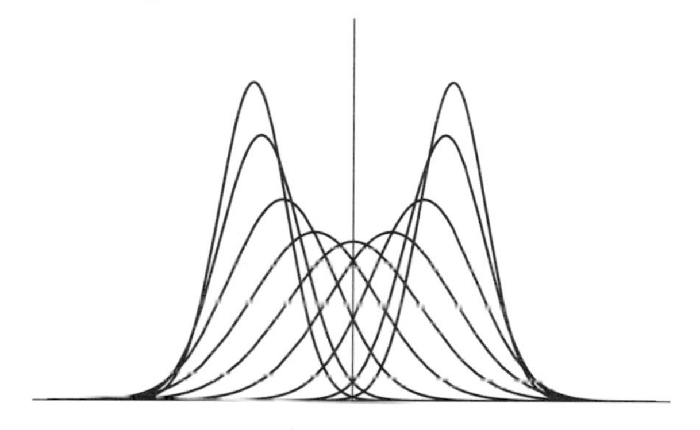

Figure 2.4.2: Exact animated packet solution for the quantum harmonic oscillator. The packet "pinches down" as it crosses the origin. However, as can be shown by running an animation for different oscillator frequency ω, in some cases the packet "peaks" at center instead.

For more general potentials and wave packets where we do not know exact solutions, consider the numerical solving schemes discussed in step (2) next.

2) Write software that numerically solves the time-dependent Schroedinger equation (2.4.4) for a wave packet $\Psi(x,t)$ impingent on a

one-dimensional potential well $V(x)$. Two important points must be made here. First, one has tremendously better success using the so-called double-time-step or Crank-Nicholson methods [Goldberg et al. 1967][Press et al. 1988]. The essential idea is that the Hamiltonian operator e^{-iHt} that propagates solutions to the Schroedinger equation is better approximated in small time step by Cayley's operator form:

$$e^{-iHdt} \sim \frac{1-\frac{1}{2}iHdt}{1+\frac{1}{2}iHdt}$$

than by, say, just $(1-iHdt)$. The reason that the Cayley form gives rise to more stable numerical iteration is that the quotient operator on the right, even though approximate, is *unitary*, thus preserving probability and so on. These observations manifest themselves in the following formulae, which assume discrete wave function values $\Psi_j^n = \Psi(j\,dx, n\,dt)$:

$$\left(1+\frac{1}{2}iHdt\right)\Psi_j^{n+1} \sim \left(1-\frac{1}{2}iHdt\right)\Psi_j^n \tag{2.4.5}$$

$$H\Psi_j^m \sim V(jdx) - \frac{1}{2(dx)^2}\left(\Psi_{j+1}^m - 2\Psi_j^m + \Psi_{j-1}^m\right) \tag{2.4.6}$$

These discrete relations comprise a neat recipe for the Schroedinger equation. The steps are:

- Choose time increment dt, space increment dx, and potential $V(x)$;
- Set up an array Ψ_j corresponding to discrete values $\Psi(j\,dx, 0)$, remembering that there needs to be a changing phase along the index j to represent initial momentum (a typical initial wave function is a real-valued Gaussian times e^{-ipx} for initial momentum p);
- Solve the tridiagonal linear equation system obtained from (2.4.5)-(2.4.6) for the elements of the new array Ψ_j^1, continuing to do this to get the evolution of the wave packet Ψ_j^n for general time indices n.

Note that it is possible to obviate the tridiagonal solution step by keeping three wave functions at any time and approximating the time derivative at any moment as the difference between wave function values *two* time steps apart, divided by $(2dt)$. This technique has proven successful for the

Schroedinger equation [Goldberg et al. 1967] and also for certain soliton equations [Crandall 1991].

The second important point is that there are some excellent test potentials for one-dimensional numerical scattering; namely, the "reflectionless potentials:"

$$V(x) \;=\; -\;\frac{N(N+1)}{2}\,\mathrm{sech}^2 x$$

which for non-negative integers N are literally reflectionless: all the matter of an impingent wave packet gets completely through the barrier; there is no back-scatter. Furthermore, in a certain sense a wave packet "reassembles" after passing through the barrier, and this should be evident in any proper software experiments [Crandall and Litt 1983]. Incidentally, such a reflectionless potential has exactly N bound states, and a magical thing happens which computations can verify: the *phase* of the reassembled final packet is either flipped or not, as the factor $(-1)^N$.

3) Here is an exploratory problem the author believes has high promise but also presents considerable computational difficulty. By performing multidimensional discrepancy integration (Project 1.3.2), numerically evaluate the following $(D-1)$-dimensional pure-geometric path integral, over N-dimensional vectors x_m:

$$F(x, x_0, E) \;=\int \cdots \int \left(\frac{\sqrt{E-W}}{C}\right)^{\nu} H_{\nu}^{(1)}\left(DC\sqrt{E-W}\right) dx_1 .. dx_{D-1}$$

where H is the Hankel function of the first kind, $\nu = DN/2-1$ and W is a potential energy average:

$$W \;=\; \frac{1}{D}\sum_{m=1}^{D-1} V(x_m)$$

while C is a kinctic term:

$$C^2 = \frac{1}{D}\sum_{m=0}^{D-1}(x_{m+1} - x_m)^2 \quad ; \quad x_D := x$$

It turns out that F is related to the Schroedinger Green's function; in fact there is reason to expect, for D sufficiently large, peaks in the magnitude of F when E is near an eigenvalue of the Schroedinger problem. The difficulty in the computation is that apparently more than $D = 10$ dimensions are required to show eigenvlaue structure even for simple problems. The high promise referred to is based on the notion that the "helium problem," quantum theory's canonical three-body dilemma, may be numerically assailable in this way.

4) Along more standard lines than step (3), investigate the standard Monte Carlo techniques for Schroedinger Green's functions and ground state energy calculations [Lee and Schmidt 1992][Koonin 1986] and related applications to quantum chemistry [Morales 1993]. An excellent introduction to quantum Monte Carlo is [Tobochnik et al. 1990]. Again the helium problem stands as a superb challenge.

5) Investigate the state-of-the-art in computational chemical physics. A good collection is [Truhlar 1988], where quantum versions of N-body problems appear and computational avenues are suggested.

6) Attempt to find symbolic power series solutions of the Schroedinger equation for various potentials. One choice is the harmonic oscillator, which after certain subtitutions allows closed-form symbolic solutions in the form of Hermite polynomials times Gaussians. Other potentials of interest are those of the reflectionless class of step (2) previous. Using for example $V(x) = -\mathrm{sech}^2 x$, one ought to be able to identify a power series solution for the ground state wave function.

Next we turn to problems relevant to structures such as crystals, large molecules, and ensembles such as liquid surfaces.

Project 2.4.3 Molecules and structure

Computation, difficulty level 2-4.

1) Investigate the tantalizing and difficult Madelung problem, which in its simplest form asks for the potential at a crystal's origin due to an infinite array of alternating charges. Let the charge at integer lattice triple (x,y,z) be as in a salt crystal, that is $(-1)^{x+y+z}$. Then the Madelung constant is the electrostatic sum:

$$M = \sum_{(x,y,z) \neq (0,0,0)} \frac{(-1)^{x+y+z}}{\sqrt{x^2+y^2+z^2}} \tag{2.4.7}$$

The very first of many difficulties is that the sum does not converge, depending on just how the summation is carried out. First, show by sample computations that when the sum for M is taken over the charge within a sphere of radius R (sphere centered at (0,0,0)) then the value does *not* converge as $R \to \infty$. It has been shown [Buhler and Crandall 1990] that the "Delord conjecture" is true; i.e., if one subtracts from the radius-R partial sum the quantity Q/R, where Q is the total integer charge lying within the sphere, then, yes, the sum does converge. The sum for M also converges, without the charge adjustment, if one simply takes partial sums over ever-expanding centered cubes. In either scenario the limit of the convergence is the same M value. The spherical partial sums are interesting in that the charge within a sphere is a complicated function of radius, as expected from the celebrated "circle problem" of analytic number theory, step (6), Project 2.1.3.

Instead of defining the Madelung constant as the limit of a properly converging partial sum, it is fashionable to define M as an analytically continued value of a multidimensional Zeta function. Consider the sum (2.4.7) but with the power s in the denominator instead of 1/2 (the root). Then the sum converges absolutely for all $Re(s) > 3/2$, and one can define the Madelung constant as the value of this $M(s)$ function if analytically continued to the value $M = M(1/2)$. Using such notions, computational chemists have used various rapidly converging sums called Ewald expansions, precisely the same in spirit as the formula (1.1.46) which expresses the Riemann Zeta function in terms of incomplete gamma functions. In modern times, more exotic and easier-to-handle summations have arisen. A landmark formula was found by [Hautot 1975] and involves only two-dimensional sums of rapidly decaying cosech functions. Hautot's sum is essentially:

$$M = -\frac{\pi}{2} + 3 \sum_{(p,q) \neq (0,0)} \frac{(-1)^q \operatorname{csch} \pi\sqrt{p^2+q^2}}{\sqrt{p^2+q^2}} \tag{2.4.8}$$

which is rapidly convergent. In [Crandall and Buhler 1987] it is shown that the Madelung potential for any sufficiently regular crystal can be written using certain triple sums over *elementary* functions. The special case of the NaCl crystal (2.4.7) is, for example:

$$M = -\pi + \sqrt{2} + \sum_{(x,y,z) \neq (0,0,0)} \frac{(-1)^{x+y+z}\left(1 - \tanh 2\sqrt{x^2+y^2+z^2}\right)}{\sqrt{x^2+y^2+z^2}} \tag{2.4.9}$$

which exhibits another rapidly converging sum.

Use the formula (2.4.8) to compute M, agreeing with:

M = `-1.74756459463318219063 6...`

Note also in step (4), Project 2.4.1 the Laplacian method of finding M, which, although inefficient, is nevertheless interesting from a physics perspective.

Note that in the expansion (2.4.9) the first terms $-\pi + \sqrt{2}$ give already $-1.727...$, which is off by only about 1 per cent! It turns out that if a certain sum can be resolved in general, then the Madelung constant M can be likewise and finally resolved. The key sum is:

$$S(x) = \sum_{u \varepsilon O^3} \frac{\operatorname{csch} \pi x|u|}{|u|}$$

where O^3 denotes all triples of odd integers. It can be shown that

$$M = \lim_{\lambda \to \infty} \left(\frac{2\pi}{\lambda} S\left(\frac{\pi}{2\lambda}\right) - \lambda \right) \tag{2.4.10}$$

so the problem of evaluating S for successively smaller arguments is an especially compelling one. A good exercise is to evaluate S numerically for the (only!) known cases of interest:

$$S(1) = -\frac{1}{8} + \frac{3 \log 2}{4\pi}$$

$$S(1/2) = \frac{1}{\sqrt{2}}$$

$$S\left(1/\sqrt{8}\right) = \sqrt{2}\, \theta_2{}^2\left(e^{-\pi\sqrt{2}}\right)$$

$$S(1/4) = \sqrt{32}\, \theta_2\left(e^{-\pi\sqrt{8}}\right) \theta_3\left(e^{-\pi\sqrt{2}}\right)$$

where Jacobi theta functions [Whittaker and Watson 1972] appear in the last two cases; and thus verify the formulae to some precision. It is amusing that, by using the above exact expression for $S(1/4)$ in the limit expression (2.4.10), one actually gets an approximation to M that is within 4×10^{-5} of the truth. There is perhap some dim hope of finding better and better identities by manipulation, preferably symbolic, of theta functions.

2) Explore computational models of DNA molecular dynamics. General theory can be found in [McCammon and Harvey 1987]. But beware of the assumption that everything must be numerical for such horrendous systems. One elementary and symbolically solvable model (of drastically simplified DNA-like structure) is found in [Powell and Crandall 1993], where the authors derive exact formulae for the acoustic speed of DNA vibrations.

3) Investigate statistical models of physics and chemistry, one example of which is the surface tension model of [Potvin 1991], in which the Potts

model is used to compute reasonable model numbers for liquid surfaces.

Roughly speaking, there are two brands of relativity: special (which deals with inertial frames and Lorentz transformations), and general (in which gravity is properly included). Below we provide a brief mix of exercises intended to sketch very briefly the spectrum of computational relativity, which spectrum runs from diagrams and visualizations in special relativity, to symbolic problems in both special and general classes, to massive computational problems involving the Einstein equations.

Project 2.4.4 Relativity

Computation, theory, possible, graphics, difficulty level 3-4.

1) Work out software to visualize a relativistically moving object. It is nontrivial to apply the correct formulae of relativity physics and to interpret these properly. One needs to use Weisskopf relations and properly set up the "photosurface" of such a moving object. Software routines are given in [Gekelman et al. 1991] and an interesting essay is [Hsiung et al. 1990].

2) Implement a finite element method for general relativistic fluid flow or relativistic collapse, which method can be thought of as means for solving the Einstein equations numerically, at least for certain situations [May and White 1967][Mann 1987].

3) A good idea is to familiarize oneself with an established general relativity package, such as *MathTensor*™. Work through the implicit exercises of a user manual for such software.

4) Attempt symbolic or partially symbolic derivations of some special- and general-relativistic effects, such as the Compton effect and the precession of the perihelion of planet Mercury [Crandall 1991].

$$\frac{x}{\sqrt{2}} \coth \frac{x}{\sqrt{2}} = \sum_{n=0}^{\infty} \frac{2^n B_{2n} x^{2n}}{(2n)!} = \frac{A_0(x) + A_1(x) + A_2(x) + A_3(x)}{A_{1/2}(x)/x}$$

$$A_k = \sum_{n=0}^{\infty} \frac{x^{8n+2k} A_{kn}}{(8n+2k+3)!} \qquad A_{k,n+1} = -136 A_{kn} - 16 A_{k,n-1}$$

$$A_{1/2,0} = 24 ; \; A_{1/2,1} = -3168$$

$$A_{00} = 6 ; \; A_{01} = -792 \; ; \; A_{10} = 20 \; ; \; A_{11} = -2704$$

$$A_{20} = -28 ; \; A_{21} = 3824 \; ; \; A_{30} = 96 \; ; \; A_{31} = -13056$$

If $(8m+2k)^{th}$ *coefficient of* $A_k(x)\left[\frac{A_{1/2}(x)}{x}\right]^{-1} \neq 0 \pmod{p}$

for $m \leq \left[\frac{p-5}{8}\right]$ *and all* $k = 0, 1, 2, 3$; *and* $B_{p-3} \neq 0 \pmod{p}$

then

$$x^p + y^p = z^p$$

is impossible in positive (x, y, z).

Sketch of numerical proof that Fermat's "Last Theorem" holds for exponents p into the millions. A fast algorithm (based on Newton's method and FFT) is used to resolve the (mod p) inverse of a multisection polynomial $A_{1/2}(x)/x$. *The regular prime exponents p that do not divide any relevant Bernoulli numbers* B_{2n} *are thus settled, with other prime exponents settled via the Vandiver criterion. This and related polynomial algorithms were used by* [Buhler et al. 1992, 1993] *to establish the theorem, and other important number-theoretic conjectures, for p to four million. Though A. Wiles recently announced a proof of the "Last Theorem," the other conjectures remain open. So it is likely that such power series calculations will be pushed further in the future.*

3 The lure of large numbers

Projects in number theory

Little [of the type of successes reported in the Guinness Book of Records] is found in the scientific domain. Yet, scientists–mathematicians in particular–also like to chat while sipping wine or drinking beer in a bar . . . Frankly, if I were to read in the Whig-Standard that a brawl in one of our pubs began with a heated dispute concerning which is the largest known pair of twin prime numbers, I would find this highly civilized.

[*Ribenboim* 1988]

The author has been drawn to the "lure" of large numbers in two different ways in two different eras. First, there was a fascination with number theory early in school. Number theory is a subject remarkable for its nobility, purity, and *civilized* character (akin to the nobility of astronomy–another field in which lures exist that will never be reached, run as we may!). Later, through channels mentioned in the Preface to this book, computational problems arose that begged for fast algorithms, parallelism, and massive storage. Thus, it became possible to compile a collection of number-theoretic projects from computationalists and theorists, many of which projects involved large integers and so came to be included in this chapter. Large-integer arithmetic is the dominant theme of this collection, primarily because many of the research problems involved *very* large numbers, some reaching even beyond one million binary bits. Printout of an integer of this kind of size (in decimal) would generate something like a book-length document. And when you need an array of these giant integers, that is like a bookshelf of digits. Integers of that magnitude will give pause even to powerful modern supercomputers: the fastest known way to multiply two "book-length" integers can consume a whole second!

To do Projects in this chapter one needs access to some multiprecision integer arithmetic package. *Mathematica* can effect large-integer operations, but often one requires the fastest possible speed. In fact, when multiplication exceeds some thousands of digits per multiplicand, it is best to use fast algorithms; faster at least than straightforward "grammar school" multiplication.

For these reasons a library "giants.c" has been included on the companion disk to the book. Its header file "giants.h" can be perused in the Appendix, to find the function calls one desires for number-theoretic problems. This software contains fast routines optimized for applications to factoring and primality testing–two topics upon which we concentrate herein.

But there are good, perhaps superior alternatives to the included "giants.c" disk library. Various big-number packages are available via network ftp. Such public software tends to come and go; one can find current information via newsgroups and announcements. One might also consider some modern ways of effecting a large-integer package, such as Objective-C, or the language C++ with its primary advantage of operator overloading.

Large-integer arithmetic

It is important that one familiarize oneself with the large-integer software at hand. In the following Project the idea is not so much to test software–although that is always a good idea–but to convey familiarity with the high-precision operations and their frank computational complexity (that is, how much CPU time they consume).

Project 3.1.1 Testing the operations

Computation, difficulty level 1-2.
Support code: Appendix "giants.[ch]."

As explained above, one may compile and use "giants.c," or use *Mathematica*, or some other equivalent large-integer package. For truly huge integers (say >10000 decimal digits), the "breakover" methods of "giants.c," which involve FFTs and so on, are essential to avoid interminable waits.

1) Time the multiplication of two integers of D digits each, in the region D = 100 to 500 digits, and verify (what should be true for virtually any software) that the time consumed per multiply behaves as $O(D^2)$. If "giants.c" is used, one simply writes a short C program having:

```
giant x = newgiant(INFINITY),
      y = newgiant(INFINITY),
      z = newgiant(INFINITY);
```

which creates three large integers having INFINITY (i.e., the largest allowed) number of 16-bit type short-integer digits. One wants at least three such integers, because intermediate saving of a giant integer may be effected with:

```
gtog(x,z);
```

which will force z to be x. Then a sequence:

```
gread(stdin, x); gread(stdin, y);
mulg(y,x);
gwriteln(stdout, x);
```

will multiply two input integers and print out the product. For larger integers than one wants to input from a file, a loop such as:

```
for(j=0; j< size; j++)
      x->n[j] = random() & 0xffff;
x->sign = size;   /* Lock in total digit count. */
```

will create a random positive integer having "size" 16-bit words, that is, $2^{16*\text{size}}$ binary bits.

2) Time, in the same way as for multiplication of the last step, division operations. If "giants.c" is used, the call divg(y,x) replaces x with the integer quotient of x/y. The remainder is obtained from modg().

3) Find, as in the grapestake problem of Project 2.1.1, the empirical probability that two large (say 100-word) integers are relatively prime, by invoking a greatest common divisor (GCD) function. It is a good option to time also the GCD, which in some elementary implementations can be even as bad as $O(D^3)$ complexity. The library "giants.c" has, luckily, a fast GCD for very large integers, compliments of [Buhler 1990], who found an efficient implementation of the fast algorithm of [Aho et al. 1974]. Such a routine in practice requires compute time no worse than $O(D \log^k D)$, for some small power $k \sim 2$. In fact, what Aho et al. show is that such a GCD can be obtained in $O(M(D) \log D)$ bit operations, where $M(D)$ is the time to multiply two D-digit numbers.

4) Use the "giants.[ch]" or equivalent software at hand to verify Fermat's little theorem, that if p is prime, and a is one of 1, 2, ..., $p-1$, then

$$a^{p-1} = 1 \pmod p$$

If this test fails, p is definitely composite, but if the test passes, one still does not know if p is prime. One may use the function powermodg() to perform an efficient powering ladder, in a sequence such as:

```
itog(1, x); gtog(p, pminus); subg(x, pminus);
itog(a, x);
powermodg(x, pminus, p);
/* x is now a^(p-1) (mod p). */
```

To test whether the result in giant x is the integer 1, there are several options. One can compare x and a giant equal to 1, via the general function gcompg(); or one can simply call the function isone(x).

Some large numbers one may investigate are the Mersenne numbers, for which a contemporary list of known prime cases is found in the next

Subject. One rapid way to create such primes is to load an integer with 1 and shift it left by q bits to create 2^q, then to subtract 1, as follows:

```
itog(1,x);
gshiftleft(521,x);
itog(1,z);
subg(z,x);    /* x is the Mersenne prime M_521. */
```

Subject 3.2 Prime numbers

Next we turn to prime numbers, a subject of continual fascination over the ages. This domain is a pleasant one, in which actual proofs (usually that a number is prime or composite) can often be effected via computer.

Project 3.2.1 Mersenne primes

Computation, difficulty level 2-4.

1) Create software to perform Mersenne primality tests, using the Lucas-Lehmer iteration, as follows. For a number $p = 2^q-1$, with $q > 2$, start with $x_0 = 4$. Then iterate:

$$x_{n+1} = x_n^2 - 2 \pmod{p}$$

The theorem is that p is prime if and only if $x_{q-2} = 0 \pmod{p}$. It should be possible in this way systematically to find all Mersenne primes up to, say, 2^{10000} in a day or so (if one's large-integer library is sufficiently fast). One notes in passing that q can be restricted to primes, so that a simple test such as that of step (4), Project 3.1.1 might be in order to rule out most non-prime exponents q prior to the Lucas-Lehmer test.

The 33 currently known Mersenne primes are:

$2^{2}-1$	$2^{3}-1$	$2^{5}-1$	$2^{7}-1$
$2^{13}-1$	$2^{17}-1$	$2^{19}-1$	$2^{31}-1$
$2^{61}-1$	$2^{89}-1$	$2^{107}-1$	$2^{127}-1$
$2^{521}-1$	$2^{607}-1$	$2^{1279}-1$	$2^{2203}-1$
$2^{2281}-1$	$2^{3217}-1$	$2^{4253}-1$	$2^{4423}-1$
$2^{9869}-1$	$2^{9941}-1$	$2^{11213}-1$	$2^{19937}-1$
$2^{21701}-1$	$2^{23209}-1$	$2^{44497}-1$	$2^{86243}-1$
$2^{110503}-1$	$2^{132049}-1$	$2^{216091}-1$	$2^{756839}-1$
$2^{859433}-1$			

These primes were gradually uncovered over the last *five hundred* years, the small ones dating back to the 1400s [Ribenboim 1988], with the last few discovered very recently. For the last few primes on the list, a program using a fast algorithm library such as "giants.c" on a typical workstation will still consume weeks or months for a single Lucas-Lehmer run. In fact, for exponents exceeding about 100000, it is best to establish custom transform-based routines specific to Mersenne arithmetic, some of which routines are discussed in Subject 3.3. In fact, the last two giant primes listed here were verified, at the request of discoverers [Gage and Slowinski 1993][Slowinski 1993] by the author and D. Smitley and J. Doenias, using fast algorithms of the type described in [Crandall and Fagin 1994]. Even under these optimized circumstances, the largest prime listed here required 3 workstation weeks to verify, which amounts to a few supercomputer hours, depending of course on the brand of computer and on program efficiency.

One should be suspicious of any Mersenne list, even the one exhibited above! This is because historically, lists of these primes, for one reason or another, have sometimes been riddled with errors. Even Mersenne's original list had numerous erroneous entries/omissions [Ribenboim 1988][Hardy and Wright 1972]. Even in modern times, investigators have "overlooked" Mersenne primes. The 29th Mersenne, namely $2^{110503}-1$, was in fact found "out of turn," meaning after some larger primes were already known [Colquitt and Welsh 1991]. It is difficult to do systematic searches when each possible Mersenne candidate takes several supercomputer hours. The author participated (with D. Smitley and others) in one such search, covering various contiguous candidate regions up to 2^{500000}. After months of computational agony and hundreds of supercomputer hours we came up empty-handed. What is more, there could have been an overlooked case due to machine or program error! The author feels that one excellent computational milestone for the year 2000 would be an *exhaustive* resolution of Mersenne candidates all the way up to one million bits.

In Mersenne searches, one is wise to sieve out exponents q and so obviate the Lucas-Lehmer test in many cases. In practice one finds that in the $q \sim 1000000$ region (a little above the largest known Mersenne prime listed above) about one-half of all the candidate prime exponents q can be removed by sieving. One uses the fact that if a prime f divides 2^q-1, then f is necessarily of the form $2kq + 1$. Another interesting theorem is that if q and $2q+1$ are both prime (i.e. q is a Sophie Germain prime), then the latter prime divides 2^q-1 [Hardy and Wright 1978].

2) Find huge primes of the related form $h2^q\pm1$ (where h is small) via a two- or three-step process. First, sieve out as many q as possible by looking for occurrences of $-2^q = \pm h^{-1} \pmod p$ for small primes p (the inverses of h can be stored in a table). Second, as an option, consider finding larger factors using an ECM program or perhaps a $(p\pm1)$ program suited for numbers of this special form. Such factorers are discussed in Project 3.4.1. To whatever prime candidates remain, one may apply known Lucas-Lehmer-like tests for primality of such numbers. One relevant theorem is as follows [Hardy and Wright 1978]. If $m \geq 2$ and $N = h2^m+1$ with $h < 2^m$, and if p is any prime mod which N is a non-residue, then N is prime if and only if:

$$p^{(N-1)/2} = -1 \pmod N$$

Note that the Pepin test for Fermat numbers is a corollary of this theorem ($h = 1$, $m = 2^n$, and take $p = 3$). Modern theorems along these lines are given by [Bosma 1993]. To this author's knowledge the largest known prime of the form $h2^q\pm1$ with $h \neq 1$ is:

$$391581 \cdot 2^{216193}-1$$

found by [Brown et al. 1989]. For a brief period this was the largest known explicit prime, being greater than the Slowinski prime $2^{216091}-1$ until the discovery of the two newest Mersenne primes.

Other exponentially growing numbers of interest for primality searches are the Cullen numbers $C_n = n2^n+1$. Try proving that infinitely many C_n are composite. Find some explicit prime C_n, other than C_{141}.

Project 3.2.2 Primes in general

Computation, theory, difficulty level 3-4.

1) Investigate the following interesting computational question: What is the best way to evaluate exactly the integer $\pi(x)$, the number of primes less than x? The value of this prime-counting function is currently known for isolated values up to $x \sim 10^{18}$. In fact M. Deleglise and J. Rivat recently obtained [Odlyzko 1993b]:

$$\pi(10^{18}) = 24739954287740860$$

What is interesting is that, though $\pi(x)$ is known for such large x, we do not have explicit knowledge (e.g., storage) of all the primes so counted. The fact is that algorithms exist which evaluate $\pi(x)$ without direct reference to all of the relevant primes. An original formula due to Legendre relates $\pi(x)$ to $\pi(\sqrt{x})$:

$$\pi(x) = -1 + \pi(\sqrt{x}) + [x] - \sum_{p \le \sqrt{x}} \left[\frac{x}{p}\right] + \sum_{p<q\le\sqrt{x}} \left[\frac{x}{pq}\right] + \ldots$$

where the sums are taken over primes p, q, ... and [ ] denotes greatest integer. A simple example would be $\pi(25) = -1 + \pi(5) + 25 - (12+8+5) + (4+2+1) = 9$, which is correct. One can therefore evaluate $\pi(x)$ recursively, using $\pi(\sqrt{x})$ and some side calculation. Though the Legendre formula is not the most efficient method, it shows elegantly the guiding principle. More advanced formulae due to Meissel, Lehmer, Mapes; and recently by [Lagaris et al. 1985][Lagaris and Odlyzko 1987] are available; and some of the fastest algorithms apparently have not yet been fully implemented [Odlyzko 1993b]. Here are some accepted values useful for testing prime-counting software [Riesel 1985][Odlyzko 1993b]:

x	$\pi(x)$
10^4	1229
10^8	5761455
10^{12}	37607912018
10^{16}	279238341033925
10^{17}	2623557157654233
10^{18}	24739954287740860

An interesting question pertains to the complexity of prime-counting. For example, the Lagaris-Miller-Odlyzko method [Lagaris et al. 1985] requires $O(x^{2/3+o(1)})$ arithmetic operations to yield $\pi(x)$; while this can be improved using newer analytic methods to $O(x^{1/2+o(1)})$ [Lagaris and Odlyzko 1987]. To reach for something much beyond $\pi(10^{20})$ it may be necessary to develop a new $O(x^a)$ algorithm for a sufficiently small a.

2) Another interesting counting problem of sorts involves the Gilbreath conjecture on iterated absolute values of consecutive primes. Let $d_0(n) = p_n$, the n-th prime for $n \geq 1$, and let $d_{k+1}(n) = | d_k(n) - d_k(n+1) |$ for $k \geq 0$, $n \geq 1$. Then the conjecture is that $d_k(1) = 1$ for all $k \geq 1$. It has been shown that the conjecture holds for all $k \leq \pi(10^{13}) \sim 3 \times 10^{11}$ [Odlyzko 1993].

3) Attempt to find Wieferich primes; viz. p such that $2^{p-1} = 1 \pmod{p^2}$. These primes figure into restricted forms of Fermat's "Last Theorem" and also into the theory of primality proving [Cohen 1993]. The only Wieferich primes below *six billion* are p = 1093 and 3511. This task is interesting because C program integer precision is often 32 bits, which is not large enough to exceed six billion. On the other hand, a large Newton-method divide as in "giants.c" has far too much overhead to be optimal on numbers in the region of 10^{10}. Thus one should probably invoke some custom mod routine intended for this scale. The proper way to test numbers p is first to check for probable primality (with a sieve and perhaps a Fermat test: $2^{p-1} = 1 \pmod{p^2}$), then check the Wieferich

relation. However, as [Vardi 1991] points out, if the (mod p^2) check is sufficiently fast, the Fermat test (mod p) is wasteful.

A similar challenge is to find Wilson primes, which would satisfy $(p-1)! = -1 \pmod{p^2}$. Though the same condition just (mod p) is necessary and sufficient for primality of p, the condition with p^2 is rarely met. Known Wilson primes are $p = 5, 13, 563$ and no others below 3×10^6 [Ribenboim 1988].

Incidentally, it is quite often stated that Wilson's criterion ($(p-1)! = -1 \pmod{p}$) is next-to-useless for primality tests. That may be so, but there are certainly some instructive and perhaps illuminating exercises that one may carry out. For example, one can actually prove primality using analytic formulae, in the following, admittedly inefficient way. The idea is to estimate the Gamma function $\Gamma(p)$ from a Sterling-Binet series (1.1.49), then to correct this asymptotic approximation rigorously using knowledge of some primes less than p. Let us prove for example that 101 is prime. We have:

$$(p-1)! \;=\; \Gamma(p) \sim \left(\frac{p}{e}\right)^p \sqrt{\frac{2\pi}{p}}\; \exp\left[\frac{1}{12z} - \frac{1}{360z^3} + \frac{1}{1260z^5} - \ldots\right]$$

If we use this asymptotic formula for $p = 101$, taking the terms within the exponential through z^{-11}, we obtain the approximation:

$$100! \sim x = 933262154439441526816992388502870069581800402\\709116103818998801362343503007700515424510555\\916524090366266959779487521623066439138868290\\1733524428994742777 3689.327\ldots$$

One would think that finding the correct *integer* value of this 100! is hopeless, but something else is known, which amounts to a strong "parity check" of sorts, and which will yield the exact value of 100!. We know that $(p-1)!$ is divisible by:

$$M(p) = \prod_{q<p} q^{\left[\frac{p-1}{q}\right]}$$

where q runs through primes and [ ] denotes greatest integer (in fact even larger powers can be used to define M but the form here suffices for the present example). Thus we align the value of our 100! approximation to equal 0 (mod $M(101)$) by computing, in this case:

$$[x] + M - ([x] \pmod M)$$

for this last expression is 0 (mod M) and furthermore is within the accuracy bound implied by our having taken six terms within the Binet exponential. The claim is that this last expression is exactly 101! Indeed, if we compute the three parts separately (or as a whole) (mod 101) the answer is 100, and by Wilson, 101 is prime.

Of course 100! (mod 101) can be computed with just a small number of rapid multiplies (mod 101). Still, interesting questions are these: how large a p can be checked for primality via Wilson's formula (and direct multiplication), and how large a p can be checked in the way we have here, via the Stirling-Binet expansion together with Chinese Remainder ideas?

4) Investigate regular primes, which are odd primes p such that p does not divide the numerator of any Bernoulli number B_{2k} for $2 < 2k \leq p-3$. These primes figure into Fermat's "Last Theorem" in the following way The equation $x^p + y^p = z^p$ has no positive (x,y,z) solutions if p is regular. In this way, prove Fermat's "Last Theorem" for $p < 37$, say. The prime 37 is irregular, and we count the "irregularity index" as the number of even integers ($2k$) that do divide the corresponding Bernoulli numerator. When p is irregular there is a Vandiver criterion, fairly easy to program [Crandall 1991], that will usually rule out that p, but one needs to know the irregularity indices to apply the criterion. It has been shown in this way, but using fast algorithms (mod p) to avoid literal computation of the radically growing Bernoulli numbers, that Fermat's "Last Theorem" holds for all exponents < 4000000 [Buhler et al. 1993][Buhler et al. 1992]. The frontispiece to this Chapter shows the basic flow of that undertaking, although for $p > 1000000$ an order-12 polynomial multisectioning algorithm (the frontispiece pictorializes order-8 multisectioning) was necessary to keep memory usage under control. The collaborators ran the

regularity computations on about 100 workstations over a one-year period to effect the proof for the stated range of exponents. Naturally, in view of the recent pending proof of the "Last Theorem" by A. Wiles of Princeton University, these numerical results may seem unnecessary, but regular primes are still elusive, important and interesting on their own. For example it is not known if there are infinitely many regular primes. Also, the results on regularity apply to still-open problems and conjectures such as Vandiver's conjecture on class numbers, Siegel's estimate on the density of regular primes, and questions pertaining to cyclotomic invariants.

Elementary experiments on regularity can use the *Mathematica* function BernoulliB[$2k$] to find primes p that do not divide any of the numerators for $2 \le 2k \le p-3$. In this way one finds that the prime $p = 37$ fails (and thus suggests further investigation via the Vandiver criterion).

Another way to proceed is to use the cot() generating function for the Bernoulli numbers (first equality of the frontispiece to this chapter). Two ways to develop the series for cot() (mod p) are as follows. One way is to generate sufficiently many (up to power x^{p-3}) terms of $\cos x$, divided by the same number of terms of $\sin x$. It is this latter division, done rapidly via a polynomial Newton method, that renders numerical attacks for large p possible. An interesting alternative is to use the continued fraction for cot(), obtained from tan() in (1.1.14), and, using the standard recurrence relations for convergents, to compute $\cot x$ (mod p) as a rational polynomial function. Also of interest, independent of (mod p) arithmetic, is to plot the resulting Pade-like approximant to cot(), and observe that in some ways it is superior to the equal-degree cos/sin form. It is not yet clear whether this continued fraction improvement carries over into (mod p) arithmetic in the guise of an enhancement.

5) A question has been asked [Guy 1980] concerning the Euler numbers E_n, which appear as coefficients in the formal expansion of sech(). The question is: Does it ever happen that a prime $p = 1$ (mod 8) divides $E_{(p-1)/2}$? It should be possible to search into the millions for such a p using techniques described in step (4) previous.

6) Find large twin primes (p, $p+2$). This task becomes difficult with modern software and equipment roughly at about the region of 2^{10000}. It is not known whether there are infinitely many such pairs. But it is

known that the sum of reciprocals of all twins *converges*. This follows as a consequence of the twin prime counting function being boundable as, for example, $\pi_2(x) < 100\ x/\log^2 x$. Another interesting problem is to evaluate Brun's constant: the sum of such reciprocals, which sum must exist. Brun's constant, taken as $B = (1/3 + 1/5) + (1/5 + 1/7) + (1/11 + 1/13) + \ldots$ has been computed by Brent [Ribenboim 1988] as:

$$B = 1.90216054\ldots$$

It is an interesting challenge to calculate B, especially because the convergence is so slow. One can conceivably use Brun's bound for π_2 (or sharper bounds established since [Ribenboim 1988]) to estimate rigorously the error in B after some point to which one has carried out the required sum. The following short table established by Brent may be used to test twin-prime counting software:

x	$\pi_2(x)$
10^4	205
10^8	440312
10^{11}	224376048

Large pairs with which to test software include the twins:

$1001 * 1691232 * 10^{4020} \pm 1$	[H. Dubner 1993]
$1706595 * 2^{11235} \pm 1$	[Brown et al. 1989]

This first of these is the largest twin prime pair of which the author is aware. Pursuant to the quote atop this chapter, I trust that such a claim is not interpreted by the reader as sounding too aggressive.

A straightforward method for finding twin prime pairs runs as follows:

- Sieve some region of integers for twin prime candidates, by cancelling whole pairs for which at least one of the pair is divisible by a small prime;
- Apply pseudoprimality tests, to further reduce the population of candidates;
- On the remaining candidates apply strict primality tests.

Because of the difficulty of the last step when the twin members have say >1000 decimal digits each, it is fashionable to choose numbers of the form $h2^q \pm 1$ as in step (2), Project 3.2.1; or of some other form such that primality proofs are not too troublesome.

7) Find long arithmetic progressions of primes. The longest of which the author is aware is the set of 21 primes $\{142072321123 + 1419763024680\ n : n = 0,\ldots,20\}$, which lists like so:

```
142072321123
1561835345803
2981598370483
4401361395163
5821124419843
7240887444523
8660650469203
10080413493883
11500176518563
12919939543243
14339702567923
15759465592603
17179228617283
18598991641963
20018754666643
21438517691323
22858280716003
24278043740683
25697806765363
27117569790043
28537332814723
```

Note that the common difference of the progression is

$$1419763024680 = 2^3 \cdot 3 \cdot 5 \cdot 7 \cdot 11 \cdot 13 \cdot 17 \cdot 19 \cdot 23 \cdot 37 \cdot 43$$

If one contemplates a search for such progressions, one should first prove that every prime not exceeding the total number of progression terms must divide the common difference. In this case that means every prime less than 21.

8) Step (5) of Project 3.1.1 was to test Fermat's little theorem, that if p is prime and p does not divide a then $a^{p-1} = 1 \pmod p$. As we said, this relation can be used to find composite p (when said relation fails even for just one such a), it cannot be used as is to prove primality. In fact, there exist Carmichael numbers C having the property that for every incidence of GCD(a, C) = 1, we have $a^{C-1} = 1 \pmod C$.

One Carmichael number is $561 = 3 \cdot 11 \cdot 17$. Find others. It has recently been proved by Alford, Granville, and Pomerance that there are infinitely many Carmichael numbers; and that up to x there are at least $O(x^c)$ for some $c \sim 1/10$ [Cohen 1993].

A converse problem is to show that some large number is indeed a Carmichael number. To this end, show that

949803513811921

is a Carmichael number and give its factorization.

9) Implement primality proving software which will determine the character (prime or composite nature) of an arbitrary integer. For relatively small (say 30-digit and below) numbers N, one may simply base an elementary prover on the prime factors of N–1 [Hardy and Wright 1978]. Obviously this requires factoring of N–1 and, though there is a similar approach if N+1 (or one of a few other forms) can be factored, or partially factored, one is probably in trouble if N has a few hundred digits. But with modern methods it is now possible to determine absolutely the character of N in the 1000 digit region and beyond. A modern example is Morain's proof of the primality of $(2^{3539}+1)/3$ [Morain 1990]. An efficient approach to determine the character of large N is first to use the Rabin-Miller probabilistic test, and if N passes that test (i.e., has not been shown composite), to invoke a rigorous algorithm such as the Adleman-Pomerance-Rumely-Cohen-Lenstra (APRCL) test. There are also modern elliptic curve-based tests due to Adleman and Morain. Old and new tests are described in the excellent course survey by [Cohen 1993].

Here are some interesting classes of numbers for primality testing:

- Mersenne numbers 2^q-1 (use Lucas-Lehmer test of Project 3.2.1);
- Fermat numbers $2^{2^n}+1$ (use Pepin test of Project 3.4.2);

- Repunit numbers, such as decimal 11111111111111111111111, generally of the form $(10^q-1)/9$. These are a kind of decimal analog to the Mersenne numbers. The largest known repunit to this author's knowledge is Williams' prime ($q = 1031$), which happens to live just about at the boundary of modern provers (although one should always be on the lookout for easier tests for numbers of such special form);
- The numbers $n! \pm 1$, and "prime factorials" $2 \cdot 3 \cdot 5 \cdot 7 \cdot ... \cdot p \pm 1$, for which factorization of one of the pair (candidate $\pm$ 1) is trivial;
- Fibonacci numbers and Lucas numbers;
- Huge random integers that have passed a sieve and then passed Fermat's test.

10) Implement (mod p) polynomial arithmetic. Such software is useful in, amongst other areas, the study of regular primes and Euler numbers respectively as in steps (4), (5) previous. These polynomial operations are interesting and instructive. For example, your software should verify that if $f(x)$ is a polynomial with all coefficients evaluated (mod p), then:

$$f(x)^p = f(x^p) \pmod p$$

Consider proving this identity. One interesting result of the relation is that partial (mod p) inverses of polynomials can be obtained without multiplication. To this end, show that if $f(x) = 1 + ax + ...$, then $f(x)^{p-1}f(x) = 1 + O(x^p) \pmod p$, so that the inverse series $1/f(x) \pmod p$ equals, up through order x^{p-1}, the power $f(x)^{p-1}$.

11) Investigate numerically the celebrated Goldbach conjecture, which states that every even number is the sum of at most two primes. This has been checked for all even numbers $\leq 4 \times 10^{11}$ [Sinisalo M K 1993]. Decades ago Vinogradov showed that every sufficiently large odd integer is the sum of *three* primes, and that it is known that "sufficiently large" can be taken to mean "$> 3^{3^{15}}$." This is a tantalizing sufficiency limit, since a somewhat smaller exponent would possibly be accessible computationally! One can imagine a scenario for future decades in which some more analytical effort, to reduce the sufficiency bound, might "meet" a computational effort coming from the ground up, as it were. Such an effort would settle once and for all the three-prime Goldbach case. As for the original Golbach case of sums of at most two primes, note that modern efforts involve interesting forms of sieving. In fact the

Goldbach problem is a sieving problem–both theoretically and computationally–*par excellence*.

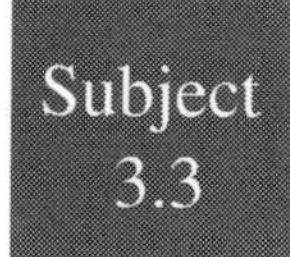

Fast algorithms

Here we turn to the application of fast algorithms to number-theoretic computation. We make use of some of the FFT concepts and results found in Chapter 4.

Project 3.3.1

Fast multiplication

Computation, theory, graphics, difficulty level 2-3.
Support code: Appendix "giants.[ch]."

1) Implement a Karatsuba recursion to multiply two integers x,y of N digits each in $O(N^{\log 3/\log 2})$ arithmetic operations. One uses the identity:

$$(a + bW)(c+dW) = ab + [(a+b)(c+d) - ab - cd]W + cd\,W^2$$

Now if W has about $N/2$ digits, and each of x, y is represented in base W, the right-hand side of the above identity thus involves three multiplications of numbers having about $N/2$ digits each. Recursion on this idea leads to the $O(N^{\log 3/\log 2})$ complexity. Note that the Karatsuba scheme is reminiscent of the Strassen matrix recursion of Chapter 1.

One possible software design option is to use Karatsuba to "fill in the gap" between $O(N^2)$ grammar-school schemes and the truly fast schemes we next discuss.

2) Here we investigate an algorithm for FFT multiply. The ideas are covered in [Crandall and Fagin 1994] and in an elementary style in

[Crandall 1991], but a self-contained overview runs as follows. The basic algorithm is to perform a convolution of two digit sequences thought of as "signals," which convolution, done properly, results in the multiplication of the signals. Assume integers x,y have digit expansions:

$$x = \sum_{j=0}^{N-1} x_j W^j \quad ; \; y = \sum_{j=0}^{N-1} y_j W^j$$

Then the convolution of signals $\{x_j\}$ and $\{y_j\}$ will be the product xy *provided* at least the upper half of the digits of each signal all vanish. The algorithm runs, then:

- Represent x,y in base W, with both representations zero-padded such that $x_j = y_j = 0$ for $j \geq N/2$;
- Compute via FFT the discrete Fourier transforms X and Y respectively. Each transform is now a floating-point (complex) signal having N elements;
- Compute the product transform $Z = \{X_k Y_k\}$;
- Take the inverse FFT of Z, call it z, which will be an approximate representation of digits for the integer product xy;
- Use a round() function to obtain the elements of z as real integers;
- Adjust the digits of z with add-and-carry to the original base representation W.

Normally one chooses N a power of two, unless of course some efficient non-power-of-two FFT is applicable. Let us run through an example. Let

$$x = 555221453977032891233$$
$$y = 442186118023559174$$

for which we shall use $N = 16$ and base $W = 1000$. Then appropriately zero-padded digit signals are:

$$x = \{233,891,\ 32,977,453,221,555,0,0,0,0,0,0,0,0,0\}$$
$$y = \{174,559,\ 23,118,186,442,\ 0,0,0,0,0,0,0,0,0,0\}$$

The elementwise product of the two FFTs turns out to be:

```
Z = {5049724., -1593431.85 - 2260339.71 i,
  -115818.41 + 90835.74 i,
  121347.57 - 251761.03 i, -85842. - 132912. i,
  -531409.48 - 309997.80 i,
  -241581.59 + 19475.74 i,
  -54078.23 + 778131.53 i,
  600576., -54078.23 - 778131.53 i,
  -241581.59 - 19475.74 i,
  -531409.48 + 309997.80 i,
  -85842. + 132912. i,
  121347.57 + 251761.03 i,
  -115818.41 - 90835.74 i,
  -1593431.86 + 2260339.71 i}
```

The inverse transform is obtained, after rounding to real integers:

```
z = {40542, 285281, 508996, 235873, 774177, 586640,
    745588, 564648, 554935, 306822, 200912, 245310,
     0, 0, 0, 0}
```

These digits have generally overflowed beyond $W = 1000$, but are patently correct in the sense that the desired product xy is the sum of $z_j 1000^j$. But it is usually necessary to perform carry-add in the following way:

```
                                            40 542
                                       285 281
                                   508 996
                               235 873
                           774 177
                       586 640
                   745 588
               564 648
           554 935
       306 822
       200 912
   245 310
___________________________________________________
245 511 219 377 500 394 175 414 413 382 281 321 542
```

Sure enough the final integer product is `555221453977032891233` * `442186118023559174` as desired.

Implement such a scheme and find a digit size at which the FFT method beats the direct, grammar-school multiply. The "giants.c" routines, for example, are designed to switch over to FFT multiply at about 2000 decimal digits for each of x and y (assumed equal-sized).

3) Investigate "balanced representation," a means by which the author and collaborators have been able to significantly increase the sizes of multiplicands before FFT floating-point errors dangerously accrue [Crandall and Fagin 1994]. The idea is that digits in an even base W no longer range from 0 through $W-1$ inclusive, but from $-W/2$ to $+(W/2-1)$. This is a currently mysterious non-linear "low-pass filter" on the "signals" of digits; which filter evidently keeps rounding errors well below the errors for standard digit representation. This is one way to control error, but another is to stay with integer arithmetic and avoid all floating-point error altogether. The next two steps describe ways to effect pure-integer fast large arithmetic. One should beware, however, that the floating-point methods are quite competitive, especially in some of the more modern supercomputers engineeered with floating-point efficiency in mind.

4) Implement a fast transform multiply using not floating-point FFTs, but some number-theoretic transform as discussed in Chapter 4. The obvious advantage to such an approach is to keep all arithmetic in integer form, obviating the need to round approximate numbers to integers. One option is to invoke a Galois transform (mod p) with all digits considered elements of $GF(p^2)$ as in step (8), Project 4.4.1. If the convolution that results in a digit of the product integer gives the digit sufficiently large range, another option for pure-integer transform is to use CRT arithmetic as in step (5) of that Project, in order to obtain sufficient precision. In such cases the Chinese remainder primes would be chosen so as to allow repsective primitive roots of unity of order equal to the desired run length.

5) Implement recursive Nussbaumer convolution [Nussbaumer 1981] to provide either cyclic or negacyclic convolution of digit signals and thus to provide another fast large-integer multiplication. This method is interesting because of its theoretical elegance (ring algebra isomorphisms), its pure-integer aspect (no floating-point requirement), and its efficient use of memory [Buhler 1993]. Ironically enough, in

actual experiments on the cataloging of regular and irregular primes [Buhler et al. 1993], it was found that inner convolutions at the bottom of recursion (where precision is not too problematic) could be conveniently effected with floating-point FFT.

We next turn to some interesting exercises pertaining to implementations of fast operations of the division class.

Project 3.3.2

Fast mod, division, and inversion

Computation, a little theory, difficulty level 2-4.
Support code: Appendix "giants.[ch]."

1) Argue that if p is a number $2^q \pm 1$ (not necessarily prime), then for $x \geq p$ the calculation $x \pmod p$ can be effected rapidly as follows. Represent x in the form $a + b2^q$, and note that $x = a - (\pm b) \pmod p$. One continues this reduction, reducing at each step the data representing $x \pmod p$ by q bits. Thus, the mod operation can be effected with shifts and add/subtracts alone. Implement this notion and observe the radically faster mod operation that should result.

These fast methods work well with Mersenne (2^q-1) or Fermat ($2^{2^n}-1$) numbers. Work out in addition the fastest way to multiply a number by 2^k, as would be required in the Mersenne- and Fermat-number transforms of Project 4.4.1.

2) Implement a Newton method divide for large integers. The procedure is intricate, but uses the same basic iteration for "divisionless divide" we encountered in Chapter 1. The method is explained in [Knuth 1981] and a detailed implementation is found in the source code "giants.c." Find empirically at what integer size a grammar-school long-division scheme is overtaken by this Newton method. Incidentally, in practice one never needs more than about 20 large-integer multiplies for such a divide, no matter how large the numerator and denominator.

3) Implement a division scheme using FFTs that applies when the

denominator, call it N, remains fixed over many divisions. For example, in a factoring run one may have to take a vast quantity of mod operations all with respect to a the fixed N to be factored. The main idea is to store the FFT of the digits of a *reciprocal* of N [Crandall and Fagin 1994]. Thus, when the division denominator is fixed, a divide time can be brought down very close to twice a multiply time, or even a little better than this.

4) Greatest common divisor (GCD) computations can be effected, as has long been known, with a straightforward Euclid algorithm. This method involves long division, but there is the advantage discovered by Lehmer that the dividend is statistically small [Knuth 1981]. On the other hand, one can effect a GCD with binary shifts alone and no long division. The functions gcdg(), binvg() in "giants.c" effect respectively the traditional and binary-shift GCDs.

5) The classical method of computing inverses (mod N) is the "extended" Euclid algorithm [Knuth 1981]. As with step (4) previous, this uses explicit long division. There is a binary-shift method for inverses as well that does not use explicit division. The routines invg(), binvg() of "giants.c" implement respectively the traditional extended method and the binary-shift method. Test these inversion methods and assess empirically their computational complexities. Note that, akin to step (3) previous, there is a parallel inversion scheme due to Montgomery, intended for finding multiple inverses (mod N) at once [Cohen 1993]. Such a scheme is useful in factoring algorithms that call for many inversions modulo the number being factored (e.g., in the ECM factorization scheme).

6) Find a method for fast inversion of x (mod p) where p is a Mersenne prime. By "fast" in this case we mean empirically faster than the traditional extended Euclid algorithm. The source "giants.c" has a routine mersenneinvg() that implements one such method.

We complete our tour of fast algorithms with two special cases that prove useful, especially in factoring problems (GCD) and parallism (CRT) for very large integers.

Project 3.3.3

Other fast algorithms

Computation, graphics, theory, difficulty level 3-4.
Support code: Appendix "giants.[ch]."

1) Implement a large-GCD along the lines of [Aho et al. 1974]. One such is found as the function ggcd() in "giants.c" as originally communicated to the author by [Buhler 1990].

2) Implement a parallel multiply scheme using the Chinese remainder theorem (CRT), in which numbers x and y are known in terms of their residues (mod p_i), where the p_i are relatively prime by pairs and fixed. Using these small residues, the products xy (mod p_i) are formed, then using all of these product residues, one can reconstruct the desired full product xy. There is even a fast pre-conditioning algorithm for the reconstruction [Aho et al. 1974][Crandall 1991]. Such a scheme has promise for parallel multiplication when many processors are available.

It is intriguing to consider combining CRT methods with FFT methods. For example, express each of the *digits* of x and y (mod p_i), and perform number-theoretic transforms (mod p_i). Then reconstruct each digit of the desired product. For this to be effective, the digits would have to be larger than just computer words. There should be some interesting optimization analysis possible here. The author attempted to work out the Pepin test for the twenty-second Fermat number (Project 3.4.2) using this parallel-digit concept. An integer x to be squared (mod F) would be split into large digits (say 1024 bits each), then various processors would compute each digit of x^2, but represent each digit (mod p_i), one i per processor. Then CRT reconstruction would be used on the digits themselves to obtain x^2. This scheme was dropped in favor of mainframe or supercomputer runs without parallelism, but only because of network overhead in shuttling digits around between machines. If network communication were infinitely fast, it is evident that powering in general could be parallelized in this fashion. The author does not yet know, however, if the speedup for N processors is linear in N.

Here is an interesting question: How can one *divide* using methods appropriate to the CRT? One attempt of interest is [Lu and Chiang 1992].

3) Implement a fast polynomial multiply (mod p) where the prime p is given. Say x and y represent polynomials (mod p) in that each of the coefficients x_i or y_i is an integer between 0 and p–1 inclusive. Then we can multiply using FFT techniques as in Project 3.3.1, except that the final rounding operation is just to reduce all coefficients (mod p) and no carry adjustment is required.

As pictorialized in the frontispiece to this chapter, such a (mod p) multiply (via Nussbaumer convolution combined with FFT in a Newton iteration) was used to find which Bernoulli numbers B_{2k} are divisible by p, for all $2 \le 2k \le p-3$.

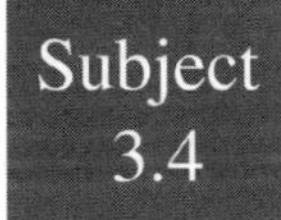

Factoring

Factoring is in a sense the "canonical hard problem" of computational number theory. Below we investigate some factoring algorithms ranging from simple-to-implement to very-recently-discovered. This chapter on large-integer arithmetic was largely motivated by problems pertaining to integers having more than 10000 digits, perhaps even 1000000 digits, but many open factoring problems involve *much* smaller numbers. For example, a recent "Most Wanted" list of numbers still not completely factored is as follows, listed in descending order of perceived importance [Wagstaff 1993]:

$2^{1024}+1$ (also known as F_{10} . . . a composite 291-digit factor remains)
$5^{256}+1$ (134-digit composite remains)
$2^{511}-1$ (123-digit composite remains)
$2^{521}+1$ (140-digit composite remains)
$3^{319}-1$ (119-digit composite remains)
$10^{149}+1$ (123-digit composite remains)
$10^{163}-1$ (138-digit composite remains)
$11^{127}+1$ (120-digit composite remains)

$7^{169}-1$ (132-digit composite remains)
$5^{206}+1$ (143-digit composite remains)

The strongest modern algorithms (quadratic sieve, elliptic-curve method, and number field sieve) evidently have been unable as yet to resolve any of these ten numbers. It seems fair to say that, these days, ~100 digits is the current effective limit of *systematic* factoring. It is still instructive and rewarding to find factors of much larger numbers. For example there are no factors known of F_{14}, F_{20} or F_{22} and even a small (say 40-digit) factor of any of these would be exciting to find.

Two important concepts that underlie many of the most efficient modern factoring algorithms are as follows. First, there is the concept of *smoothness*. We say that an integer x is P-smooth if the prime factorization of x involves no primes $> P$. Many of the asymptotic complexity results for factoring can be derived from the heuristic approximation:

$$Prob\{x \text{ is } P-\text{smooth}\} \sim \left(\frac{\log x}{\log P}\right)^{-\frac{\log x}{\log P}} \tag{3.4.1}$$

Thus, for example, the probability that, for a random integer of order 10^{100}, each prime factor is 25 digits or less is about 4^{-4}. When a factoring algorithm involves a choice of smoothness limit P and a basic scale size x (which may or not be known *a priori*), and if factorization of N requires # occurrences of P-smooth integers of that size x, each occurrence costing C, the expected cost to factor is approximately:

$$T \sim C \left(\frac{\log x}{\log P}\right)^{\frac{\log x}{\log P}} \# \tag{3.4.2}$$

If N is a number to be factored, both x and P should be optimized if possible to minimize the total cost T.

The second important concept is that, to factor N, one could find a non-trivial congruence (non-trivial meaning $x \neq \pm y \pmod N$):

$$x^2 = y^2 \pmod{N} \qquad (3.4.3)$$

then $GCD(x \pm y, N)$ can be tested for factors of N. It turns out that such non-trivial square relations can be found by first finding enough random *smooth* squares; i.e., x^2 = (smooth residue) (mod N), with fixed smoothness limit P. In fact, one needs no more than about $\pi(P)$ such independent congruences. In this way the complexity analysis of most –but not all–modern factoring methods depends on the connection between smoothness and small squares (mod N).

Project 3.4.1

Factoring algorithms

Computation, theory, difficulty level 3-4.
Support code: Appendix "ContFract.ma."

1) Implement a Pollard rho factoring algorithm, elementary discussions of which are found in [Vardi 1991][Crandall 1991]. The basic idea, given N to be factored, is to start with integers a (such as, say, $a = 3$) and x_0 and iterate:

$$x_{n+1} = x_n^2 + a \pmod{N}$$

Say that some mystery p divides N. Eventually there will be a completion of a cycle, i.e., an occurrence of $x_j = x_k$ (mod p) for $j > k$. This p can generally be uncovered using greatest common divisor. In fact GCD(N, $x_j - x_k$) will likely produce the solitary p. But how does one find the cycle? One answer lies in the Floyd cycle-finding algorithm, which says that eventually $x_{2m} = x_m$ (mod p) for some m. So here is the basic algorithm:

- Choose a, x_0 (e.g., $a = 2$, $x_0 = 3$);
- Set an initial $w_0 = x_0$ and iterate three assignments every loop pass:

$$\begin{aligned} x &:= x^2 + a \pmod{N} \\ w &:= w^2 + a \pmod{N} \\ w &:= w^2 + a \pmod{N} \end{aligned}$$

- Accumulate (mod N) the product Q of all the (x–w) obtained after each set of three iterations;

• From time to time check $GCD(N, Q)$, hoping for a factor of N.

The success of the Pollard rho scheme can be thought of as a consequence of the "birthday paradox," which says that in a room of, say, 23 people, the probability of a common birthday is about 1/2. The point is that one may imagine dropping 23 objects into 365 boxes, so in the object-box picture it is not so intuitively shocking that a box is fairly likely to have >1 object. Likewise, the iteration $x^2+a \pmod p$, where p is an unknown factor, is a "drop" into the residues (mod p). If the iteration lands on a previously struck position, this will be detected by the algorithm. Many speed enhancements of this basic algorithm are possible, as explained in [Vardi 1991]. The basic heuristic complexity estimate is that the time to find a factor p of N is:

$$O(p^{1/2})$$

whose worst-case is $O(N^{1/4})$. This method is very effective for 10- to 30-digit numbers N, beyond which one is hard pressed to find the CPU cycles necessary to perform $N^{1/4}$ iterations of the Pollard loop.

When N to be factored has special form, there are theoretical enhancements. In fact if factors of N are known to be of the form $kM+1$, it makes sense to iterate instead $x := x^M + 1 \pmod N$. In fact the original factorization of F_8 used $M = 1024$. The expected time to factor is increased by $\sqrt{M}$, while the extra time to power up to the M-th power only grows logarithmically with M. It is even possible to "hope for" a mystery factor = 1 (mod M) and just guess a fairly composite power M.

The author does not know, but would like to know the answers to questions such as: What is the speedup factor if K parallel machines each start with a different x_0? A different a? The answer is presumably $\sqrt{K}$, on statistical grounds. But here is a harder question: what if various *powers* M are given to various machines?

2) Implement a Pollard $(p-1)$ factoring method. This is one of the easiest to program. The idea is to choose an integer a and raise it to a highly composite power $M = 2^b 3^c \ldots$, and then check:

$$GCD\left(a^M - 1, N\right)$$

and hope for this to be a factor of N. The method works spectacularly well when the (unknown) factor p of N happens to enjoy the property that $p-1$ has no large primes in its own factorization. Furthermore, there are radical enhancements possible when $p-1$ is comprised of small factors plus a *solitary* final large factor. One then implements a "second stage" to the algorithm [Montgomery 1987].

A software implementation can be tested by, for example, re-enacting Gostin's discovery of a factor $p = 2327042503868417$ of the 15-th Fermat number $2^{2^{15}} + 1$. The point is,

$$p-1 = 2^{17} \cdot 14431 \cdot 1230263$$

so the solitary final prime factor is conveniently small.

There is an elegant FFT-based extension to the $(p-1)$ method due to [Montgomery and Silverman 1990]. This method involves a fast algorithm for accumulating special cases of the general form:

$$\prod_{j \neq k} (x_j - x_k)$$

It should be no surprise that there exist fast algorithms for computation of this product. For one thing, the multiplicands are far from independent; e.g. $(x_a - x_b) - (x_c - x_b) = (x_a - x_c)$. When the integers involved are sufficiently large, there is also the additional enhancement of accumulating such products by storing the FFTs of the stated x_j and forming the product from these FFTs [Crandall and Fagin 1994]..

3) Implement a continued fraction factoring method. The basic idea is to generate small squares (mod N) (which will, by being small, often be smooth) using the fact that if p_n/q_n is the n-th convergent approximation for the simple continued fraction (the form (1.1.10) but with all b numerators = 1) of $\sqrt{(kN)}$ for positive integer k, then

$$\left| \frac{p_n}{q_n} - \sqrt{kN} \right| < \frac{1}{q_n^2}$$

from which it follows that the least positive residue $p_n^2 \pmod N$ is typically of order $N^{1/2}$. It is instructive to note at this point that a naive random square would typically be of order N itself.

Let us work through a concrete example: the factorization of $N = 2^{29}-1$. It should be noted that pure-integer arithmetic can in principle be used to find continued fractions for irrational reals $\sqrt{(kN)}$ [Cohen 1993], but for this small example N one can easily solve the factoring problem via numerical floating-point fraction routines. Take $k = 1$, and calculate (using, for example, Appendix code "ContFract.ma"):

$$\sqrt{2^{29}-1} = 23170 + \cfrac{1}{2+\cfrac{1}{9+\cfrac{1}{2+\cfrac{1}{39+\ldots}}}}$$

From a list of the p_n one finds three useful cases where $p_n^2 \pmod N$ is relatively smooth:

$$258883717^2 = -2 \cdot 3 \cdot 5 \cdot 29^2 \pmod N$$
$$230618656999^2 = -3 \cdot 5 \cdot 11 \cdot 79 \pmod N$$
$$1175120510305245312^2 = 2 \cdot 3^2 \cdot 11 \cdot 79 \pmod N$$

The goal in such analyses is to end up with a composite square that factors into primes of all even powers. The above three cases were chosen so that the composite square is merely the product of the three squares:

$$(258883717 \cdot 230618656999 \cdot 1175120510305245312)^2 =$$

$$2^2 \cdot 3^4 \cdot 5^2 \cdot 11^2 \cdot 29^2 \cdot 79^2 \pmod N$$

Now we have a congruence $x^2 = y^2 \pmod N$ and proceed to test $GCD(x \pm y, N)$, hoping for a factor of N. In the present case $GCD(x+y,N) =$ 1103, which leads to the factorization $2^{29}-1 = 233 \cdot 1103 \cdot 2089$.

For larger numbers, such as the seventh Fermat number factored by Brillhart and Morrison in 1974 (see Project 3.4.2), one usually cannot find *visually* a product of squared convergents that yields all prime factors with even powers. In such cases one must store "vectors" of primes, one for each of the squared convergents (mod N), together with these primes' powers (even/odd flag will do) and pursue a Gaussian reduction technique on the "vectors" of binary power flags, to end up with a product representation $x^2 = y^2 \pmod N$. The common practice is to establish a smoothness bound P, and only look at convergents p_n whose squares (mod N) are P-smooth. One then hopes for a total of $O(\pi(P))$ such smooth convergents, and proceeds to reduce into a final square relation.

When square-relation reduction is done sufficiently well, the continued fraction method evidently enjoys complexity (expected number of arithmetic operations to factor N) [Riesel 1985]:

$$e^{(1+o(1))\sqrt{\log N \log\log N}} \tag{3.4.4}$$

which can be derived from the heuristics associated with the smoothness and complexity estimates prior to this Project. In formula (3.4.2) one may estimate $x \sim \sqrt{N}$, $C = 1$, and $\# \sim P$, say, then optimize the compute time T with respect to smoothness limit P. It should be noted, however, that methods with the same formal heuristic complexity can be much faster in practice. For example the quadratic sieve of step (5) below essentially generates *pre-factored* small squares (mod N), while in the continued fraction method we are obliged to factor the small squares. Thus, when the number N to be factored is much more than about 50 digits, one runs into trouble because ≥ 25 digit "small" square residues need be factored. One is essentially forced, then, to adopt an "early abort" strategy, in which residues that do not factor into a fixed base (or factor into said base with at most one prime extant) are ignored.

4) Implement an elliptic curve (ECM) factoring method. For the theory of elliptic curves, see the very readable account [Koblitz 1987]. A good factoring algorithm reference which also contains enhancements and optimization information is [Montgomery 1987]. The present author prefers the Montgomery parametrization of ECM in that no inversions are required, at the expense of more multiplies. There is also a way to use FFTs to speed up the ECM [Montgomery 1992], and a different FFT

method to effect more rapid elliptic curve algebra [Crandall and Fagin 1994]. This, the continued fraction method previous, the quadratic sieve method next, and several other methods all enjoy worst-case complexity to factor N as (3.4.4). But the ECM method's special efficacy depends on the magnitude of the smallest prime factor of N, so in terms of the smallest prime factor p a more refined estimate is:

$$e^{(1+o(1))\sqrt{2\log p \log\log p}}$$

whose worst case ($p^2 \sim N$) gives the previous complexity form. Thus, the ECM method is especially effective when there is an unknown but relatively small prime factor, or when N has no particular algebraic form (since in such cases some other methods can sometimes be enhanced by exploiting that form).

A straightforward implementation of ECM by I. Vardi is found as the package NumberTheory/FactorIntegerECM in most v2.0 versions of *Mathematica*. A very elementary program is exhibited in [Crandall 1991]. Like the Pollard (p–1) method, there is a second-stage enhancement for ECM which markedly improves efficiency [Montgomery 1987].

There are also enhancements by which one chooses elliptic curves adroitly, as opposed to just randomly [Atken and Morain 1993].

5) Implement a quadratic sieve (QS) method [Pomerance et al. 1988][Riesel 1985]. This method class, together with the ECM class of the previous step, is responsible for many of the "most wanted" factorizations of the 1980s and 1990s [Wagstaff 1993]. The theoretical worst-case complexity of the QS methods is essentially the same as the ECM complexity previous; but these methods vary in practical effectiveness according to the precise nature of the factors of N. Furthermore, QS remains superior to ECM for numbers having something like two 50-digit factors (ECM rarely finds factors with more than 35 digits, but for 30 digits and below ECM is a healthy choice).

One approach to state-of-the-art systematic factoring is, given N:

- Hit N with sieves, or Pollard rho, or Pollard $(p\text{–}1)$ methods for a little time; say, until factors of $\leq 10^{10}$ are ruled out;
- Invoke ECM until factors $\leq 10^{30}$ are ruled out;
- If N has roughly 80-150 digits, invoke QS, MPQS, etc.;
- Perhaps invoke NFS, discussed in step (6) next.

Note that some investigators might disagree with this overall recipe; for example, it is often said that the $(p\text{–}1)$ method is akin to just one ECM pass, and so one may as well spend any extra valuable computation time on the ECM. However, if N has specific form, one may elect to stick with one of the more elementary options. For example, if it is suspected that unknown prime factors are of the form $C+1$ where C is a relatively smooth composite, the $(p\text{–}1)$ method is especially efficient. Note that any factors of Mersenne or Fermat numbers enjoy this property. For another example, only ECM was used (for many weeks on hundreds of machines) to find new factors of F_{13} as mentioned in the next Project–the QS class and NFS currently being hopeless at that size of N–while Pollard rho and $(p\text{–}1)$ methods have been, in practical terms, pretty much exhausted for F_n with $n < 18$.

6) Implement a number field sieve (NFS) factoring method. Good references are [Lenstra and Lenstra 1993][Cohen 1993]. The asymptotic complexity of the basic algorithm represents a breakthrough over the handful of pre-NFS methods:

$$e^{(c+o(1))\log^{1/3} N\,(\log\log N)^{2/3}}$$

An implementation of NFS finally conquered the ninth Fermat number as mentioned in the next Project. The NFS is more effective than any method for numbers exceeding roughly 120-150 digits, and is also competitive for numbers smaller than this, in certain cases. The lucrative case for the NFS is when a multiple of the number N to be factored can be written as a polynomial in powers of $[N^{1/d}]$, with d being a small integer such as 4,5; and furthermore when the coefficients of this polynomial are small. The Cunningham numbers such as $2^{428}+1 = ((2^{86})^5+4)/4$ admit of such representations convenient for NFS algorithms, which is how R. Silverman and others have had success factoring certain Cunningham numbers >100 digits in this way [Wagstaff 1993].

One might consider as an initial implementation the factorization of $F_7 = 2^{128} + 1$. One might take, for example, the polynomial representation $F_7 = M^4 + 1$, $M = 2^{32}$. Then the primary NFS task, as the references will explain, is to set a smoothness limit P and then to sieve for relatively prime pairs (a, b) such that both

$$a + bM$$
$$a^4 + b^4$$

are P-smooth. One needs to find about $2\pi(P)$ such pairs in order to proceed with the NFS algorithm after the sieve. A typical workstation with efficient sieve program can generate several hundred such pairs per hour, for $P \sim 1000$. In this way the NFS brings the F_7 factorization task down to a "one sitting" problem.

To close this chapter we turn to a topic which, as the author and collaborators have found, is a superb testing ground for large-integer arithmetic, new fast algorithms, and number-theoretical software in general.

Project 3.4.2 Status of Fermat numbers

Computation, theory, difficulty level 2-4.
Support code: Appendix "FermatConvolution.ma."

Investigate, using available software and aforementioned (or not mentioned!) techniques, the status of Fermat numbers:

$$F_N = 2^{\left(2^N\right)} + 1$$

Fermat conjectured that these are all primes, but in fact only the first few (F_0 through F_4) are prime and no other Fermat primes are known. It is furthermore known that for $N \geq 2$ any factor of F_N must have the form:

$$f = k\,2^{N+2} + 1$$

which is certainly useful as one attempts to "sieve out" possible factors. A rigorous primality test known as Pepin's test says that F_N is prime if and only if:

$$3^{(F_N - 1)/2} \; = \; -1 \pmod{F_N}$$

Evidently one may compute the Pepin residue via successive squarings. Like the Lucas-Lehmer test for Mersenne primes, the number of squarings required for a proof is roughly the number of bits in the number whose character is under test.

Data on the character of F_m for various m reaching into the thousands is available [Riesel 1985][Keller 1992]. The present author's status table below represents current knowledge for all $m < 23$, but many other interesting data are known. For example, the monstrous number

$$F_{23471}$$

(which has how many decimal digits?) was shown to admit the factor $5 \times 2^{23473} + 1$, by Keller in 1984 [Riesel 1985]. It is instructive to prove this in software, and perhaps to find even more impressive monsters.

The Appendix code "FermatConvolution.ma" implements various DWT run-length reduction techniques for transform-based arithmetic modulo Fermat numbers. Note that these fast algorithms apply not only to primality testing via the Pepin test, but also to the factoring problem for F_N, since for any known factoring method, many multiplications (mod F_N) are required.

The fast multiplication algorithms we have outlined have additional enhancements when arithmetic is to proceed modulo a Fermat number. Specifically, the multiplication xy can be effected precisely as the negacyclic convolution of the digit signals, thus reducing the required FFT run-lengths by 2 or even 4. These algorithms are special cases of the discrete weighted transform (DWT) [Crandall and Fagin 1994], also mentioned in step (5), Project 4.2.7. Such methods have enabled the author to find in 1991, with an ECM factoring method based on DWT arithmetic, two new factors of F_{13}; and collaborators J. Doenias, C. Norrie, and J. Young to establish small pieces of the following Fermat number status list. In particular, Norrie implemented some DWT code, written by the present author for the Pepin test, on an Amdahl mainframe during much of the 1993 calendar year, in an assault on the character (prime or composite) of F_{22}. After perhaps 10^{16} arithmetic operations, this Fermat number has finally been proven composite.

A striking modern factorization achievement is the Lenstra-Menasse final factorization of F_9, obtained via an implementation of the number field sieve (NFS). The account of this adventure [Lenstra et al. 1993] is good reading.

It is instructive to re-enact the historical methods used at various places in the legend. Methods of interest are Pollard rho method (to resolve F_8), Pollard (p–1) method, and elliptic curve method (ECM) (to resolve Brent's factors of F_{11}). Straightforward sieving, using the special form for possible factors explained above, is responsible for many of the initial factors of the larger F_N.

Concerning Fermat numbers there are many entertaining theoretical questions; some of which are assailable, while some seem impossible. Here is one solvable problem: show that no Fermat number can be a prime power p^k, $k > 1$; by proving that the Diophantine equation:

$$b^k - 4^m = 1$$

has no solutions for $k > 1$ and odd b. But a seemingly impossible problem is: show that there are finitely many prime Fermat numbers. Not only does this problem seem beyond the reach of current mathematics, but the computational results (such as the complicated factorizations and population of composite cofactors) displayed in the following status table suggest that, conceivably, there may be, beyond $F_4 = 65537$, *no more* prime Fermat numbers at all!

Status of Fermat Numbers F_N, $N \leq 22$

```
P = a proven prime;  C = a proven composite

F0, F1, F2, F3, F4 = P
F5  = 641 * 6700417           [Euler 1732]
F6  = 274177 * 67280421310721 [Landry, Le Lasseur 1880]
F7  = 59649589127497217 * 5704689200685129054721
                              [Morrison, Brillhart 1974]
F8  = 1238926361552897 * P    [Brent, Pollard 1980]
F9  = 2424833 *               [Western 1903]
      7455602825647884208337395736200454918783366342657 * P
                              [A. Lenstra, Manasse 1990]
F10 = 45592577 *              [Selfridge 1953]
      6487031809 * C          [Brillhart 1962]
F11 = 319489 * 974849 *       [Cunningham 1899]
      167988556341760475137 * 3560841906445833920513 *
                              [Brent 1988]
      P                       [Morain 1988]
F12 = 114689 *                [Pervouchine, Lucas 1877]
      26017793 * 63766529 *   [Western 1903]
      190274191361 *          [Hallyburton, Brillhart 1974]
      1256132134125569 * C    [Brent 1987]
F13 = 2710954639361 *         [Hallyburton, Brillhart 1974]
      2663848877152141313 * 3603109844542291969 * C
                              [Crandall 1991]
F14 = C                       [Selfridge and Hurwitz 1963]
F15 = 1214251009 *            [Kraitchik 1925]
      2327042503868417 *      [Gostin 1987]
      C                       [Suyama 1989]
F16 = 825753601 *             [Selfridge 1953]
      C                       [Baillie 1989]
F17 = 31065037602817 *        [Gostin 1980]
      C                       [Baillie 1989]
F18 = 13631489 *              [Western 1903]
      C                       [D. and G. Chudnovsky 1988]
F19 = 70525124609 *           [Reisel 1962]
      646730219521 *          [Wrathall 1963]
      C                       [Crandall,Doenias,Norrie,Young
                               (CDNY) 1993]
F20 = C                       [Young and Buell 1988]
F21 = 4485296422913 *         [Wrathall 1963]
      C                       [CDNY 1993]
F22 = C                       [CDNY 1993]
```

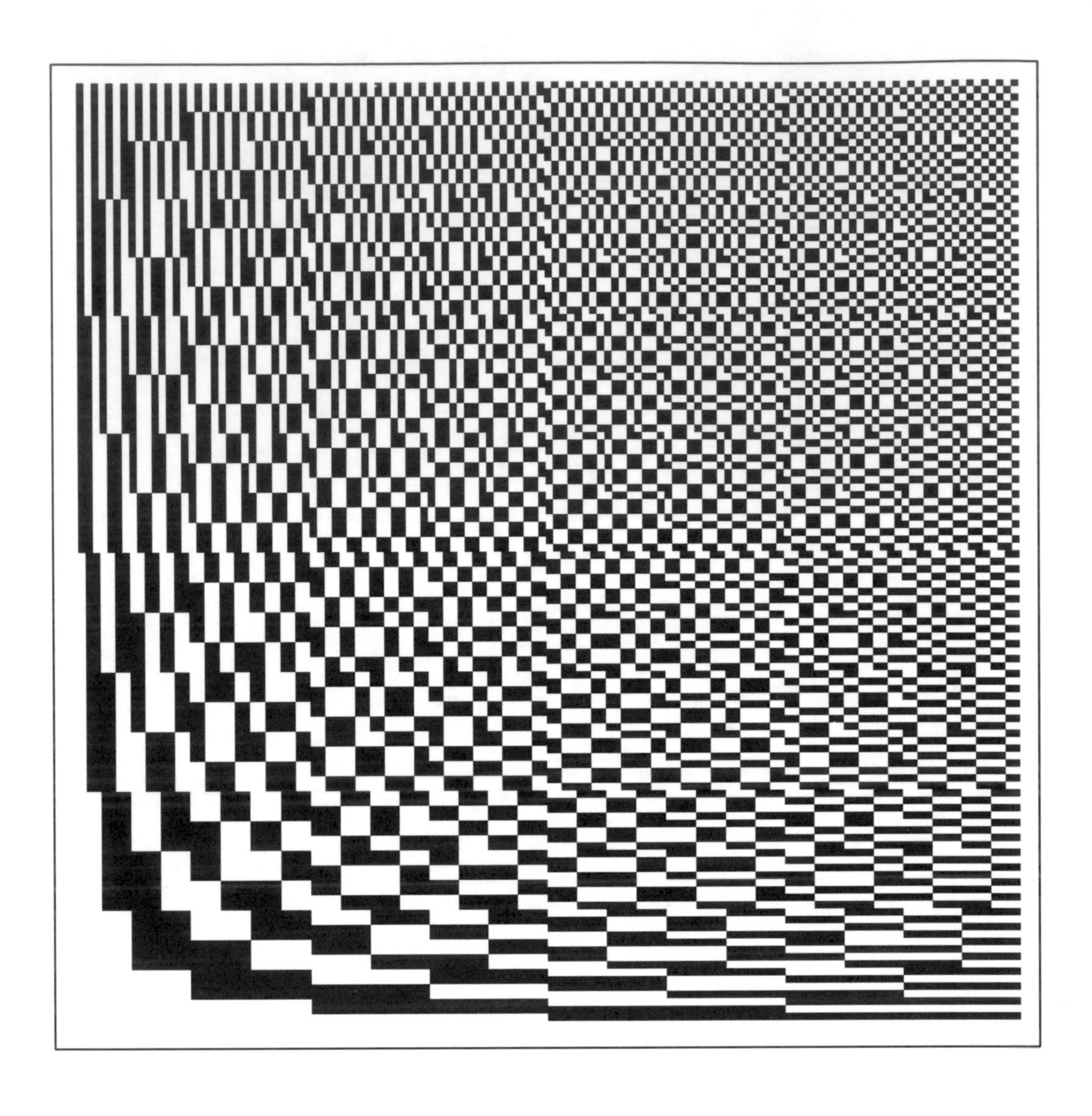

Two-dimensional pictorial of the first 128 Walsh-Hadamard basis functions reveals various symmetries and the concept of sequency. The symmetry about a diagonal is a manifestation of Wal(m,n) = Wal(n,m). Sequency, the count of zero crossings, appears as the number of color transitions as one traces across a typical horizontal section; with no transitions for the lowest row, rising consecutively to 127 transitions for the top row.

4 The FFT forest

The ubiquitous FFT and its relatives

Perhaps the most interesting [contribution to the modern theory of the FFT] is the discovery that the Cooley-Tukey FFT was described by Gauss in 1805. That gives some indication of the age of the topic. . .the fact that a recently compiled bibliography contains over 3000 entries indicates its volume.

[*Burrus* 1990]

Beyond the impressive age and ubiquity of the fast Fourier transform (FFT) topic, the researcher will also sense scope and depth. The FFT forest is a thick, gnarled one. One might say dark, too, because various algorithmic issues remain unresolved. To this day FFT algorithms have not been optimized in all circumstances–there are just so many cases. For example, data is purely real-valued, data is time-symmetric, data is largely zero, data is >1 dimensional, only a few FFT components are needed, or combinations of these. Even new brands of computing machinery can give rise to new FFT structures, especially in this age of vectorization and parallelism. It seems that new optimizations arise in the literature at a healthy average rate.

Projects herein will overview some of these special cases, with emphasis on performance. Related transforms are overviewed as well: Hartley transform, Walsh-Hadamard transform, square-wave transform, real-valued FFT, discrete cosine transform, and number-theoretic transforms. The new, attractive, and powerful wavelet transforms are actually covered separately, Chapter 5. This segregation of wavelet transforms makes sense in one respect: all the other transforms discussed in the present chapter, when applied to signals of run-length *N*, are of the "divide-and-conquer" $O(N \log N)$ complexity class in at least one direction, forward or inverse; whereas wavelet algorithm complexity can be $O(N)$ in both directions.

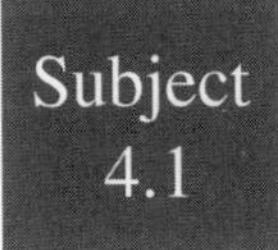

Discrete Fourier transform

By FFT we refer to the fast Fourier transform algorithm. Perhaps the first FFT lesson to be learned is that "FFT" is a misnomer of sorts. The FFT is not of itself a mathematical construct; rather it is an algorithm for computation of the discrete Fourier transform (DFT). This DFT in turn is defined on the basis of a signal vector

$$x = \{x_0, x_1, x_2, \ldots, x_{N-1}\}$$

possessed of N complex values. We denote the DFT of x by another, generally complex vector X also having N elements, the k-th of which is:

$$X_k = \sum_{j=0}^{N-1} x_j \, e^{-2\pi ijk/N} \tag{4.1.1}$$

On the other hand, if X is known, the elements of x always can be reconstructed according to the inverse DFT:

$$x_j = \frac{1}{N} \sum_{k=0}^{N-1} X_k \, e^{+2\pi ijk/N} \tag{4.1.2}$$

We have used what might be called "engineer's" nomenclature, although several research disciplines have by now adopted this notational scheme. It is an easy mnemonic to remember that the inverse (4.1.2) involves a change in sign for the exponent, together with division by N. There are alternative DFT definitions. For example some treatments have sign reversals, or a prefactor of $N^{-1/2}$ for each of the two transformation formulae; the common prefactor giving rise to an even easier mnemonic for the formulae. However, overall multiplication by a square root often consumes noticeable redundant machine time. Computations are more efficient in the forward DFT direction with the prefactorless definition (4.1.1).

For computational purposes there is an important observation about the inverse DFT. If we denote a computer function that, given a signal x, creates (or modifies) an array to end up with transform values X by:

$$X = \mathrm{DFT}(x)$$

then in principle we do not need any extra significant software to compute the inverse transform. Indeed, the original signal can be reconstructed via:

$$x = (1/N)\,(\mathrm{DFT}(X^*)\,)^*$$

where the conjugation (*) applies to each element of the indicated complex vector; likewise multiplication by $(1/N)$ applies elementwise. Thus, it is not uncommon to see program test sequences such as:

```
 (* Assume input signal x is ready. *)
    dft(x, X);    (* X becomes transform (4.1.1). *)
    conjugate(X);(* Conjugate every element *)
    dft(X, x);
    conjugate(x);
    div(x,N)    (*Divide each element by N. *)
(* x should now be the original signal. *)
```

This program sequence is an identity test for one's DFT function. Note that the definition (4.1.2) is never used explicitly. However, in the applications of modern FFTs, sometimes the researcher prefers to use an actual inverse algorithm with distinct structure for reasons of speed in regard to manipulation of element indices. Often a simple sign change within an execution loop will change a forward DFT to an inverse one. We discuss later in Subject 4.2 explicit forms of inverse FFTs. But the present notion is that one does not have to repeat all of the forward transform programming work in order to forge an inverse, for the latter is virtually free.

Incidentally, it is a trivial but valuable point that, because of the usual sluggishness of division in a custom function such as `div(x,N)` above, one should generally avoid dividing each element of a transform by N, opting instead to create once a floating-point reciprocal $1.0/N$ and do elementwise multiplication.

Systems and even languages can have built-in transform functions. As always, one must carefully look over the transform definition used by the original programmer(s). For example, in *Mathematica* 2.1, our forward DFT (4.1.1) would, because of sign and prefactor discrepancies, have the equivalent call:

```
Sqrt[n] * InverseFourier[x]
```

where x is actually a List, while our inverse DFT (4.1.2) would have an equivalent:

```
Sqrt[1/n] * Fourier[X]
```

However, as described previously, one could in principle live with either one of these forward-inverse *Mathematica* functions because the function `Conjugate[]` is also available.

The elements $\{X_k\}$ of the transform signal X are sometimes called Fourier components, or spectral values, while X itself can be called the spectrum. When real-world signals are sampled at discrete time intervals τ, meaning that measurements x_m and x_{m+1} generally occur a time τ apart on the time-line, the index k means that X_k is a measure of the signal content at frequency $k/(N\tau)$. When τ is in seconds, the frequency corresponding to k is in Hertz (cycles/sec).

The motivation for a fast algorithm is the following. Say that one desires the complete DFT, meaning the set of components X_k. If one proceeds by computing each X_k from (4.1.1) by summing all summands (x_j $*$ (exp factor)) directly, then the entire spectrum ($k = 0,\ldots,N-1$) takes $O(N^2)$ multiplications ($*$). As is well known, this complexity can be brought down to $O(N \log N)$ by taking into account certain redundancies in the transform arithmetic; the manner of this reduction is the FFT algorithm.

There are special situations, however, when other complexity reduction techniques apply, especially when the signal "run-length" N is small, as we see in the following project.

Project 4.1.1 Fundamental DFT manipulations

Theory, difficulty level 1.

This problem amounts to a tour of various algebraic manipulations that familiarize one with various mathematical aspects of the DFT. The pathway we take to the answer is somewhat arbitrary, but shows what can sometimes be done without explicit numerical computation. We assume a signal $x = \{1,1,1,1,-1,-1,-1,-1\}$ of length $N = 8$, amounting to the discretization of one cycle of a square wave. There will be eight components of the spectrum X. To find all of these exactly we start by assuming a kind of signal power identity (which we generalize later):

$$\sum_{k=0}^{7} |X_k|^2 = 64$$

which is a special case of Parseval's rule, and that

$$X_{8-k}^* = X_k \quad ; \quad k = 1,\ldots,7$$

which is a statement of Hermitian symmetry. The next steps are:

1) For the given eight square-wave values x_j, argue on the basis of the definition (4.1.1) that X_k is zero for even $k = 0, 2, 4, 6$. Complex plane drawings, where each summand of (4.1.1) is a complex vector, may be useful for this proof. This leaves only the evaluation of the odd-indexed spectral components.

2) Argue from the Hermitian symmetry that knowing X_k for $k = 1,3$ would

completely settle all eight spectral values.

3) Writing, then, $X_1 = a + bi$, $X_3 = c + di$, deduce from the inverse transform (4.1.2) for signal element x_0 that $a + c = 4$; and deduce from the inverse transform for x_2 that $d = b + 4$.

4) From the Parseval rule deduce a value for $a^2 + b^2 + c^2 + d^2$ and using step (3) previous, obtain a relation between a and b.

5) From a complex plane diagram as suggested in step (1) previous, but this time for the evaluation of component X_1, deduce that $a = 2$, and from step (4) and the diagram obtain b.

6) Put all this analysis together to deduce the exact spectrum (transform):

$$X = \{0,\ 2-(2+2\sqrt{2})\,i,\ 0,\ 2+(2-2\sqrt{2})\,i,\ 0,\ 2-(2-2\sqrt{2})\,i,\ 0,\ 2+(2+2\sqrt{2})\,i\}$$

It is of interest that this transform (spectrum) is obtained here via observed symmetries and algebra, without numerical (meaning floating-point) computation.

Having seen some elementary DFT manipulations, we turn to some formulae relevant to more general values of the run-length N.

Project 4.1.2

Algebraic aspects of the DFT

Theory, difficulty level 2.

1) For the DFT (4.1.1) and arbitrary N, prove the Hermitian symmetry for pure-real signals; that is, if each element of x is real, then

$$X^*_{N-k} = X_k \quad ; \quad k = 0, \ldots, N-1 \tag{4.1.3}$$

where all indices are interpreted (mod N); for example X_N means X_0. Is the converse true, that is, if (4.1.3) holds for the transform, must all original signal values be real? What analogous theorem holds for pure-imaginary signals?

2) Argue on the basis of (1) that for N even and pure-real x, both X_0 and $X_{N/2}$ must be real. Then answer quantitatively this possible classroom query: "On an information-theoretic basis, shouldn't the number of bits required to represent X be the same as for x? I thought the transform was complex and therefore, for real signals x, won't X have more data?"

3) Prove the Convolution Theorem which applies to any complex signals x, y:

$$\sum_{j+k = n \,(mod\, N)} x_j\, y_k^{\,*} \;=\; \frac{1}{N} \sum_{k=0}^{N-1} X_k\, Y_k^{\,*}\, e^{2\pi ikn/N} \qquad (4.1.4)$$

The left-hand summation is the convolution of x and y; the right-hand side reveals that said convolution is an inverse DFT of a product signal formed as an elementwise product of X and Y^*.

4) From the Convolution Theorem, establish for general N a Parseval Rule (as exemplified in Project 4.1.1) involving the sum of the absolute squares of a single signal's elements. What does this say about the power in a signal and the power in its spectrum?

5) Show that the DFT (4.1.1) can be written as the sum of an "even part" and an "odd part;" that is, exhibit a complex number W such that:

$$X_k \;=\; \sum_{j=0}^{N/2-1} x_{2j}\, e^{-2\pi ijk/(N/2)} \;+\; W^k \sum_{j=0}^{N/2-1} x_{2j+1}\, e^{-2\pi ijk/(N/2)} \qquad (4.1.5)$$

Note that, though k runs from 0 through $N-1$, each summation has period $N/2$ in k; that is, each summation starts repeating its values as k crosses its midrange. This is the celebrated Danielson-Lanczos identity, which reduces the problem of DFT computation down to a little more than two problems each involving halved run-length. In this way the identity forms the basis of one class of FFT algorithms–the "decimation in time" class, so named because the time indices of the original signal are decimated in an even/odd fashion. See step (10) below for a complementary approach.

6) Clearly from step (5) previous we see that the DFT for a signal with N even and all odd-indexed elements equal to zero can be obtained via a length $N/2$ DFT. Show a somewhat deeper fact, that for a symmetrical signal (i.e., such that again N is even and $x_j = x_{N-j}$), the DFT can likewise be obtained via a length $N/2$ DFT [Rabiner 1979] and some few extra operations.

7) It is possible to obtain the DFTs of two independent pure-real signals by combining these signals into a complex signal, then performing the DFT on that. More precisely, if x and y are such signals and the DFT of the complex signal $x + iy$ is denoted Z, then the desired transforms X and Y can be recovered from Z. Show this recovery is possible by proving:

$$X_k = \frac{(Z_k + Z^*_{N-k})}{2}$$

and a similar formula for Y_k. Assume that in actual practice the divisions by 2 would be multiplications by 0.5, and answer this question: If the run-length of each of the original x, y is N, then the DFT of $x + iy$ is taken, exactly how many multiplications by 0.5 are required to reconstruct X, Y? (Take care to apply the known Hermitian relations from step (1) previous, for these relations significantly reduce the required multiply count.) The double-DFT technique can be used when two pure-real signals enter naturally into a problem, such as the problem of computing the convolution of x and y as embodied in (4.1.4).

8) Similar to the double DFT notion of step (7) previous is that of halving the (assumed even) run-length N when a signal x is pure-real, by putting exactly half of the elements of x into a separate signal and performing the double DFT. One way to effect this reduction in DFT run-length is as follows. Create "even" and "odd" signals denoted xe, xo respectively:

$$\begin{aligned} xe &= \{x_0, x_2, x_4, \ldots, x_{N-2}\} \\ xo &= \{x_1, x_3, x_5, \ldots, x_{N-1}\} \end{aligned}$$

each having run-length $N/2$. Now let Z be the (length $N/2$) DFT of the complex signal $xe + i\, xo$. Show that:

$$X_k = \frac{Z_k + Z^*_{N/2-k}}{2} - \frac{i\left(Z_k - Z^*_{N/2-k}\right)}{2} e^{-2\pi ik/N} \tag{4.1.6}$$

hence the desired transform may be recovered from knowledge of Z. As usual, it is unnecessary to do the arithmetic implicit in (4.1.6) for $k > N/2$, due to the Hermitian symmetry (4.1.3).

9) Apply the ideas derived in step (8) previous to obtain the DFT of the $N = 8$ square wave of Project 4.1.1, establishing first that $Z = \{0, 4, 0, 4i\}$. Then use identity (4.1.6) to establish the final explicit X for that Project.

10) Given a signal x of even length N, show that if only the even-indexed components of the DFT X of x are desired, then it is enough to compute the DFT of the length $N/2$ sequence:

$$y = \{x_j + x_{j+N/2} : j = 0, 1, ..., N/2 - 1\}$$

Then find another length $N/2$ sequence whose DFT yields all the odd-indexed components of X. In this way one obtains a formula analogous to the Danielson-Lanczos identity (4.1.5), and so has yet another way to establish an FFT algorithm. What arises from this partitioning of the transform indices into even/odd sets is the "decimation-in-frequency" class of FFTs.

11) The Danielson-Lanczos identity of step (5) previous dates back to the 1940s. In modern times [Burrus 1988] complicated but computationally more efficient reduction identities have arisen. Establish the following identities for N divisible by four [Duhamel and Hullman 1984]:

$$X_{2k} = \sum_{j=0}^{N/2-1} (x_j + x_{j+N/2})\, e^{-2\pi ijk/(N/2)} \tag{4.1.7}$$

$$X_{2k+1} = \sum_{j=0}^{N/2-1} \left((x_j - x_{j+N/2}) - (-1)^k i (x_{j+N/4} - x_{j+3N/4}) \right) \times$$
$$\times\, e^{-2\pi i j/(N/4)}\, e^{-2\pi i j m/N}$$

where $m = 2k + 1 \pmod 4$. Though these identities form the core of the modern split-radix FFT algorithm class we discuss later, the manipulation of DFT indices, starting from the defining equation (4.1.1), without any foreknowledge of fast algorithms, is instructive.

Project 4.1.3 DFT test signals

Theory, difficulty level 2.
Support code: Appendix "DFTgraphs.ma."

1) Consider the general square wave for N even:

$$x_j = \begin{cases} 1, & j \le N/2 \\ -1, & \text{otherwise} \end{cases} \tag{4.1.8}$$

The transform X of this square wave should vanish at even-indexed components, and, for $k << N$, decay asymptotically as $1/k$ in amplitude at odd indices. Derive these two facts by treating the defining sum (4.1.1) as the sum of two geometric series. In signal processing, it is an old and useful idea that a square wave is possessed solely of odd harmonics, and that these possess power as the squared reciprocal of frequency.

2) Consider for N even the "chirp wave:"

$$x_j = e^{\frac{\pi i j^2}{N}} \tag{4.1.9}$$

The transform X of the chirp wave enjoys the useful property of constant absolute amplitude; i.e., $|X_k|$ is independent of k. From the defining sum (4.1.1) prove this constancy, and give the constant. Hint: attempt to "complete a square" in the exponent of the general summand. Then find an analogous chirp wave for the case N is odd.

3) Find a closed-form expression for the DFT of a ramp signal, for which $x_j = j$ and length N is arbitrary. In the spirit of the square wave analysis in step (1) previous, find, for small k/N, the k-dependence of amplitude.

4) Given the DFT X of a signal x of length N, show how immediately to obtain the DFT of the discrete difference signal:

$$\Delta x = \{x_j - x_{j-1} : j = 0, 1, ..., N-1\}$$

As always, indices are evaluated (mod N), so index -1 means index $N-1$. Argue that if the DFT of a difference signal is known, and also X_0 is known, then X can be reconstructed. Note that the requirement for an extra constant, such as the zero-th component of X, is analogous to the requirement of a constant of integration in calculus.

5) As discussed in Project 1.1.1, white noise is a kind of derivative of Brownian motion. An interesting test signal is a white noise signal for which each x_j is an independent Gaussian random variable of unit variance and zero mean. Argue that a Brownian motion signal, for which the white noise is the discrete difference signal, should have spectral power behaving as $1/k^2$ for $k << N$., and is thus a model for "$1/f^2$ noise."

The fractional Brownian motion concept is useful for creating other test signals, such as simulations of the ubiquitous "$1/f$ noise" that nature seems to favor. In this context, $1/f$ would mean that spectral amplitude falls off as $1/\sqrt{k}$ for $k << N$. These studies for two-dimensional settings give rise to models for fractal mountains, as discussed in Chapter 6.

Regardless of the profundity of FFT algorithm studies, it often turns out that at some point the researcher needs some pedestrian, direct DFT software, if for no other reason than to test by comparison the integrity of any new FFT attempt. Naturally, this software is not useful for large run-lengths N, but it will certainly be good for small N, and perhaps more importantly for N that are not highly composite (consisting of "small" primes to "high" powers) as is usually stipulated for the fast algorithms.

Project 4.1.4 Direct DFT software

Computation, graphics, difficulty level 1.
Support code: Appendix "fft.c," function DFT().

1) Write a DFT program that computes, directly according to the sum (4.1.1), and for arbitrary positive integers N via floating-point arithmetic, the complete transform $\{X_k : k = 0,1,...,N-1\}$. One approach is to create an auxiliary array of complex numbers (i.e., pairs of reals) corresponding to X, and fill said array one component at a time, each single filling requiring N complex multiplications. Start first with evaluation of sin/cos terms (components of the complex exponentials) computed every time they are needed. This is quite slow but serves as an instructive starting efficiency. Work out some timing software that reports the time per DFT. (For small N one should perform some known number of DFTs and divide by that number to get the time per DFT.) For sufficiently large N it will be possible with refinements below, and eventually via FFT algorithms, to achieve stunning improvements in the speed of the overall calculation.

2) Write an inverse DFT that uses the simple conjugation method of the text. Alternatively, pass a flag to the DFT routine that, by indicating forward or inverse, determines the sign of all sin() terms and also determines whether there is an eventual divide by N.

3) Ensure the integrity of the DFT software by applying it to some test signals from Project 4.1.3. It is a good idea to establish some form of error measure which indicates whether the DFT is valid. One measure is to compare the original signal with a forward-inverse transform (two calls to the single DFT if the flag method of step (2) previous is used), by summing all terms:

$$|x_j - x_{fi,j}|^2$$

where x_{fi} is the doubly transformed signal (and therefore should agree exactly with x in theory). If this sum of squares is divided by N and an overall square-root is taken, one has a form of Root-Mean-Square-Error (RMSE) measure, so commonly used in signal processing. Some investigators prefer to use just the maximum deviation

$$max_j \left| x_j - x_{fi,j} \right|$$

as the measure, and so on. A word of caution: there are many ways to harbor DFT bugs for which the forward-inverse transform pair *appears* correct. One might do well to consider a numerical test of the Parseval identity, as in step (4) of Project 4.1.2. This would test fairly stringently the intermediate transform in a forward-inverse operation.

4) To speed up the software, evaluate all necessary sin/cos terms you will need just once, and store these in a table. Then within the DFT loop, access table elements as needed. Compare the table-lookup DFT time with that of the first approach. Note that this comparison is ambiguous: do we include the table creation time in the total time of this second method? In practice, some DFT/FFT projects involve vast numbers of repeated transforms, so in such cases the table creation time is negligible. The point of this project part is the comparison of as-needed computation to table lookup of the trigonometric functions. The good news about this table creation experiment is that the same table software can be used again in sophisticated FFT implementations.

5) Investigate performance of a Goertzel algorithm which avoids table lookup and almost all of the trigonometric evaluation. The idea is to use an identity such as:

$$e^{2\pi i(a+b)/N} = e^{2\pi ia/N}\left(c + i\sqrt{1-c^2}\right)$$

where $b \le N/2$ and $c = \cos 2\pi b/N$. If one sets $b = 1$, say, and evaluates c just once, then complex exponentials for large angles can be evaluated by "ratcheting up" with multiplies alone.

6) Develop means to graph the DFTs of test signals. Figure 4.1.1 shows a plot of the DFT of the length-128 one-cycle square wave $x = \{-1, -1, ..., 1, 1\}$. The Hermitian symmetry (around $k = 64$) is evident, as is the harmonic reciprocal damping. It is often revealing to make decibel plots,

that is to plot

$$(db)_k = -20 \log_{10} (| X_k |/M)$$

where the constant M (often just the maximum value of all the $|X_k|$) is chosen to force some convenient zero-decibel reference. Furthermore it is sometimes convenient to plot a logarithm of the index k, as is often done in filter studies. Figure 4.1.2 shows a db-per-octave plot of the odd harmonics of the square wave's transform.

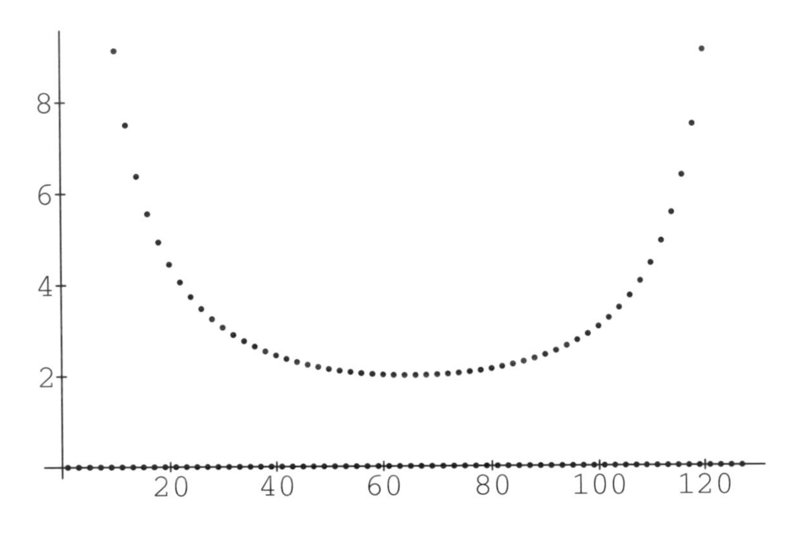

Figure 4.1.1: Amplitude plot of the DFT of an $N = 128$ square wave showing several aspects of DFT algebra, such as vanishing even-indexed elements, Hermitian symmetry, and odd harmonic reciprocal damping.

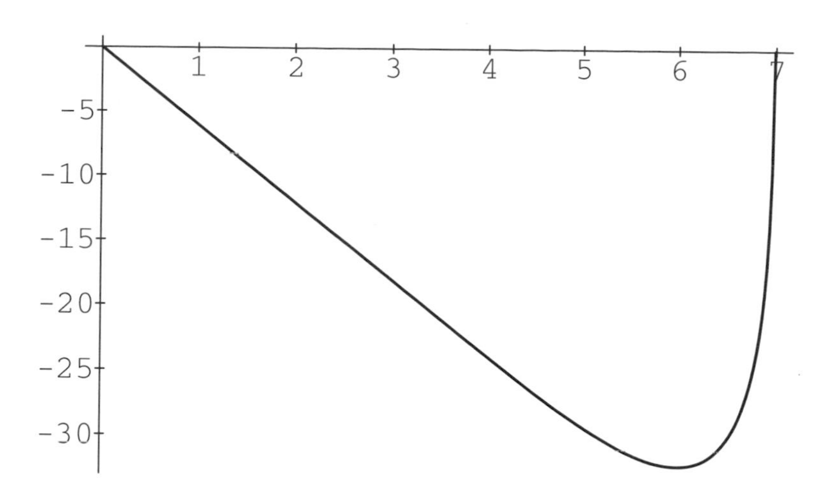

Figure 4.1.2: Decibel plot of the odd-k components of the square wave of Figure 4.1.1. Vertical axis is decibels, horizontal is $\log_2 k$. The 6 db/octave (one horizontal unit is an octave) roll-off is evident, up to the Nyquist index $k = 2^6 = 64$.

Subject 4.2 FFT algorithms

The fast Fourier transform algorithms are divide-and-conquer schemes *par excellence*. The Danielson-Lanczos identity (4.1.5) told us that a DFT for N even can be cast as little more than two DFTs each of length $N/2$. When N is a power of 2, then evidently we can recurse on this idea down to very small lengths. In fact, it is possible to stop only at length 1, and a length-1 DFT is merely an identity transform. Though not the most speed-efficient approach, an instructive and elegant exercise is to perform the Danielson-Lanczos recursion all the way down to trivial length. The following is a *Mathematica* sequence that does the recursion for N (the length of the List x) any power of 2:

(4.2.1)

```
fft[x_] :=
   Block[{n, xeven, xodd, xsplit, g, w},
    If[(n=Length[x])==1,Return[x]];
```

```
    xsplit = Transpose[Partition[x,2]];
    xeven = fft[xsplit[[1]]]; xodd = fft[xsplit[[2]]];
    g = Exp[-2 Pi I/n]; w = Table[g^m,{m,0,n-1}];
    Return[Join[xeven,xeven] + w * Join[xodd,xodd]];
  ]
```

Whether this manner of coding is transparent or not, one can easily see at least two things: the recursive nature (`fft[]` calls itself), and the relationship to the fundamental identity (4.1.5). The Appendix source "DanielsonLanczos.ma" shows alternative ways of coding the recursion.

Project 4.2.1

Recursive FFTs

Computation, a little theory, difficulty level 1.
Support code: Appendix "DanielsonLanczos.ma."

1) Implement a Danielson-Lancsoz recursion in any language of choice (that will support recursion, arrays, and so on). Verify that the transform result of Project 4.1.1 is obtained for signal $x = \{1,1,1,1,-1,-1,-1,-1\}$. Chances are good that if the recursive program agrees with the Project on that eight-element square wave then everything is working. Still, it is a good idea to try other signals, such as the test signals of Project 4.1.3.

2) If the results of step (10), Project 4.1.2 are available, write a recursive program that uses the ideas of that Project step, instead of the Danielson-Lanczos identity. Test this program as in step (1) previous.

3) For one of these recursive programs, attempt speed optimization of various kinds. One improvement is to use a single, precalculated sin/cos table. Another is to exit the recursion not at final length 1, but at length 2 or even 4, at which time the final DFTs are relatively trivial.

4) Work out the numbers of (complex) multiplies and adds required for a full recursion. For large N the theory should give asymptotically some multiple of $N \log N$, certainly radically better than the DFT's complexity $O(N^2)$.

5) Insert custom counter variables to add up the number of multiplies and

adds for the full recursion. Compare these results to the theoretical $O(N \log N)$ for the recursion, for example by creating a table of operation count vs. $N \log N$.

The previous recursion study is illuminating, but for real computational efficiency it is desirable to perform the FFT calculation without creating new arrays, not evaluating sin/cos on every pass, and so on. Along these lines, a highly pragmatic development is the modern FFT [Cooley and Tukey 1965]. (The author is aware of the controversy surrounding who really did discover the FFT, which controversy is implied in the opening quote atop this chapter.) The essential contributions of Cooley and Tukey were to provide an "in-place" scheme for carrying out the recursion, and to publish the idea in a widely read journal. From adroit re-indexing of the original signal, a cascade of pairwise complex computations, or "butterflies," result in the desired transform, and all of this happens within the original array: the signal x is eventually replaced with the transform X.

Project 4.2.2 FFT indexing and butterflies

Theory, difficulty level 2.

Consider a length-4 signal x and its transform X to be column vectors. Observe that for $\omega = e^{-2\pi i/N} = -i$, the DFT can be cast algebraically in terms of a matrix operation:

$$X = Fx$$

or, explicitly:

$$\begin{bmatrix} X_0 \\ X_1 \\ X_2 \\ X_3 \end{bmatrix} = \begin{bmatrix} 1 & 1 & 1 & 1 \\ 1 & \omega & \omega^2 & \omega^3 \\ 1 & \omega^2 & \omega^4 & \omega^6 \\ 1 & \omega^3 & \omega^6 & \omega^9 \end{bmatrix} \begin{bmatrix} x_0 \\ x_1 \\ x_2 \\ x_3 \end{bmatrix}$$

1) Find a matrix P such that FP has the block matrix form

$$FP = \begin{bmatrix} A & BA \\ A & -BA \end{bmatrix}$$

where each of A, B is a 2-by-2 matrix, explicitly:

$$A = \begin{bmatrix} 1 & 1 \\ 1 & -1 \end{bmatrix}, \quad B = \begin{bmatrix} 1 & 0 \\ 0 & -i \end{bmatrix}$$

This P should be a permutation matrix; that is, upon acting on a column it shuffles (but does not otherwise affect) the elements.

2) Clearly P^{-1} is also a permutation matrix, and furthermore

$$X = (FP)\,(P^{-1}x)$$

so that the DFT can be obtained by applying the block matrix form to a permuted signal. Comment on the validity of the following claim: "The application of the block matrix FP to the permuted signal is a statement of the Danielson-Lanczos identity (4.1.5) for the case $N = 4$."

3) We see that a permutation can simplify the required DFT matrix. A second lesson is that the simplified matrix FP can be factored into two matrices *each of whose rows has only two non-zero entries.* Find a matrix C such that

$$FP = \begin{bmatrix} A & BA \\ A & -BA \end{bmatrix} = C\begin{bmatrix} A & 0 \\ 0 & A \end{bmatrix}$$

and note that C, as well as the matrix with the off-diagonal 0 blocks, has the property that *each row has only two non-zero entries*. (There are other ways to factor FP to this end.) The point is that now we can establish a butterfly diagram for the full order-4 DFT, by inspection of the formula:

$$X = C\begin{bmatrix} A & 0 \\ 0 & A \end{bmatrix} P^{-1}\, x$$

In order to create the full diagram, assume that a fundamental butterfly, which we draw like so:

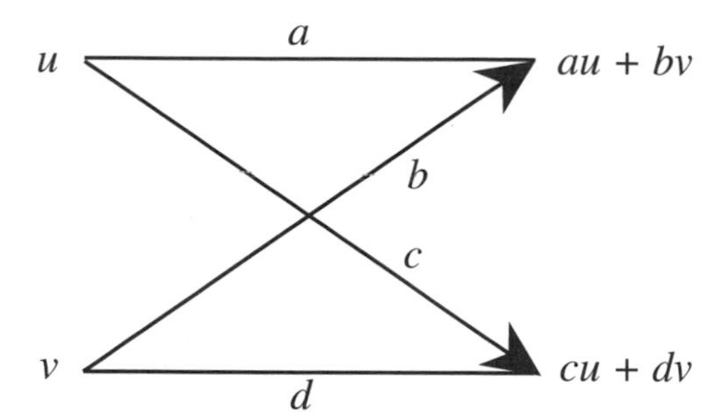

embodies the matrix operation

$$\begin{bmatrix} a & b \\ c & d \end{bmatrix} \begin{bmatrix} u \\ v \end{bmatrix}$$

Note that in actual practice, the butterfly can be done "in-place" with a program sequence:

```
t = a*u + b*v;
v = c*u + d*v;
u = t;
```

where t is a temporary variable. It is interesting that dedicated integrated circuits have been designed to perform a butterfly as if it were a single vector operation. Now complete the diagram, partially drawn below, by labeling the correct factors and indices not shown.

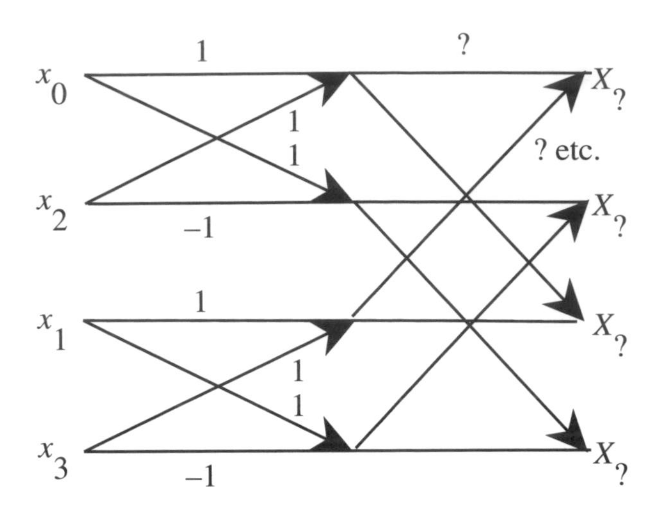

4) Generalize the order-4 matrix analysis, by showing that, for N even, the DFT matrix can be factored:

$$F_N = C_N \begin{bmatrix} F_{N/2} & 0 \\ 0 & F_{N/2} \end{bmatrix} P_N^{-1}$$

where $F_{N/2}$ is the $N/2$-by-$N/2$ DFT matrix and $P^{-1}{}_N$ is a permutation matrix that puts the even indices in the top half, and the odd indices in the bottom half of a column vector. The key to the computational butterfly structure is that C_N, corresponding to the last (right-hand) stage of an order-N diagram, should have only two non-zero elements per row, thus giving rise to the pairwise butterflies. See [Van Loan 1992] for the rigorous mathematics of such FFT decompositions.

5) Now we explore the Cooley-Tukey radix-2 factorization scheme for N greater than 4. This scheme accumulates a more complicated permutation than $P_4{}^{-1}$, with a simpler interior matrix. Let us focus upon the simple case $N = 8$. Find the matrix M such that

$$F_8 = C_8 M \Pi_8$$

where Π_8 is the permutation matrix which, when applied to an arbitrary order-8 column, produces:

$$\begin{bmatrix} x_0 \\ x_4 \\ x_2 \\ x_6 \\ x_1 \\ x_5 \\ x_3 \\ x_7 \end{bmatrix}$$

corresponding to "bit-reversed" ordering of indices. The permuted indices are obtained by writing the original indices j as 3-bit binary 000 through 111, and reversing the order of bits.

If done correctly, this task produces an M which is clearly block structured, and further factorable along the lines of previous Project steps. Draw out, then, a complete order-8 FFT butterfly diagram, starting on the left with the permuted x column exhibited above. Use this diagram to compute manually the exact order-8 DFT of the square wave of Project 4.1.1.

6) One way to understand the advantage of bit-reversal is to realize that the Danielson-Lanczos identity involves "hopping over" indices in order to create lower-length even/odd signals and their DFTs. At deeper recursive levels the even/odd hopping continues, so the binary expansion of an original signal index determines where it ends up according to the permutation. The Cooley-Tukey idea is to bit-reverse prior to the execution of the butterflies, although there are other FFTs, as we see later, that require bit-reversal after the butterflies. An intuitive discussion of bit-reversal can be found in [Press et al. 1988]. For the present, modify a recursive FFT from Project 4.2.1 to print out a bit-reversal table, by tracking what happens to various original signal indices.

7) Consider deeper study and research into FFT structures. One lofty goal is to end up with a completely general Cooley-Tukey matrix factorization, of which we studied the order-8 case in step (5) previous. Such an

undertaking can benefit from some very fine treatments, such as [Van Loan 1992][Burrus 1988], and references therein.

The rationale for all of the permutation, in-place butterflies, matrix factorization, and so on is to relieve ourselves of the burden that FFT recursion presents in the matter of performance and memory. Using the Cooley-Tukey ideas, and variants of same, we can arrive at two basic FFT classes having radix-2 architecure (N is assumed a power of 2).

Decimation-in-time FFTs require, as in the previous Project, pre-scrambling according to the bit-reversal algorithm. The decimation-in-frequency FFTs on the other hand require scrambing *after* the butterfly execution.

Samples of the two FFT classes follow, as *Mathematica* code. These cases are exhibited in *Mathematica* for two reasons: First, it is possible to do symbolic FFTs in this way, using exact complex numbers; and second, it is easier to see how calling sequences work for such problems as convolution.

```
(* Gentleman-Sande, decimation in frequency radix-2
complex FFT. Scramble bits *after* this call. *)

fftfreq[x_] :=
     Block[{y,n,m,i,j,a,ii,im},
        y = x;
        n = Length[x];
        For[m = Floor[n/2] , m > 0 , m = Floor[m/2],
          For[j = 0, j<m, j++,
            For[ i = j, i < n, i += 2m,
              ii = i+1; im = i+m+1; (* List indices. *)
              a = Exp[-2 Pi I j/(2*m)];
              {y[[ii]],y[[im]]} =
                {y[[ii]]+y[[im]],a*(y[[ii]]-y[[im]])};
            ];
          ];
        ];
        Return[y];
     ];
```

```
(* Cooley-Tukey, decimation in time radix-2 complex
FFT.  Scramble bits *before* this call. *)

ffttime[x_] :=
     Block[{y, n,m,i,j,a,ii,im},
        y = x;
        n = Length[x];
        For[m = 1, m < n , m += m,
          For[j = 0, j<m, j++,
            For[ i = j, i < n, i += 2m,
              ii = i+1; im = i+m+1; (* List indices. *)
              a = Exp[-2 Pi I j/(2*m)];
              {y[[ii]],y[[im]]} =
                {y[[ii]]+a*y[[im]],y[[ii]]-a*y[[im]]};
            ];
          ];
        ];
        Return[y];
     ];
```

There are at least three cautions to heed in regard to these fundamental FFT functions. First, the *Mathematica* indices `ii,im` can be removed for fast C programs, say, in which case `i` and `i+m` are the appropriate `y`-array indices. The +1 offset is due to the fact that *Mathematica* Lists are indexed from 1 on. It was deemed best to live with only this one indexing change for ease in porting between language systems (similar indexing discrepancy occurs in FORTRAN translations). Second, the implicit butterfly algebra of the form { , } = { , } must be handled, in C programs, by explicit use of real and imaginary floating-point parts, and one or two temporary variables to effect the butterfly.

```
xfreq = scramble[fftfreq[z]];
xtime = ffttime[scramble[z]];
xtest = Sqrt[Length[z]] InverseFourier[z];
```

Usable–not necessarily optimal–scrambing functions are found in the Appendix code: "FFTs.ma," "fft.c," and "fft_real.c."

With this *Mathematica* style of FFT one can write quite transparent FFT-based algorithms. For example, here is the generation of a numerical cyclic convolution of two Lists `a,b` in one line:

```
N[ffttime[Conjugate[fftfreq[a]*fftfreq[b]]]/Length[a]]
```

which makes explicit reference to the Convolution Theorem (4.1.4). Note that the scrambling is not required (see step (2) in the following Project).

Project 4.2.3 Complex FFTs, N a power of 2

Computation, difficulty level 1.
Support code: Appendix "FFTs.ma", "fft.c."

1) In the fastest computer system on which it makes sense to develop software, work out an FFT package that has a forward FFT as a decimation-in-frequency type, and an inverse decimation-in-time FFT. Recall the observation of the previous subject, that a mere complex conjugation will allow the same code to be used for inverse transforms. But there is perhaps a more efficient approach, which is to use a flag that indicates whether to invert, and this will simply affect sin() calls with an overall conjugation sign.

Note: Appendix code "fft.c," function FFT2raw(), is a decimation-in-frequency implementation with a "skip" option for traversing arithmetic progressions of the signal data (a useful option for multidimensional FFTs).

The reason we are considering here a decimation-in-frequency first, is that such a forward call will leave the output scrambled. *But this is the way that the decimation-in-time FFT expects it.* Therefore, test the package by noting that the intentional scrambling calls (one for each type of FFT) can be omitted and yet we still achieve a forward-inverse identity transform.

2) Use the two FFTs to perform convolutions according to the theorem (4.1.4). Again, scrambling can be dropped for such problems because of the way that both sides of the equation are indexed. Such observations are of interest, but in practice the scrambling time is usually an order of magnitude smaller than the actual FFT time.

3) Write convolution software for two pure-real signals, using the double-DFT idea in step (7), Project 4.1.2, for simultaneous calculation of the two forward DFTs X, Y of equation (4.1.4). The final inverse FFT to get

the convolution can be just the decimation-in-time variety, but one should be aware that the foreknowledge of a pure-real convolution result makes it possible to speed up the inverse transform markedly. See Project 4.2.4.

4) Implement a split-radix FFT that uses the Duhamel-Hullman relation (4.1.7). [Burrus 1988] describes a FORTRAN implementation but one should beware of a misprint bug in one of the loops. Appendix code "fft.c," function FFT2splitraw(), is a tested way to do the split-radix transform. Compare performance with standard Cooley-Tukey or Gentleman-Sande transforms (the latter is embodied in function FFT2raw(), but the skipping option should be optimized out for a fair test).

5) Optimize one or more of the transforms using one or more of the following ideas, or ideas not listed here.

- Remove trivial multiplications by $1, -1, i, -i$; not with "if" statements or the like but by inspecting the logic. Often it happens that nested FFT loops have trivial arithmetic at one end or the other of their loop ranges.
- Explore the fact that complex multiplication can be done in three real multiplies rather than four, at the expense of a few adds. Specifically, the typical butterfly performs:

$$u = x * \cos t + y * \sin t$$
$$v = y * \cos t - x * \sin t$$

  which has four multiplies and two adds. The algebra can be rewritten in convolution form, by setting $a = y * \cos t$, $b = x * \sin t$:

$$u = (x + y) * (\cos t + \sin t) - a - b$$
$$v = a - b$$

  amounting to three multiplies and four adds if the trigonmetric sum is stored. A little algebra shows that if multiplies are more than twice as slow as adds, the three-multiply scheme is advantageous.
- Establish sin/cos arrays that are initialized just once. Take care to know whether interleaving the sin/cos terms is advantageous in terms of the memory speed.

- Depending on machine type, beware of operations such as n << 1, which may be better rendered as n += n.
- Again depending on machine and compiler, invoke += and –= where appropriate.
- Avoid divisions that occur often, instead defining just once a floating-point reciprocal.
- Unroll short-written code loops known to have many passes, to reduce loop overhead.

6) By inserting custom operation counters in your decimation-in-time FFT, compare with the claims in the literature [Burrus 1988], that:

Length $N = 1024$ requires 20480 multiplies and 30720 adds,

which, after optimization work, can supposedly be improved to:

13324 multiplies and 27652 adds

(These operation counts are for real floating-point operations.)

7) The scrambler functions in the Appendix code "fft.c" are not too efficient. Write a scrambler based on the following idea: a table is created just once, that says what swaps need to be done for a complete scramble. Then when a scramble is required, one traverses the table and just swaps the stated pairs. For advanced scrambling techniques see [Evans 1987, 1989].

The next algorithms we investigate are the class of real-signal FFTs. When the original data x is pure-real, a good deal of the computational effort of the classic complex FFTs can be obviated. Indeed, on information-theoretic grounds, it seems reasonable that one should be able to obtain the DFT components for pure-real signals in half the time (see step(2) of Project 4.1.2). Conversely, an inverse DFT that is known to produce a pure-real signal should be more efficient than a general inverse.

Project 4.2.4 Real-signal FFTs

Computation, difficulty level 2.
Support code: Appendix "fft.c," "fft_real.c"

By "real-signal FFT" is meant either a forward FFT on a pure-real signal, or an inverse FFT whose result is known *a priori* to be pure-real.

1) Write real-signal FFT software using the complex packing idea of step (8), Project 4.1.2, in which a complex signal of length $N/2$ is formed from even/odd halfs of the original signal. Endeavor to do the final evaluations of relation (4.1.6) "in-place;" that is, replace the FFT's result array Z with the final transform X. Compare performance with previous complex FFTs for which you would merely have input signals whose N elements were each of the form (real) + $0i$.

2) Implement a split-radix real-signal FFT and real-result inverse FFT as in [Sorenson et al. 1987]. A tested implementation appears in Appendix code "fft_real.c." For such a project it is convenient to express the transform, after a forward FFT call, say, in Hermitian sequence:

$$\{Re(X_0),\ Re(X_1),\dots,Re(X_{N/2}),Im(X_{N/2-1}),\ \dots,Im(X_1)\}$$

and conversely to have the inverse FFT expect this format and yield a pure-real result. Compare the performance of these split-radix, real-signal transforms with other, more traditional transforms.

3) Write convolution software that uses transforms of step (2) on the assumption of pure-real original signals x, y. The author knows of no faster way to effect such (floating-point) convolution over a certain range of run-lengths. For very short N there are often special tricks, and for very large N there is Nussbaumer convolution discussed in Chapter 3, but it is hard to compete with the split-radix FFT approach for $N \sim 10^2$ to 10^5.

Besides the radix-2 and split-radix FFT formulations previously discussed, it is also possible to write FFTs of radix-4, radix-8, and prime-power radices, depending of course on the divisors of N. There are also prime-factor algorithms (PFAs) and the Winograd Fourier transform algorithms (WFTAs) that are more modern than the Cooley-Tukey indexing approach, and especially suitable for certain N not a power of 2.

Project 4.2.5

FFTs for other radices

Computation, difficulty level 2.
Appendix code: "fft.c," function FFTarb().

1) Write an FFT for arbitrary index N, using the prime factorization of N and some indexing scheme, such as the scheme in [Burrus 1988], which uses theorems concerning relatively prime factors of N. One implementation–not optimized in any sense, but functional–is function FFTarb() in Appendix code "fft.c." This approach is only efficient when N is highly composite, i.e., factors into a "few" small primes to "high" powers.

2) Implement a prime-length FFT, observing that the DFT for N a prime number can be expressed in terms of convolutions of two power-of-two length signals [McClellan and Rader 1979]. This means that in principle the "least composite" N, that is prime N, can still be handled in no more than $O(N \log N)$ operations.

3) Implement a Winograd Fourier transform (WFTA) for length $N = 15$, and verify the claim that only 34 floating-point multiplies are required [Burrus 1988].

4) Implement a Prime Factor Algorithm (PFA) for length $N = 15$ and verify that 50 multiplies are required.

5) Implement a radix-4 FFT (assuming N is a power of 4) and compare its performance with a standard radix-2 FFT.

6) Implement ideas of Bailey and Carlson [Carlson 1992], who refined FFTs on Cray supercomputers. One of their ideas is to insert a matrix transpose operation between Stockham decimation-in-frequency blocks, thereby reducing the power-of-two array strides.

Every FFT has a generalization to higher-dimensional signal spaces. The canonical example of a two-dimensional signal is an image. In that case one uses an FFT of length equal to the width of the data, and (possibly a different) FFT whose length is the height. In the case of an M-by-N signal $x = \{x_{jk} : j = 0...M-1;\ k = 0...N-1\}$ the two-dimensional DFT is taken to be:

$$X_{mn} = \sum_{j=0}^{M-1} \sum_{k=0}^{N-1} x_{jk}\, e^{-2\pi i\left(\frac{jm}{M} + \frac{kn}{N}\right)} \qquad (4.2.2)$$

which admits of an inverse:

$$x_{jk} = \frac{1}{MN} \sum_{m=0}^{M-1} \sum_{n=0}^{N-1} X_{mn}\, e^{+2\pi i\left(\frac{jm}{M} + \frac{kn}{N}\right)} \qquad (4.2.3)$$

Project 4.2.6

FFTs in higher dimensions

Computation, theory, difficulty level 2.
Appendix code: "fft2D_real.c," "fft.c."

1) Investigate the following folk theorem (which happens to be true). Think of a two-dimensional M-by-N signal x as having N rows and M columns. Then the theorem goes: "The the-dimensional FFT of x can be obtained by doing a length M, *in place* FFT on every row, followed by a length-N FFT, also in-place, on every column, leaving an M-by-N transform in place of x." Argue further that the entire 2-dimensional transform can be done in $O(MN \log MN)$ operations.

2) For an M-by-N signal, argue that the two-dimensional DFT can be written as a one-dimensional DFT of length MN, provided M and N are relatively prime [McClellan and Rader 1979].

3) Describe analytically the two-dimensional DFT of an N-by-N signal $x = \{x_{jk}\}$ consisting of all zeros except for a straight line of the form $j + k = a \pmod N$, where a is a constant integer.

4) State and prove a two-dimensional convolution theorem whose one-dimensional form is (4.1.4). State the natural Parseval corollary.

5) Implement the notion of step (1) previous, using for convenience say, power-of-two run-lengths for each dimension. Appendix code "fft.c," function FFT2dim() does this, though the performance could be much improved.

6) Prove, as with one-dimensional pure-real signals, that half the information in a two-dimensional transform of a two-dimensional signal is redundant; i.e., derive a Hermitian relation similar to (4.1.3). Describe geometrically what this symmetry means (thinking of the original signal as an image, say).

7) Work out software using the idea of step (6) previous. A working example is found in Appendix code "fft2D_real.c."

8) Implement ideas of [Press and Teukolsky 1989b] for higher-dimensional FFTs. (Note the slight difference in their DFT definition.)

The true depth of the FFT forest is by no means sounded by the previous Projects. We end this Subject with an assortment of FFT applications that reveal the research possibilities that arise from this forest of algorithms.

Project 4.2.7 Applications of the FFT

Computation, theory, difficulty level 2-4.

1) Study and implement a "very long signal" FFT as described by [Hocking 1989].

2) Study the problem of optimal FFT in the case that only some of the elements of the transform X are required. One approach is a split-radix method of [Roche 1992], which exhibits marked speed improvement when the number of required elements is $\ll \log_2 N$.

3) Study FFT techniques for the numerical approximation of integrals of the form:

$$\int_a^b e^{ikt}\, x(t)\, dt$$

Some methods are discussed in [Press and Teukolsky 1989a], in which the authors give some valuable caution, and bolster this with beautiful correction formulae.

4) Investigate the research domain of vectorized or parallel FFTs, for which tensor algebra looks promising [Johnson and Tolimieri 1989]. An actual, very efficient implementation has been outlined by [Hertz P 1990], suitable for Connection Machines or comparable parallel devices. This and other parallelizations are discussed in [Hillis and Boghosian 1993].

5) Study discrete weighted transforms, which are DFTs for which the signal $\{x_j\}$ is replaced by $\{a_j x_j\}$, and which apply naturally to certain computational problems. Here is one example problem: We have seen that when x is pure-real, about half the work suffices to perform the FFT, so show that even when x is replaced by:

$$\{e^{-\pi ij} x_j\}$$

i.e. every real element is given a certain phase, the FFT can still be obtained in half the usual work [Crandall and Fagin 1994].

6) Investigate the interesting Fractional Fourier transform (FRFT) of [Bailey and Swartztrauber 1991], defined by:

$$F_k = \sum_{j=0}^{N-1} x_j \, e^{-2\pi ijk\alpha}$$

where α is any complex number. It turns out to be possible, using ideas of Bluestein regarding filter convolutions, to forge a fast algorithm that resolves once and for all that for *any* α (even irrational values!), it is possible to construct an $O(N \log N)$ algorithm. This implies in turn the trivial corollary that any run-length N admits of a fast algorithm. Remarkably enough, it turns out that, independent of α, about 20 $N \log_2 N$ operations suffice.

Real-valued transforms

We have seen interesting FFT optimizations in the case of pure-real signals–an important study for the obvious reason that nature loves reality, at least in regard to signal processing. A further notion is to establish transforms that also, like the original signals, have real elements. We shall henceforth in this subject assume real signals x and describe several transforms that enjoy the reality property.

We begin with the Hartley transform, with its attendant algorithms we can call FHTs. The transform is defined:

$$H_k = \sum_{j=0}^{N-1} x_j \, cas \frac{2\pi jk}{N} \qquad (4.3.1)$$

where we define $cas\ z = \cos z + \sin z$. The inverse transform is:

$$x_j = \frac{1}{N} \sum_{k=0}^{N-1} H_k \, cas \, \frac{2\pi jk}{N} \qquad (4.3.2)$$

Good treatments of the Hartley transform concept can be found in [Bracewell 1986][Le-Ngoc and Vo 1989] and references therein.

Project 4.3.1 Hartley transform

Computation, theory, difficulty level 2.
Support code: Appendix "RecursiveFHT.ma."

1) Prove that if (4.3.2) holds, then (4.3.1) does; and vice-versa.

2) Show that the DFT of x can be obtained immediately from H according to:

$$X_k = \frac{H_k + i H_{N-k}}{1+i} \qquad (4.3.3)$$

and find a converse relation that expresses H_k in terms of elements of X. These relations are quite useful for testing FHT programs, assuming one already has FFT testers.

3) Show that Fourier spectral power can be obtained according to:

$$2\,|X_k|^2 = |H_k|^2 + |H_{N-k}|^2 \qquad (4.3.4)$$

4) Find a convolution theorem of the brand (4.1.4) that relates the convolution of two real sequences (the left-hand side of that formula) to an inverse Hartley transform of some construct involving the Hartley transforms of x and y.

5) It is possible to define an "odd" Hartley transform by:

$$G_k = \sum_{j=0}^{N-1} x_j \, cas \, \frac{2\pi j\left(k+\frac{1}{2}\right)}{N} \tag{4.3.5}$$

Establish various properties, such as the correct form of the inverse transform, relationship of G to H and X, and so on. Work out a fast algorithm for computation of G. The so-called negacyclic convolution of two real sequences, as defined in Chapter 3, is expressible in terms of G in a natural way [Hu et al. 1992].

6) Establish an analog of the Danielson-Lanczos identity (4.1.5) which relates the order-N DHT to two order-$N/2$ DHTs. (One should end up with three terms on the right-hand side, one of which involves a simple permutation of one of the half-length DHTs.) Then implement a recursive Fast Hartley transform (FHT) algorithm, an example of which is shown in "RecursiveFHT.ma." If the recursion has been implemented as an efficiently compiled program, compare its execution speed with that for the recursive complex FFT. (Naturally the sin/cos values in either recursion should be stored in an overall lookup table, or at least computed with precisely the same delay, to allow fair timings.)

7) Design a non-recursive FHT algorithm and verify that it has just about twice the speed of a standard radix-2 complex FFT. In this regard see [Le-Ngoc and Vo 1989] who observe that FFT butterflies can be "Hartley-ized" in a systematic way. Note, however, that claims to the effect that the FHT is an optimal real-signal transform are suspect: in practice it appears difficult, if not impossible, to beat the best split-radix real-signal FFTs. Still, an FHT might bring dividends aside from execution speed, such as a convenient pathway to the closely related fast discrete cosine transform algorithms.

8) Show how a length-8 FHT can be constructed to require only two floating point multiplies. How many adds are required? Compare these results with analogous ones for various real-signal FFTs.

Next we overview the discrete cosine transform (DCT) which has found serious application in signal and image processing. The DCT strikes a good balance between transform speed and resiliency to endpoint and edge aliasing.

For a given signal x define, the DCT by a vector D whose elements are:

$$D_k = c(k) \sum_{j=0}^{N-1} x_j \cos \frac{\pi(2j+1)k}{2N} \qquad (4.3.6)$$

where $c(k) = 1/\sqrt{N}$ for $k = 0$ and otherwise $\sqrt{(2/N)}$. The inverse transform is defined by:

$$x_j = \sum_{k=0}^{N-1} c(k)\, D_k \cos \frac{\pi(2j+1)k}{2N} \qquad (4.3.7)$$

which is to say that the transform matrix is unitary: the inverse equals the transpose.

Project 4.3.2

Discrete cosine transform

Computation, graphics, theory, difficulty level 2.

1) Show that indeed (4.3.7) holds when (4.3.6) does and vice-versa.

2) Study a remark from the literature [Rabbani and Jones 1991]: "...the DCT has a higher compression efficiency [than the DFT] since it avoids the generation of spurious components." In particular, sketch or plot some discretely sampled smooth signals, together with plots of such signals *reconstructed* via inverse transform, from their DCT and DFT. Clearly the reconstructed plots should agree, and should also agree with the original signal. But now perform quantization of the DFT and DCT, for example, choose a positive integer Q and force every transform element to be assigned merely a code from 0 through $Q-1$ depending on its value between, say, the minimum and maximum transform values. Then compare how the reconstructions appear on the basis of these

quantized, or "noisy" transforms, with special regard for the values at the *endpoints* of the signals. The fact of more spurious qualitative errors for the Fourier quantization can be traced to the property of the DCT that, as indices "wrap around" the natural boundary $N-1$, discontinuity effects do not appear as with the DFT. When two-dimensional DCTs are used for image processing, the advantage is that "blocking" effects are relatively suppressed, as blocks tend to agree on mutual edges.

3) Can a DCT of length N be determined directly from components of some DFT, and what is the appropriate length of such a DFT? One approach to the answer (though not the only approach) is to attempt to "fix up" the spurious DFT behavior described in step (2) previous, and in so doing accidentally "discover" the DCT [Rabbani and Jones 1991].

4) Given a length-N Hartley transform H of x, give formulae that will allow direct computation of the length-N DCT of x. See [Chakrabarti and JaJa 1990] [Malvar 1987].

5) Argue that, just as with the two-dimensional DFT, one may perform a two-dimensional DCT by doing rows in place, then columns in place. The two-dimensional DCT is defined as having components D_{mn}, obtained by double summing

$$x_{jk} \cos \frac{\pi(2j+1)m}{2N} \cos \frac{\pi(2k+1)n}{2N}$$

and then multiplying by $c(m)\,c(n)$. As in step (1), Project 4.2.6, argue that the M-by-N, two-dimensional DCT therefore requires $O(MN \log MN)$ operations. Now consider the image processing trick of breaking a two-dimensional signal into rectangular blocks, say K equal-sized blocks in each direction, so there are K^2 total blocks. The trick is simply to perform a DCT on each block independently. In particular, argue that the work required to do all the two-dimensional DCTs, one such per block, is *asymptotically less than* the work required for the full-size two-dimensional DCT.

6) Work out a fast DCT transform for $N = 8$, which is a popular choice especially in image processing, where images are often tesselated into 8x8 pixel blocks. Usually the industrial applications require a fast inverse

DCT, but either direction is equally difficult. In fact, there is no need to worry about the normalizer $c(k)$ because its computational effect amounts to a modification of only the zero-th term of the transform. How many multiplies and adds are required for this case $N = 8$? Powerful tricks have been developed for reduction of the work in order-8 and 8x8 fast DCTs. One trick is embodied in step (4) previous, in connection with step (7), Project 4.3.1. The next two steps cover yet other means of speeding up a fast DCT.

7) Consider the real-valued FFT (RVFFT) defined by:

$$R_k = \sum_{j=0}^{N-1} x_j \, cs_{kN}\left(\frac{2\pi jk}{N}\right) \qquad (4.3.8)$$

where the function cs_{kN} is the cosine function for $0 \le k \le N/2$, and the sine function for $k > N/2$. Show how to compute the DCT from knowledge of R. Then see if any new optimizations for step (6) previous are thereby suggested. See the work of [Vetterli 1988].

8) Find some matrices that can be used to pre-scale the signal x so that the DCT involves fewer operations after the scaling. This is important in software that already requires some scaling scaling, so that a new scaling matrix can be completely absorbed into the old, required scaling. The general idea is that if Δ denotes the DCT matrix, and for some matrix M the matrix product ΔM is easier than Δ to compute, then signals x can be scaled by the inverse of M; in other words

$$\Delta M M^{-1} x$$

will still be the DCT. One way to start this task is to assume order-8 and find 8-by-8 matrices M such that ΔM has relatively simple visual form.

The Walsh-Hadamard transform (WHT) is of a distinct class of transforms that enjoy an attractive property: no multiplications are required. One merely uses addition (and subtraction) because the fundamental basis set, rather than sinusoidal, is "square wave like."

The WHT is defined in terms of Walsh functions. We adopt here the discretized Walsh functions:

$$Wal_n(t) = Par(n \,\&\, t) \tag{4.3.9}$$

where a length N (a power of 2) is understood, both n and t run from 0 through $N-1$, the "&" is a bitwise "and" operation, and Par function is $(-1)^{\#}$ where # is the number of 1's in the binary expansion of its argument. Then we define the WHT by:

$$W_k = \sum_{j=0}^{N-1} x_j \, Wal_k(j) \tag{4.3.10}$$

The index k on these components W_k is difficult to interpret physically in the case that x is a real-world signal. A partial solution to this problem is to focus on the number of zero crossings of Wal, which number has come to be called "sequency."

Project 4.3.3 Walsh-Hadamard transform

Computation, graphics, theory, difficulty level 2.
Suuport code: Appendix "walsh.c."

1) Show that the Walsh function (4.3.9) enjoys the following properties.

$$Wal_m \, Wal_n = Wal_{m \wedge n} \quad \text{where } \wedge \text{ means "exclusive-or;"}$$

$$Wal_n(t) = Wal_t(n)$$

2) Find the inverse WHT in the form of the usual sum complementary to (4.3.10).

3) Show that if a Walsh function Wal_n has sequency s, which is the number of zero-crossings, then the index is given by:

$$n = \sim(s \wedge (s >> 1))$$

where ">>" denotes right-shift by one binary bit (destroying the original right-hand bit) and " ~ " denotes binary complement.

4) Show that the inverse of the sequency-to-index operation can be written:

$$\sim s = n \wedge (n >> 1) \wedge (n >> 2) \wedge \ldots$$

5) Sketch the Walsh functions, say for $N = 16$, in sequency order.

6) Develop an analog to the Danielson-Lanczos FFT recursion formula, and write a fast WHT algorithm in recursive style.

7) Work out a fast algorithm for the WHT. Appendix "walsh.c" shows how relatively easy the Walsh-Hadamard butterflies can be computed. Time the fast software against other transforms.

8) Work out means to plot the celebrated two-dimensional Walsh density, which would be a gray-level or colored plot of the product

$$Wal_{Nx}(Ny)$$

over the unit x-y square. An example plot is the frontispiece for the present chapter.

9) Explain the relationship between the order $N = 2^k$ WHT and the k-dimensional DFT having run-length 2 in each dimension.

10) By inspecting for various $N = 2^m$ the N-by-N matrix that represents the WHT of order N, work out a recursive algorithm for construction of these matrices, called Hadamard matrices.

An interesting new transform, called by [Pender and Covey 1992] a square-wave transform (SWT), has some potentially powerful properties. Like the Walsh-Hadamard transform, only additions and subtractions are required; no multiplies. Furthermore, and unlike the WHT, there is an $O(N)$ algorithm for the forward SWT, the inverse being of complexity

$O(N \log N)$. The SWT is defined in terms of the Rademacher functions, discrete versions of which are:

$$R_j(t) = (-1)^{\left[2^j t/N\right]}$$

where j runs through $0,1,2,...,N/2-1$; t runs through $0,1,2,...,N-1$; and [ ] denotes greatest integer. It is convenient first to define the *inverse* SWT, as follows. For length-N signal x and SWT transform $S = Qx$, where Q is thought of as an N-by-N matrix, we have:

$$x = Q^{-1} S$$

$$Q^{-1} = \begin{bmatrix} \vdots & \vdots & \vdots & & \vdots & \vdots \\ V_1 & V_2 & V_3 & \cdots & V_{N-1} & V_0 \\ \vdots & \vdots & \vdots & & \vdots & \vdots \end{bmatrix}$$

where the V_k are column vectors whose t-th row component (with t taking values from 0 through $N-1$ inclusive) is a Rademacher evaluation according to the following groupings:

$$V_0 = \{R_0(t)\}$$

$$V_1, \ldots, V_{N/2} = \left\{R_1\left(t+\frac{N}{2}-1\right)\right\}, \left\{R_1\left(t+\frac{N}{2}-2\right)\right\}, \ldots, \{R_1(t)\}$$

$$V_{N/2+1}, \ldots, V_{N/2+N/4} = \left\{R_2\left(t+\frac{N}{4}-1\right)\right\}, \left\{R_2\left(t+\frac{N}{4}-2\right)\right\}, \ldots, \{R_2(t)\}$$

and so on until V_{N-1} is defined as $\{R_{N/2-1}(t)\}$. An example of a forward SWT is given by the matrix Q for $N = 16$:

$$16\ Q \ = \ \begin{matrix}
4 & -4 & 0 & 0 & 0 & 0 & 0 & 0 & -4 & 4 & 0 & 0 & 0 & 0 & 0 & 0 \\
0 & 4 & -4 & 0 & 0 & 0 & 0 & 0 & 0 & -4 & 4 & 0 & 0 & 0 & 0 & 0 \\
0 & 0 & 4 & -4 & 0 & 0 & 0 & 0 & 0 & 0 & -4 & 4 & 0 & 0 & 0 & 0 \\
0 & 0 & 0 & 4 & -4 & 0 & 0 & 0 & 0 & 0 & 0 & -4 & 4 & 0 & 0 & 0 \\
0 & 0 & 0 & 0 & 4 & -4 & 0 & 0 & 0 & 0 & 0 & 0 & -4 & 4 & 0 & 0 \\
0 & 0 & 0 & 0 & 0 & 4 & -4 & 0 & 0 & 0 & 0 & 0 & 0 & -4 & 4 & 0 \\
0 & 0 & 0 & 0 & 0 & 0 & 4 & -4 & 0 & 0 & 0 & 0 & 0 & 0 & -4 & 4 \\
4 & 0 & 0 & 0 & 0 & 0 & 0 & 4 & -4 & 0 & 0 & 0 & 0 & 0 & 0 & -4 \\
2 & -2 & 0 & 0 & -2 & 2 & 0 & 0 & 2 & -2 & 0 & 0 & -2 & 2 & 0 & 0 \\
0 & 2 & -2 & 0 & 0 & -2 & 2 & 0 & 0 & 2 & -2 & 0 & 0 & -2 & 2 & 0 \\
0 & 0 & 2 & -2 & 0 & 0 & -2 & 2 & 0 & 0 & 2 & -2 & 0 & 0 & -2 & 2 \\
2 & 0 & 0 & 2 & -2 & 0 & 0 & -2 & 2 & 0 & 0 & 2 & -2 & 0 & 0 & -2 \\
1 & -1 & -1 & 1 & 1 & -1 & -1 & 1 & 1 & -1 & -1 & 1 & 1 & -1 & -1 & 1 \\
1 & 1 & -1 & -1 & 1 & 1 & -1 & -1 & 1 & 1 & -1 & -1 & 1 & 1 & -1 & -1 \\
1 & -1 & 1 & -1 & 1 & -1 & 1 & -1 & 1 & -1 & 1 & -1 & 1 & -1 & 1 & -1 \\
1 & 1 & 1 & 1 & 1 & 1 & 1 & 1 & 1 & 1 & 1 & 1 & 1 & 1 & 1 & 1
\end{matrix}$$

It is of interest that this forward transform can be performed in $O(N)$ additions, while the inverse transform apparently requires $O(N \log N)$.

Project 4.3.4 Square-wave transform

Computation, graphics, theory, difficulty level 2-3.
Support code: Appendix "FSWT.ma."

1) For the $N = 16$ SWT matrix above, find the inverse matrix and, merely by reading off some columns, make plots of various of Pender and Covey's square wave basis functions $V_k(t)$. See step (6) below.

2) Develop a fast square-wave transform (FSWT) and associated butterfly diagrams (the latter for small N such as 8 or 16). The Appendix code example ignores the power-of-2 scalings; e.g. it assumes the forward Q matrix in the above text example has only 0, +1, –1 as possible entries.

3) As in step (2) previous, develop a fast inverse SWT algorithm.

4) Argue that the forward transform can be done generally in $O(N)$ total

additions (and perhaps some scalings by powers of two). Argue that the inverse SWT can be done in $O(N \log N)$ additions.

5) Investigate algebraic properties of the basis vectors V_k. For example, are these basis vectors orthonormal? Are they linearly independent?

6) As in step (8), Project 4.3.3, make a two-dimensional plot of the basis functions $V_k(t)$, where t runs horizontally and k runs vertically. An example plot is shown in Figure 4.3.1, for $N = 16$ basis functions:

Figure 4.3.1: The order-16 Pender-Covey square wave basis functions, plotted with indices running from $k = 1$ (bottom row) to $k = 15$ and finally $k = 0$ (top row) White means +1, black means −1. One can see that the basis functions are not orthogonal, but they are linearly independent.

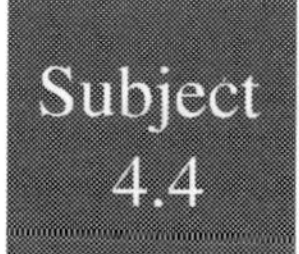

Number-theoretic transforms

The transforms so far covered in this chapter are suitable for application to signals consisting of floating-point data. But when data are known to be integer-valued, or at least to consist of elements in an algebraic ring, a host of number-theoretic transforms can be brought to bear, especially on problems of convolution and multiplication.

Let a signal x consist of N ring elements and let g be an N-th root of unity in the ring. Then a useful transform is

$$T_k = \sum_{j=0}^{N-1} x_j \, g^{-jk} \tag{4.4.1}$$

where the implied summation and multiplication are ring operations. Under certain further conditions on g and N, a valid inverse transform is

$$x_j = N^{-1} \sum_{k=0}^{N-1} T_k \, g^{+jk} \tag{4.4.2}$$

One advantage of such transforms lies in their numerical exactitude: one may handle a given problem with integer arithmetic throughout. Another advantage that often matters is speed, especially when the ring arithmetic allows simplifications, such as binary shifting or other logical operations.

A primary disadvantage of number-theoretic transforms is that run-lengths are tightly constrained. Much research has gone, and will go, into the problem of more general (especially longer) possible run-lengths for the transforms.

Project 4.4.1 Exploring Number-theoretic transforms

Computation, theory, difficulty level 3-4.

1) Give general conditions on g and N such that the inverse transform (4.4.2) exists.

2) One common algebraic domain for such transforms is the finite field $GF(p)$ where p is a prime (that is, just the integers (mod p)). One assumes that all elements of x are integers (mod p), and that g is a primitive N-th root of unity (mod p). Show that under these conditions the inverse (4.4.2) exists.

3) If the algebraic ring is Z_m, where m is not necessarily prime, give conditions that ensure the existence of the inverse transform.

4) State an analog to the Convolution Theorem (4.1.4), together with any extra necessary conditions beyond those of step (1) previous.

5) Write software that will allow a prime p, run-length N, and a primitive N-th root of unity to be passed; and performs a fast transform (mod p) as described in step (2) previous. This technique is useful for high-precision arithmetic, because one may take the fast transform (mod p) for various distinct small p, then reconstruct the (possibly large) convolution elements using the Chinese remainder theorem (CRT). In writing such transforms, the butterfly style of the classical FFTs in Appendix code "FFTs.ma" should be easy to mimic, where one must take care to generate the required powers (mod p) of the primitive root somehow. It is sometimes better to precompute a table of all powers of the primitive root, just as tables of sin/cos values have been contemplated in previous FFT projects.

6) One useful example is the Mersenne transform, defined for a Mersenne prime $p = 2^q - 1$. One may consider:

$$T_k = \sum_{j=0}^{q-1} x_j \, 2^{-jk} \ (mod\, p)$$

Implement a fast Mersenne transform by following the butterfly style of a classical FFT. Attempt to perform a maximum amount of the arithmetic using binary shifts [McClellan and Rader 1979][Nussbaumer 1981]. See step (1), Project 3.3.2, for certain considerations involving arithmetic (mod p) when p has special form.

7) Another interesting case is the Fermat number transform (FNT), defined over integers (mod F_n) where $F_n = 2^{2^n} + 1$. Clearly for run-length $N = 2^n$ one can take $g = 4$. Find a root of order 2^{n+1}(mod F_n), which allows for a longer run-length. Implement a fast FNT by following the butterfly style of a classical FFT. Attempt to optimize by using shifts as discussed in step (1), Project 3.3.2.

8) It turns out that if one is willing to do arithmetic with Gaussian (complex) integers, there are transforms that will offer the reward of enormous run-lengths. Investigate what might be called Galois transforms, for which the field is taken to be $GF(p^2)$. It is convenient to take $p = 3$ (mod 4); for example, p can be a Mersenne prime $2^q–1$. All arithmetic is performed (mod p), but signal elements and the root g are generally represented as complex Gaussian integers $a+bi$, and at all stages both real and imaginary parts are reduced independently (mod p). These transforms support large power-of-two run-lengths [Crandall and Fagin 1994]. Taking $p = 2^{61}–1$, $N = 2^{16}$, show that

$$g = 1510466207055935382 + 1200425448497313353\, i$$

is a primitive N-th root of unity (mod p). The resulting length 65536 transform could be used, for example, to assail number-theoretic problems involving numbers even larger than the largest known explicit primes of today (such as the great Mersenne primes listed in Chapter 3). One problem whose resolution is unknown to the author is this: Can these complex Galois transforms be given a split-radix fast algorithm?

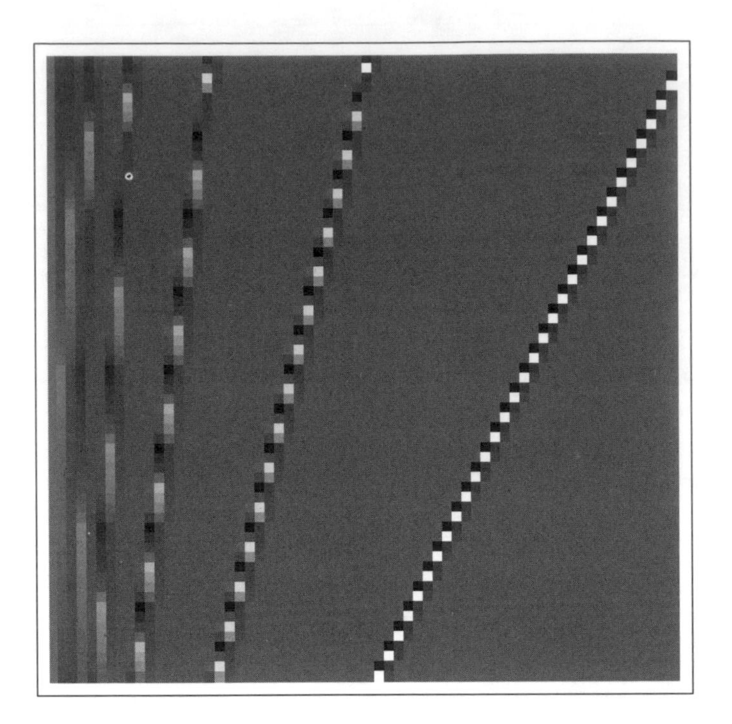

Portrait of an inverse fast wavelet transform matrix, together with a discrete Daubechies wavelet. The gray level in the 4096-element matrix corresponds to element value. Trains of scaled and translated wavelets traverse the matrix. Each matrix column is a discrete representation of a member of the orthonormal order-4 Daubechies wavelet basis. The ribbon was created by extracting and replotting the eighth matrix column.

5 Wavelets

Young arrivals in the transform family

Wavelets are based on translation . . . and above all on dilation. . . It is remarkable how long it has taken for "dilation equations" to be mentioned beside differential equations and difference equations. True, they are hardly in the same league. But ideas about wavelets are coming fast . . .

[Strang 1989]

It is a beautiful thing when a new field of thought grows out of practical necessity, eventually to become a focus for intense, renewed research in pure mathematics. The modern wavelet field (and even the name "wavelet") sprang from seismic analysis in the early 1980s, although many of the basic notions have long been known to engineers and mathematicians. Some such notions arose in the 1950s and 1960s with the disparate works of Gabor on signal processing and Calderon on harmonic analysis; going back even to the beginning of this century, for example to the pioneering work of Haar on orthogonal bases. But the modern direction started not only from an interest in analyzing particular wave phenomena; there has for decades been a consternation relating to the non-locality of Fourier bases. A perturbation–intentional or otherwise–in a traditional, non-local basis expansion can cause ringing, aliasing, and generally, repercussions all over the relevant domain. But with modern wavelets we can use basis functions that are contained neatly in finite regions, thus lending themselves handily to certain problems of signal processing and compression. With wavelets, one can perform multiresolution analysis, literally sorting signal components by their location and resolution scale. Thus, whereas classical Fourier methods sort signals into their spectra, the wavelet transforms sort signal details into a locale-scale collection. Transients (as in seismic disturbance!) are in some sense the opposite of pure tone waves, so are perhaps better approached this way on grounds of detail and spatio-temporal structure.

It is difficult responsibly to reference the origins of wavelet theory, not to mention the current state of the field. Still, one must associate with the development of the modern theory the names of [Morlet et al. 1982][Grossman, in [Combes. et al. 1987]][Coifman, in [Meyer 1991]][Mallat 1989][Daubechies 1988, 1992] and many others. Specific references appear later within Projects. Wavelets already enjoy connection with many fields, which we exhibit here possibly to intrigue the reader who wishes to go beyond the present chapter (note also that several of the references just named are excellent survey works). The following lists are nowhere near exhaustive.

Aspects of pure mathematics:

- vector spaces
- orthonormal bases
- splines
- frames
- matrix algebra
- functional spaces
- coherent states
- renormalization group
- phase space analysis
- elliptic operators

Applied fields:

- compression
- acoustics
- nuclear engineering
- sub-band coding
- signal and image processing
- neurophysiology
- music
- magnetic resonance imaging
- speech discrimination
- optics
- fractal analysis

The theory of wavelets is deemed so compelling that within this chapter a fair amount of theoretical background is presented, finally ending up in computational exercise. Perhaps the single most important finding on the computational side is that fast wavelet algorithms exist that require, for signals of length N, only $O(N)$ operations. It is the implementation of such algorithms that amounts to the primary goal of this chapter.

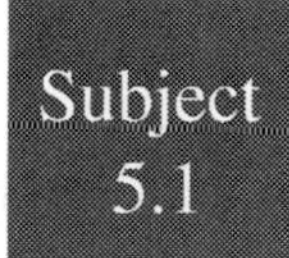

Chords, notes, and little waves

We begin our tour of wavelet theory with consideration of some classical continuous transforms that give rise to the notions of musical "chords" and "notes." We end up with the notion of continuous, and ultimately discrete wavelet transforms. We start with the continuous Fourier transform whose discrete counterpart was the central theme of Chapter 4. For a signal–now a continuous function $f(t)$ in L_2, meaning f is square-integrable–define the Fourier transform

$$f^{\sim}(\omega) = \int_{-\infty}^{\infty} f(t)\, e^{-i\omega t}\, dt \tag{5.1.1}$$

whose inverse transform is

$$f(t) = \frac{1}{2\pi}\int_{-\infty}^{\infty} f^{\sim}(\omega)\, e^{i\omega t}\, d\omega \tag{5.1.2}$$

As has been well known and celebrated for many decades, the transform $f^{\sim}(\omega)$ provides a measure of signal content at frequency ω. A good metaphorical exercise is to think of the function $f^{\sim}$ as defining a "chord" containing just the right continuous mixture of amplitudes $|f^{\sim}(\omega)|$ such that the playing of same (with the precise phases dictated by the complex value of $f^{\sim}$) will precisely reconstruct the original signal $f(t)$. On the other hand, there is the notion of "musical notes" that arrive at different times; actually at approximate times, for it is the nature of tone bursts–an uncertainty principle of acoustics–that they cannot be given precise arrival times. In order to model this notion of serial notes, we consider a windowed Fourier transform that depends on a suitable window function $g(t)$:

$$T(\omega, s) = \int_{-\infty}^{\infty} f(t)\, g^*(t-s) e^{-i\omega t}\, dt \tag{5.1.3}$$

The function g is often chosen to be a pulse localized in time, so that only those phenomena occurring approximately at time s are selected out by the integral. In this way the transform T indeed measures the signal f's frequency content, but only that content occurring near the specific time s. It is a useful observation that a correlation (similar to a convolution) of two signals f and g, namely:

$$(f \otimes g)(s) = \int_{-\infty}^{\infty} f(t)\, g^*(t-s)\, dt \tag{5.1.4}$$

is simply the special evaluation $T(0,s)$. If g is an even function, as often happens, this correlation is equivalent to the convolution.

There are various ways to invert this windowed transform to recover the original signal f. Sometimes one may resconstruct f from the ω-dependence of T, and sometimes from the s dependence. It will be interesting later when, for wavelet transforms, we typically use a full two-dimensional dependence to recover signals.

Project 5.1.1 Windowed Fourier transform

Theory, possible computation, difficulty level 1-3.

1) Derive a formal inversion of (5.1.3); namely:

$$f(t) = \frac{1}{2\pi\, g^*(0)} \int_{-\infty}^{\infty} T(\omega, t) e^{i\omega t}\, d\omega \tag{5.1.5}$$

Note that the sampling pulse value $g(0)$ must not vanish in order for this inversion to work.

2) If one knows g and the correlation (5.1.4) of f and g, the task of decorrelation (to find f) is sometimes possible. Show that if the Fourier transform $g^{\sim}(\lambda)$ has no zeros for λ real, then

$$f^{\sim}(\lambda) = \frac{1}{g^{\sim *}(\lambda)} \int_{-\infty}^{\infty} T(0, s) e^{-i\lambda s} \, ds \qquad (5.1.6)$$

which amounts to a continuous reminiscence of the discrete Convolution Theorem (4.1.4). In turn, $f(t)$ may then be recovered by inverse Fourier transform on $f^{\sim}$.

3) Here we do an explicit decorrelation. Assume a Laplacian sampling pulse $g(t) = e^{-|t|}$ and a correlation:

$$(f \otimes g)(s) = (1 + |s|)\, e^{-|s|}$$

and from the decorrelation formula (5.1.6) recover the original signal $f(t)$. After finding this f, verify the correlation claim above by direct integration as in (5.1.4).

4) Assume a Gaussian sampling pulse $g(t) = e^{-t^2}$, and assume a frequency modulated (FM) signal $f(t) = e^{iat^2}$. Argue first that the instantaneous angular frequency of f (the derivative of the phase) rises linearly in time. Obtain an exact expression for the windowed Fourier transform and show that its magnitude is:

$$|T(\omega, s)| = \frac{\sqrt{\pi}}{(1 + a^2)^{1/4}} \, e^{-\left(\frac{\omega}{2} - a s\right)^2 / \left(1 + a^2\right)}$$

From this formula, explain the sense in which T "picks out" frequency phenomena that occur near time s. The windowed Fourier transform for such a Gaussian is known as a Gabor transform, dating back to the 1940-1950s [Chui 1992 I].

5) Develop software that calculates, via FFT techniques, numerical approximations to (5.1.3) and (5.1.5). Test the software on exactly solvable cases such as in step (4) previous. Note the remarks of step (3),

Project 4.2.7 in regard to integration via FFT.

Now we exhibit the continuous wavelet transform as a certain generalization of the windowed transform concept. The basic form of a continuous wavelet transform, again for square-integrable signals f, is:

$$F(a,b) \;=\; \frac{1}{\sqrt{|a|}} \int_{-\infty}^{\infty} f(t)\, W^*\left(\frac{t-b}{a}\right) dt \qquad (5.1.7)$$

where W is the wavelet, literally a "little wave," generally a time-localized disturbance. $F(a, b)$ essentially measures the content of signal f occurring at time locale b, at scale a. Here is how one can visualize the fundamental behavior of this wavelet transform. As an example, imagine W to be a wiggle of duration about unity, whose integral is zero (in engineering terms, W is an "AC pulse"). If f in the neighborhood of time b has no appreciable detail at scale a (i.e., if f is relatively smooth over the interval $(b–a, b+a)$), then this interval's contribution to the integral (5.1.7) will be accordingly small. In such cases F is revealing to us that "at time b there is little detail at scale a." On the other hand, if f possesses tight, high-resolution phenomena, then there may be appreciable contributions to the integral when a is about the duration of these tight phenomena. In this way the transform F reveals, on a two-dimensional a-b plane, a breakdown of where and at what scales live the details of f.

Remarkably, under certain admissibility conditions on the wavelet W, an inverse of (5.1.7) can be written down. This inverse transform involves an integral over the locale-scale plane:

$$f(t) \;=\; \frac{1}{C[W]} \int_{-\infty}^{\infty} \int_{-\infty}^{\infty} \frac{da\, db}{|a|^{5/2}}\, F(a,b)\, W\left(\frac{t-b}{a}\right) \qquad (5.1.8)$$

where $C[W]$ is a constant depending only on the W function. As we see in the following Project, an implication of the admissibility condition on W is its "AC" nature, i.e., W should have vanishing integral.

Project 5.1.2

Continuous wavelet transform

Theory, possible computation, possible graphics, difficulty level 1-3.

1) Show that (5.1.8) is a formal inverse of the transform (5.1.7), with:

$$C[W] = \int_{-\infty}^{\infty} \frac{d\omega}{|\omega|} |W^{\sim}(\omega)|^2 \qquad (5.1.9)$$

where the Fourier transform $W^{\sim}$ is defined as in (5.1.1). Hint: insert the defining integral for F into the inverse formula (5.1.8), but with W removed in two places in favor of $W^{\sim}$. This shows that the existence of a non-vanishing C integral is an admissibility condition. Now conclude that the wavelet W is indeed an "AC pulse;" i.e., it satisfies:

$$\int_{-\infty}^{\infty} W(t)\, dt = 0$$

2) For given positive real A, find B such that the wavelet:

$$W(t) = (1 - Bt^2)\, e^{-At^2}$$

is admissible in the sense of step (1) previous, in so doing also giving the exact value for $C[W]$. Verify that for such a B the integral of W indeed vanishes. Because of its graphical appearance, this wavelet is sometimes called the "Mexican hat" wavelet [Daubechies 1992].

3) For the Mexican hat wavelet found in step (2) previous, with $A = 1$, find the exact wavelet transform $F(a,b)$, when f is the sum of two arbitrary real Gaussians:

$$f(t) = Ce^{-(t-u)^2/r^2} + De^{-(t-v)^2/s^2}$$

Explain how the wavelet transform thus picks out the scale and position of detail.

4) Exhibit graphically an exact continuous wavelet transform, by plotting

$F(a,b)$, from step (3) previous, as the height of a surface on the locale-scale (a-b) plane.

5) Do the graphics of step (4) previous, but instead of obtaining an exact F, create software that numerically performs the integral (5.1.7). In this regard see [Schiff 1992][Combes et al. 1987].

6) Investigate, in the manner of previous steps in this Project, the Littlewood-Paley wavelet:

$$W(t) = \frac{1}{\pi t}(\sin 2\pi t - \sin \pi t)$$

7) Investigate, in the manner of previous steps in this Project, the Morlet wavelet:

$$W(t) = e^{iat-t^2/2} - \sqrt{2}\, e^{iat-t^2-a^2/4}$$

8) Work out a Parseval identity for continuous wavelet transforms, by relating the signal power

$$\int_{-\infty}^{\infty} |f(t)|^2 \, dt$$

to a certain integral over the a-b plane. Test this theorem on special cases for f and g.

9) Investigate more of the vast array of theoretical aspects on continuous wavelets already to be found in the literature [Daubechies 1992][Chui 1992][Combes et al. 1987].

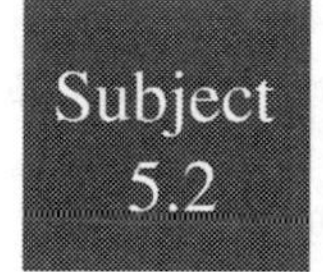

Discrete wavelet bases

By analogy with the continuous transform (5.1.7), we consider a discretized version:

$$F_{mn} = 2^{m/2} \int_{-\infty}^{\infty} f(t) W(2^m t - n) dt \qquad (5.2.1)$$

Instead of a locale-scale, or "*a-b*" plane, we now have a discrete lattice of integers m (so that scale is 2^{-m}) and n (so that the locale is $n/2^m$) that together index the transform. It is remarkable that wavelets W exist such that this use of only binary powers for scale, together with only integer translates, results in a discrete wavelet basis for square-integrable functions, and thus an invertible transform. In fact, if one can establish a general orthonormality condition for such a wavelet basis:

$$\int_{-\infty}^{\infty} W(2^a t - b)\, W(2^c t - d) dt = 2^{-a} \delta_{ac}\, \delta_{bd} \qquad (5.2.2)$$

then one expects a formal inverse:

$$f(t) = \sum_{m \in Z} \sum_{n \in Z} 2^{m/2} F_{mn} W(2^m t - n) \qquad (5.2.3)$$

which is a discrete version of the continuous inverse (5.1.8). The expansion coefficients thus tell us signal content at scale-locale pairs m, n. What one might call the "canonical" wavelet basis is generated by the Haar wavelet, defined by:

$$W(t) = \begin{array}{rl} 1 & , \ 0 \le t < 1/2 \\ -1 & , \ 1/2 \le t < 1 \\ 0 & , \ otherwise \end{array} \qquad (5.2.4)$$

amounting to a solitary cycle of a square wave of period unity. Though this Haar basis dates back to the early twentieth century, other localized wavelets comprise a genuinely new class. It is the intellectual leap that (5.2.2) may be satisfied by more complicated, but still localized waves that leads to modern wavelet theory.

Project 5.2.1 Example wavelet expansions

Theory, a little computation, difficulty level 1.

1) For the Haar wavelet (5.2.4), show that general orthonormality (5.2.2) is achieved.

2) For Laplacian signal $f(t) = e^{-k|t|}$, and the Haar wavelet as W, find the transform F_{mn} in closed form. Then, by evaluating the inverse transform for $f(0)$, establish the attractive identity

$$\sum_{m \in Z} 2^m \left(1 - x^{-2^{-m-1}}\right)^2 = \log x$$

valid at least for for $0 < x \leq 1$. Verify cases of this identity numerically.

3) Work out a Parseval identity for discretized wavelet transforms, by relating the signal power

$$\int_{-\infty}^{\infty} |f(t)|^2 \, dt$$

to a certain sum over the m-n (i.e., the scale-locale) lattice. Apply this theorem to special scenarios, such as that of step (2) previous.

A key observation in the 1980s was that suitable wavelets W can be constructed from certain "mother functions" Φ [Daubechies 1988]. Assume that Φ exists and satisfies a certain dilation equation with constant coefficients $\{h_k : k = 0,1,...,M-1\}$ where M is even. Note that M simply counts the number of coefficients. There is valid wavelet theory for which M can be infinite [Resnikoff 1989], but for the present treatment we consider finite even M. The dilation equation is taken to be:

$$\Phi(t) = \sum_{k=0} h_k \, \Phi(2t - k) \qquad ; k = 0,\ldots,M-1 \tag{5.2.5}$$

Assume also an orthonormality condition for integers a, b:

$$\int_{-\infty}^{\infty} \Phi(t-a)\,\Phi(t-b)\,dt = \delta_{ab} \tag{5.2.6}$$

Now we define the wavelet function explicitly as:

$$W(t) = \sum_{k=0} (-1)^k h_{M-1-k} \, \Phi(2t - k) \qquad ; k = 0,\ldots,M-1 \tag{5.2.7}$$

where here and elsewhere h_j with $j < 0$ or $j \geq M$ is taken to be zero. It is a beautiful thing that now an extra set of conditions on the h coefficients, namely:

$$\sum_{k=0} h_k h_{k+2q} = 2\,\delta_{0q} \tag{5.2.8}$$

for $q = 0,1,2,3,\ldots$, results in the W-orthonormality condition (5.2.2). As an example, the Haar system having wavelet (5.2.4) can be generated by $M = 2$, $h_0 = h_1 = 1$; and:

$$\Phi(t) = \begin{cases} 1 & , 0 \leq t < 1 \\ 0 & , \textit{otherwise} \end{cases} \tag{5.2.9}$$

Not only do the definition (5.2.7) and the condition (5.2.8) give rise to orthonormal wavelets, they also allow us to establish more than one basis representation of a square-integrable function. From (5.2.7) it is clear by induction that for any *negative* integer m, $W(2^m t - n)$ can be written as a linear superposition of integer translates $\Phi(t - k)$. But this means tha,t instead of the original discretized transform F_{mn}, we may as well define a transform as a collection $\{s_k, d_{mn}\}$, where:

$$s_k = \int_{-\infty}^{\infty} f(t)\, \Phi(t-k) \qquad (5.2.10)$$

$$d_{mn} = 2^{m/2} F_{mn} \quad ; m \geq 0 \qquad (5.2.11)$$

The normalization for the d terms with respect to F terms is for future convenience. Accordingly, f may be recovered via the inverse transform:

(5.2.12)

$$f(t) = \sum_{k\,\varepsilon Z} s_k\, \Phi(t-k) + \sum_{m\,\geq 0} \sum_{n\,\varepsilon Z} d_{mn}\, W(2^m t - n)$$

Note the essential difference between expansion (5.2.3) and its alternative (5.2.12): the latter has a truncated W sum. This split expansion is an example of the interesting algebra that obtains from the formal wavelet theory.

We shall see later how the splitting of expansion coefficients into these "s" and "d" sets gives rise to a recursive procedure known as "multiresolution analysis." For some sets $\{h_k\}$ this analysis can be performed efficiently via a fast wavelet transform. Indeed, we shall see that the "s" part of the expansion (5.2.12) can be broken down systematically and elegantly into the missing part of the m-sum. In the meantime, however, we next investigate properties of $\{\Phi, W\}$ basis collections.

Project 5.2.2 Mother function and its wavelet

Theory, a little computation, difficulty level 2.

1) Assume a Φ exists such that the dilation equation (5.2.5) holds. Show that if the integral of Φ does not vanish, then

$$\sum_{k=0} h_k = 2$$

The converse question, whether this simple identity is enough to ensure the existence of a non-trivial solution to the dilation equation, has been answered, essentially in the positive [Daubechies 1988][Meyer 1992].

2) For the Haar system starting with mother function (5.2.9), show that the dilation equation (5.2.5) is satisfied, that Φ-orthonormality (5.2.6) holds, and that the definition (5.2.7) indeed gives the Haar wavelet (5.2.4).

3) For the Laplacian signal of step (2), Project 5.2.1, expanded in the Haar wavelet system, use the split expansion (5.2.12) to establish the identity

$$\sum_{m=0}^{\infty} 2^m \left(1 - x^{-2^{-m-1}}\right)^2 = \frac{1}{x} + \log x - 1$$

and verify cases numerically.

4) Here we derive mutual orthonormality results for M even, using the hypotheses (5.2.5)-(5.2.8). Show first that W is orthogonal to every integer translate of Φ:

$$\int_{-\infty}^{\infty} \Phi(t-k)W(t)\,dt = 0$$

To find the underlying symmetry for this proof, it helps to write out some special cases of this integral in terms of h coefficients, for say $M = 4$ or 6. Argue from this result that for integers n, k and integer $m \geq 0$

$$\int_{-\infty}^{\infty} \Phi(t-k)W(2^m t - n)\,dt = 0$$

so that any second sum in the split expansion (5.2.12) is orthogonal to any first sum. Next, argue that for integers $m \neq n$:

$$\int_{-\infty}^{\infty} W(2^m t - a)W(2^n t - b)\,dt = 0$$

Finally–and now we definitely need the wavelet condition (5.2.8)–show that orthonormality (5.2.2) holds. Thus, the basis collection $\{\Phi, W\}$

appearing in the split expansion (5.2.12) enjoys mutual orthonormality.

5) Assuming the dilation equation (5.2.5), derive the following formal expression for the Fourier transform (defined as 5.1.1) of the mother function Φ:

$$\Phi^{\sim}(\omega) = \prod_{j=1}^{\infty}\left(\sum_{k}\frac{h_k}{2}\, e^{-i\omega k/2^j}\right) \tag{5.2.13}$$

For the Haar system, derive thus the identity:

$$\frac{\sin x}{x} = \prod_{j=1}^{\infty} \cos\left(\frac{x}{2^j}\right)$$

and test this identity numerically.

6) Let $H(\omega)$ denote the parenthesized sum in (5.2.13) but for $j = 0$. This is a digital filter function for the wavelet system. Show that wavelet conditions (5.2.8) are equivalent to the filter condition:

$$|H(\omega)|^2 + |H(\omega+\pi)|^2 = 1$$

which says in essence that the filter in question passes a constant energy 1 which in turn arises from low- and high-bands. Such filters can allow perfect reconstruction, as opposed to many filters that remove some information from a signal. Note the remarks of step (6), Project 5.3.2.

7) Show that orthornormality (5.2.6) is equivalent to:

$$\sum_{n \,\varepsilon\, Z} |\Phi^{\sim}(\omega + 2\pi n)|^2 = 1$$

and verify, theoretically or numerically or both, this identity for the Haar system.

8) On the normalization assumption

$$\int_{-\infty}^{\infty} \Phi(t)\, dt = 1$$

together with orthonormality (5.2.6), show that Φ provides a "partition of unity," that is, for all t,

$$\sum_{m \,\varepsilon Z} \Phi(t-m) = 1$$

One approach is to obtain the value $\Phi^{\sim}(0)$ and use the orthonormality equivalence in step (7) previous. The partition of unity provides a sharp test of numerical iteration schemes that compute mother functions.

A great discovery in modern wavelet theory is that one can actually find wavelets of compact support, meaning essentially that such wavelets are really and truly "localized;" in fact they vanish outside of a *finite* interval. For $M \geq 2$ even it turns out that if we posit wavelet conditions:

$$\sum_k h_{2k} = \sum_k h_{2k+1} = 1 \qquad (5.2.14)$$

$$\sum_k h_k \, h_{k+2q} = 2\delta_{q0}$$

then a mouthful of successes follow: a mother function of compact support exists, can be orthonormalized, satisfies a dilation equation (5.2.5), and gives rise to a wavelet of compact support, which is in turn orthonormalized and forms a basis for square-integrable signals in the sense of (5.2.3) [Daubechies 1988, 1992]. Armed with these compactly supported wavelets, we can use the split expansion (5.2.12) and a recursive formula to appear later, to engage in multiresolution analysis.

An order $M = 4$ Daubechies wavelet arises from the assignment:

$$\{h_k\} = \left\{\frac{1+\sqrt{3}}{4}, \frac{3+\sqrt{3}}{4}, \frac{3-\sqrt{3}}{4}, \frac{1-\sqrt{3}}{4}\right\} \qquad (5.2.15)$$

whose mother function $\Phi(t)$, and hence $W(t)$, turn out to vanish outside of the interval $t \in [0,3)$. Generally speaking, a compactly supported wavelet will vanish outside $[0,M-1)$. This is not the only wavelet of compact support of order 4, but it has a certain smoothness property unique to order-4 wavelets.

Project 5.2.3 Wavelets of compact support

Theory, computation, graphics, difficulty level 1-4.
Support code: Appendix "Daubechies.ma," "PhiHat.ma"

1) Show symbolically that the Daubechies coefficients (5.2.15) satisfy the wavelet conditions (5.2.14), which we write out explicitly here for clarity:

$$h_0 + h_2 = h_1 + h_3 = 1$$
$$h_0^2 + h_1^2 + h_2^2 + h_3^2 = 2$$
$$h_0\, h_2 + h_1\, h_3 = 0$$

Show further that a certain moment condition

$$0\,h_0 - 1h_1 + 2h_2 - 3h_3 = 0$$

is satisfied. This is a smoothness criterion which renders the Daubechies system unique up to trivial permutation, an important observation in signal processing.

2) Here we compute and plot the mother function and wavelet of compact support. The idea is to initialize an array–representing the mother function's values at discrete points–with all elements equal and non-zero. The array is thought of as spanning $[0, M-1]$. By iterating the dilation equation (5.2.5) on this array, one generally converges to an approximate shape for Φ . To this end, start with the Daubechies system (5.2.15) and

create Φ and W plots, which should appear essentially like so:

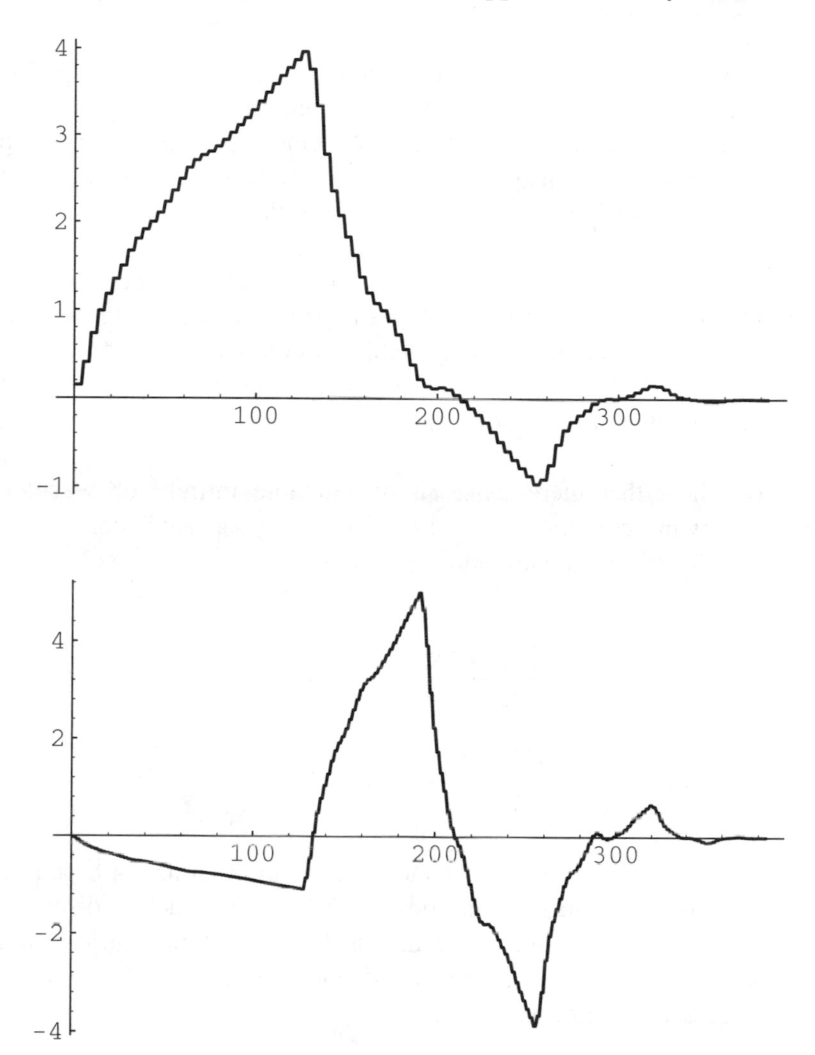

Figure 5.2.1: Daubechies mother function and wavelet, respectively obtained by array iteration of the dilation equation (the vertical axis is arbitrary and the horizontal axis corresponds to 128*t). The plots can be taken to much finer scale; the rough discretization shown is for purpose of illustration. In fact, these plots represent only the fifth iteration of the dilation equation on an array of 3*128 elements.

3) Using the product representation (5.2.13) plot the Fourier transform (actually one may plot just absolute value) of the Daubechies mother

function. Explain how the orthonormality equivalence in step (7), Project 5.2.2 is qualitatively evident.

4) Make again a plot of the Daubechies functions, as in step (2) previous, but this time use the product representation (5.2.13) for $\Phi^{\sim}$, and use an FFT to recover a discretization of Φ. One might think that this procedure would introduce unsightly errors, but one really can obtain plots that are indistinguishable from the finest limit of those of Figure 5.2.1.

5) Verify numerically for some computer-stored approximation to the Daubechies mother function that the partition-of-unity theorem of step(8), Project 5.2.2 holds to a reasonable approximation. In fact, a measure of departure from the partition property may be a good way to assess numerical approximations of mother functions.

6) Show that there exist an uncountable infinity of wavelet systems satisfying conditions (5.2.14), by verifying said conditions for the following general parametric solution:

$$h_0 = \frac{1+\sqrt{2}\,cos\,\theta}{2} = 1 - h_2$$

$$h_1 = \frac{1+\sqrt{2}\,sin\,\theta}{2} = 1 - h_3$$

7) Find some nontrivial wavelet systems of order $M = 4$ based on *rational* solutions $\{h_k\}$ to the conditions (5.2.14). One such is discussed in step (8) next. These are especially useful in computer applications of fast wavelet transforms, because of the potential simplifications in the necessary arithmetic.

8) Here is a hard but fascinating exercise in integral calculus; furthermore, calculus where one never knows the explicit function being integrated! Consider the mother function Φ for the system:

$$\{h_k\} = \left\{\frac{3}{5}, \frac{6}{5}, \frac{2}{5}, -\frac{1}{5}\right\}$$

which is easily seen to fulfill the wavelet conditions (5.2.14). We assume the usual normalizations:

$$\int_{-\infty}^{\infty} \Phi(t)\,dt \;=\; \int_{-\infty}^{\infty} \Phi^2(t)\,dt \;=\; 1$$

Prove that the cubic integral is not 1; rather:

$$\int_{-\infty}^{\infty} \Phi^3(t)\,dt \;=\; \frac{2492903}{2531945}$$

Similarly, the integral of the Daubechies mother function cubed (using 5.2.15) is 2321/2352, and one can even give a (monstrous) closed form expression for the cubic integral in terms of general h coefficients. In some ways these are the most maddening integrals the author has ever seen, because they come tantalizingly close to fulfilling "Fermi's Zero-th Rule," which in physics folklore states that a dimensionless integral that shows up in a random derivation tends to be 1. The author conjectures that there is an order-4 wavelet system having *maximum* cubic integral, for which h_0 is an algebraic number 0.62864... and for which the cubic integral is 1.38625... Another conjecture is that the cubic integral is exactly 1 only for degenerate h sets (that provide just a Haar system). Integrals of any higher powers of the mother function are unknown to the author. Then there is the forbidding problem: for constant k, what is the integral of $\exp(k\,\Phi)$?

9) For an $M = 4$ wavelet system, show that the conditions (5.2.14) imply yet more conditions, namely for the alternating sum:

$$h_0 \,-\, h_1 \,+\, h_2 \,-\, h_3 \;=\; 0$$

the cyclic product:

$$h_0 h_1 \,+\, h_1 h_2 \,+\, h_2 h_3 \,+\, h_3 h_0 \;=\; 1$$

and the end terms:

$$|h_0 h_3| \;\le\; 1$$

These conditions prove useful in fields such as signal processing and statistical applications of wavelets.

10) Study the fractal properties of the Daubechies wavelet system defined by (5.2.15). In particular, exactly where is the mother function differentiable? See [Daubechies 1988] on the notion that whether the function is differentiable at t depends on t's binary expansion!

11) Find an order $M = 6$ wavelet system by assuming an *ansatz*:

$$\{h_k\} \;=\; \{a+x,\ b+3x,\ c+2x,\ c-2x,\ b-3x,\ a-x\}$$

and finding a solution to conditions (5.2.14) in a, b, c, x. Plot a mother function as in step (2) or step (4) previous.

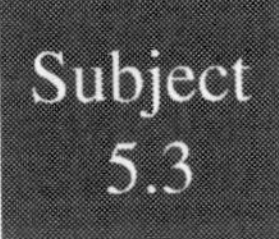

Subject 5.3 Discrete wavelet transform

Assume a wavelet system with even M and conditions (5.2.14) as in the previous Subject. Consider a discrete signal:

$$s \;=\; \{s_0,\ s_1,\ s_2,\ \ldots,\ s_{N-1}\} \tag{5.3.1}$$

where we shall assume, as one often does for discrete Fourier transforms, that N is a power of 2. Now imagine a "phantom" continuous signal function

$$f(t) \;=\; \sum_{j=0}^{N-1} s_j\, \Phi(t-j) \tag{5.3.2}$$

that is, a signal whose mother function components as in the split expansion (5.2.12) are just the components of signal s, and the d coefficients are zero. But, as we have seen before, it is possible to represent f as a superposition of W's alone, as in (5.2.3). Multiresolution analysis will be the procedure of casting the phantom signal (5.3.2) as successively scaled superpositions that accordingly involve more and more d terms. A recursive relation that ignites the whole procedure is the following:

$$\Phi(t-k) = \sum_{k} c_{ik}\, \Phi\left(\frac{t}{2}-i\right) + \sum_{m} d_{mk}\, W\left(\frac{t}{2}-m\right) \tag{5.3.3}$$

where the coefficients are the constants:

$$c_{ik} = \frac{1}{2} h_{k-2i} \qquad ; \; d_{mk} = \frac{1}{2}(-1)^k h_{M-1-k+2m} \tag{5.3.4}$$

Clearly we can insert the identity (5.3.3) into (5.3.2) to end up with a representation of the phantom signal that has some new mother function terms and some new W terms. This is the first stage of a discrete wavelet transform, which can be applied again and again, always to the prevailing mother function terms, using (5.3.3).

To avoid infinite-dimensional matrix algebra, it is convenient to periodicize the original discrete signal, much as one does in discrete Fourier analysis. One way to effect signal replication is to replace the phantom function definition (5.3.2) with:

$$f(t) = \sum_{j=0}^{N-1} s_j\, F_N(t-j) \tag{5.3.5}$$

where F_N is a periodic train of mother functions, which we define together with a periodic train of W functions, as:

$$F_N(t) = \sum_{a \varepsilon Z} \Phi(t-aN) \; ; \qquad G_N(t) = \sum_{a \varepsilon Z} W(t-aN) \tag{5.3.6}$$

Now the phantom signal (5.3.5) is is a replication, at intervals spaced apart by N, of the originally defined phantom function. This adjustment will make the discrete wavelet transform especially simple. For the periodicized functions, the fundamental multiresolution identity (5.3.3) can be written in terms of N-dimensional column vectors:

$$\begin{pmatrix} F_N(t) \\ F_N(t-1) \\ \vdots \\ F_N(t-N+1) \end{pmatrix} = M_N \begin{pmatrix} F_{N/2}\left(\frac{t}{2}\right) \\ \vdots \\ F_{N/2}\left(\frac{t}{2}-\frac{N}{2}+1\right) \\ G_{N/2}\left(\frac{t}{2}\right) \\ \vdots \\ G_{N/2}\left(\frac{t}{2}-\frac{N}{2}+1\right) \end{pmatrix} \tag{5.3.7}$$

where M_N is an N-by-N matrix whose entries are:

$$(M_N)_{ki} = \begin{cases} \dfrac{1}{2} \displaystyle\sum_{b \varepsilon Z} h_{k-bN-2i} & ; \; i < N/2 \\ \dfrac{1}{2} \displaystyle\sum_{b \varepsilon Z} (-1)^k h_{M-1-k+bN+2i} & ; \; i \geq N/2 \end{cases} \tag{5.3.8}$$

where all matrix indices start at 0 and run through N–1. Observe now that the phantom signal (5.3.5) is simply the dot product of the original signal vector s with the left-hand column of (5.3.7). But this means that the phantom signal is also the dot-product of ${M_N}^T s$ with the right-hand column of (5.3.7), where T denotes transpose. We are moved to think of ${M_N}^T s$ as a discrete wavelet transform of s. Actually this will be one stage of a full wavelet transform, and furthermore, for reasons of convenience in later work, we choose a certain normalization for the wavelet transform matrix. Such a choice of normalization is analogous to the choice for prefactors in forward- and inverse-DFTs. Putting all of this together, we define:

Single-stage discrete wavelet transform matrix: (5.3.9)

$$(T_N)_{ki} = \begin{cases} \dfrac{1}{\sqrt{2}} \sum\limits_{b \varepsilon Z} h_{k-bN-2i} & ; \; i < N/2 \\[2ex] \dfrac{1}{\sqrt{2}} \sum\limits_{b \varepsilon Z} (-1)^k h_{M-1-k+bN+2i} & ; \; i \geq N/2 \end{cases}$$

where again all matrix indices run from 0 through N–1. The T_N matrix thus defined is a unitary matrix that acts on column vectors of discrete data much like the standard DFT matrix does. The full wavelet transform is defined as a product of matrices derived from the T matrices:

Full wavelet transform matrix: (5.3.10)

$$W_N = T'_2 T'_4 \cdots T'_{N/2} T'_N$$

Where T' matrices are understood always to be N-by-N, obtained by filling out a sufficient number of 1's on the main diagonal for $T_{N/2}$, $T_{N/4}$, and so on. In other words, successive application of T'_k matrices are to affect only the top k terms of the current column. The count of these "s" terms will continue to be halved, leaving more "d" terms, until after the application of T_2 we end up with a solitary "s" term at the very top of the column, and N–1 "d" terms below it. The full transform is to be applied to original signal columns s of length N, so that $W_N\, s$ is another column we define as the discrete wavelet transform of s.

Here is one pictorialization of the discrete wavelet transform procedure, for which we use the example $N = 8$ for the length of the original signal:

(5.3.11)

$$s = \begin{pmatrix} s_0 \\ s_1 \\ s_2 \\ s_3 \\ s_4 \\ s_5 \\ s_6 \\ s_7 \end{pmatrix} \underset{T'_8}{\rightarrow} \begin{pmatrix} s \\ s \\ s \\ s \\ d \\ d \\ d \\ d \end{pmatrix} \underset{T'_4}{\rightarrow} \begin{pmatrix} s \\ s \\ d \\ d \\ d \\ d \\ d \\ d \end{pmatrix} \underset{T'_2}{\rightarrow} \begin{pmatrix} s \\ d \\ d \\ d \\ d \\ d \\ d \\ d \end{pmatrix} = W_8 s$$

From an algorithmic point of view, the important realization is that, once a "d" term is created, it is never again affected. This leads, as we shall see, to $O(N)$ fast algorithms. The successive columns provide a multiresolution analysis by exhibiting detail in the original data, but at changing scale and locale. In the above pictorial, the four "d" terms resulting from application of T'_8 represent the greatest signal detail; i.e., the smallest scale. Generally a slowly changing signal will have these four terms all small. After application of T'_4 there are two new d terms created, and these represent a larger scale; in fact, twice as large as before.

We next look at some special cases, to further illustrate the manner in which h coefficients enter into the transform operation. For the Haar system, the length-4 transform is:

(5.3.12)

$$W_4 = \frac{1}{\sqrt{2}} \begin{bmatrix} 1 & 1 & 0 & 0 \\ 1 & -1 & 0 & 0 \\ 0 & 0 & 1 & 0 \\ 0 & 0 & 0 & 1 \end{bmatrix} \frac{1}{\sqrt{2}} \begin{bmatrix} 1 & 1 & 0 & 0 \\ 0 & 0 & 1 & 1 \\ 1 & -1 & 0 & 0 \\ 0 & 0 & 1 & -1 \end{bmatrix}$$

The T matrix definition (5.3.9) involves a kind of "wrap-around" effect for the general wavelet matrix, although this is not yet visible for the Haar system. For an order $M = 4$ wavelet, the length-8 wavelet transform starts, at its first stage, with application to a signal of eight data the matrix:

$$T'_8 = \frac{1}{\sqrt{2}} \begin{bmatrix} h_0 & h_1 & h_2 & h_3 & 0 & 0 & 0 & 0 \\ 0 & 0 & h_0 & h_1 & h_2 & h_3 & 0 & 0 \\ 0 & 0 & 0 & 0 & h_0 & h_1 & h_2 & h_3 \\ h_2 & h_3 & 0 & 0 & 0 & 0 & h_0 & h_1 \\ h_3 & -h_2 & h_1 & -h_0 & 0 & 0 & 0 & 0 \\ 0 & 0 & h_3 & -h_2 & h_1 & -h_0 & 0 & 0 \\ 0 & 0 & 0 & 0 & h_3 & -h_2 & h_1 & -h_0 \\ h_1 & -h_0 & 0 & 0 & 0 & 0 & h_3 & -h_2 \end{bmatrix} \tag{5.3.13}$$

in which wrap-around is evident. Then one applies to the resulting column the matrix:

$$T'_4 = \frac{1}{\sqrt{2}} \begin{bmatrix} h_0 & h_1 & h_2 & h_3 & 0 & 0 & 0 & 0 \\ h_2 & h_3 & h_0 & h_1 & 0 & 0 & 0 & 0 \\ h_3 & -h_2 & h_1 & -h_0 & 0 & 0 & 0 & 0 \\ h_1 & -h_0 & h_3 & -h_2 & 0 & 0 & 0 & 0 \\ 0 & 0 & 0 & 0 & 1 & 0 & 0 & 0 \\ 0 & 0 & 0 & 0 & 0 & 1 & 0 & 0 \\ 0 & 0 & 0 & 0 & 0 & 0 & 1 & 0 \\ 0 & 0 & 0 & 0 & 0 & 0 & 0 & 1 \end{bmatrix} \tag{5.3.14}$$

Finally one applies the matrix:

$$T'_2 = \frac{1}{\sqrt{2}} \begin{bmatrix} h_0 + h_2 & h_1 + h_3 & 0 & 0 & 0 & 0 & 0 & 0 \\ h_3 + h_1 & -h_0 - h_2 & 0 & 0 & 0 & 0 & 0 & 0 \\ 0 & 0 & 1 & 0 & 0 & 0 & 0 & 0 \\ 0 & 0 & 0 & 1 & 0 & 0 & 0 & 0 \\ 0 & 0 & 0 & 0 & 1 & 0 & 0 & 0 \\ 0 & 0 & 0 & 0 & 0 & 1 & 0 & 0 \\ 0 & 0 & 0 & 0 & 0 & 0 & 1 & 0 \\ 0 & 0 & 0 & 0 & 0 & 0 & 0 & 1 \end{bmatrix} \tag{5.3.15}$$

which exhibits the rule–ultimately based on (5.3.9)–that wrap-around be additive.

Project 5.3.1

Fast wavelet transform algorithms

Computation, theory, graphics, difficulty level 2-3.
Support code: Appendix "fwt.[ch]," "FWTInPlace.ma," "FWTNotInPlace.ma," "WaveletMatrix.ma."

1) From the wavelet conditions (5.2.14) show that T_N is unitary (i.e., its inverse equals its transpose), and that the full matrix W_N is likewise unitary. Thus, *the inverse wavelet transform is obtained simply as* W_N *transpose.*

2) From unitarity as in step (1) previous, argue that at every stage of the full wavelet transform, a Parseval relation holds; viz., the sum of the squares of the column elements is an invariant, for all stages.

3) For an orthonormal basis collection $\{\Phi, W\}$ derive the fundamental multiresolution identity (5.3.3) with coefficients (5.3.4).

4) Here we investigate an explicit example of multiresolution. Consider the staircase function $f(t)$ shown below:

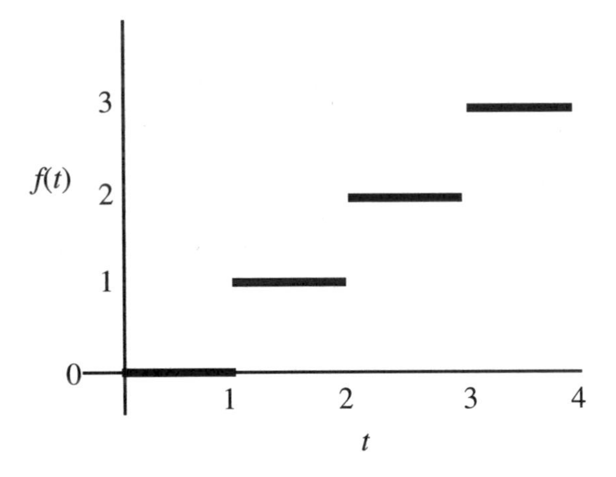

Here $f(t)$ is the greatest integer in t, except that f vanishes outside $[0,4)$. Clearly, using the Haar mother function, we have an expansion:

$$f(t) = 0\Phi(t) + 1\Phi(t-1) + 2\Phi(t-2) + 3\Phi(t-3)$$

in the style of (5.3.2). Show graphically how the successive application of the T' matrices in (5.3.12) corresponds to alternative expansion of f into functions $\Phi(t/2-k)$, $W(t/2-k)$ (for the first matrix application), then (for the second matrix application) the functions $\Phi(t/4)$, $W(t/4-k)$, and $W(t/2-k)$.

5) Draw butterfly diagrams in the style of Project 4.2.2, for the full Haar wavelet transform, say for $N = 8$. Notice that the diagram is "pruned" for d terms; that is, every time one subtracts instead of adds, the resulting term is never touched again. Optionally, make butterflies for higher-order wavelets as well. Argue on this basis that a fast $O(N)$ Haar transform is possible. Argue further that $O(N)$ is the complexity for any of the compactly supported wavelets we have been considering.

6) Implement software that constructs full wavelet matrices as products of T' matrices. The Appendix code "WaveletMatrix.ma" implements the arithmetic implicit in the definition (5.3.9). Test symbolically or numerically or both, the fact of unitarity of W_N as in step (1) previous.

7) Implement a fast wavelet transform by inferring the general pattern from (5.3.13)-(5.3.15). The Appendix code has *Mathematica* examples, both in-place and not-in-place, together with much faster C code. Verify that a forward-inverse transform results in a numerically reasonable identity. Verify also that a Parseval theorem (stability of the sum of squares of the data column upon any transform in any direction) is holding.

8) Provide yet another set of plots for the Daubechies system, by taking the inverse wavelet transform of some adroitly chosen special data. (How would you create pure sine or cosine waves, starting with an inverse FFT?) One can use either a fast wavelet transform algorithm, or explicit matrix contruction as in step (6) previous. Note that, as in step (1) previous, columns of the inverse transform matrix are rows of the forward matrix–a realization that underlies the frontispiece of this chapter.

Applications of wavelets abound, and several are found in other chapters of this book. To close this chapter, we investigate some general ideas for applications, leaving more field-specific applications for elsewhere.

Project 5.3.2

Applications of fast wavelet transforms

Computation, graphics, theory, difficulty level 2.
Support code: Appendix "WaveletOnSignal.ma, " "fwt.[ch]."

1) Apply a fast wavelet transform to some synthetic data, such as is done in the Appendix code "WaveletOnSignal.ma." Try several different signals until it becomes clear how high- and low-resolution details are sorted out by the transform. The beginnings of compression science occur when one tries to "throw out" small wavelet coefficients after the transform, and notices that the reconstruction is relatively good.

2) Here we investigate a fascinating connection of wavelets with statistics. Assume that random signal data has mean m and covariance given by a Markovian power law:

$$< s_j > \; = m$$

$$< (s_j - m)(s_k - m) > \; = \begin{cases} \sigma^2 & ; j = k \\ \rho^{2|j-k|} & ; j \neq k \end{cases}$$

where $\rho < 1$ determines how correlated are adjacent data. Consider order $M = 4$ wavelet systems and show that when a T_N matrix is applied to a column of N such data $\{s_j\}$, one has:

$$\textit{mean of resulting } (N/2) \text{ "}s\text{" } \textit{terms} = m\sqrt{2}$$
$$\textit{mean of resulting } (N/2) \text{ "}d\text{" } \textit{terms} = 0$$

which is interesting enough; but also derive the attractive theorem:

$$\textit{variance of resulting } (N/2) \text{ "}s\text{" } \textit{terms} \geq \sigma^2$$
$$\textit{variance of resulting } (N/2) \text{ "}d\text{" } \textit{terms} \leq \sigma^2$$

The results of step (9), Project 5.2.3 are useful in this analysis. This all means that the wavelet transform literally sorts variances, by depositing random variables of smaller variance into the high-detail components of the transform. This phenomenon can be thought of as a major reason why wavelets have application for sound and image compression: When data are sufficiently correlated, wavelets show what parts of the data require more bits of representation. As an option, consider testing the theorems of this step on pseudorandom data which you force to be correlated as hypothesized.

In addition, describe what happens when, for the Daubechies wavelet defined by (5.2.15), T_N is applied to a data column whose elements lie in arithmetic progression; i.e., $s_j = a + bj$. This has implications for signals possessed of linear changes. The conditions discuessed in step(1) and step(9), Project 5.2.3 are relevant here.

3) Investigate matrix algebra applications of wavelets. For example, consider an 8-by-8 matrix U whose entries are:

$$U_{ij} = 2 - 2\delta ij$$

By applying an order $M = 4$ wavelet transform in two-dimensional style; i.e. by computing symbolically:

$$W_8{}^T U W_8$$

then taking the determinant, show that det $U = -1792$. The idea is that a highly correlated matrix is often given large regions of zero–or nearly zero–elements by such a two-sided wavelet operation. The situation is analogous to that of image processing, as we discuss later in the book, with highly spatially correlated images having happily skewed distributions of wavelet components after the transform.

Other matrices interesting to study include, say, $U_{ij} = 1/(1 + |i-j|)$. A sufficiently large U of this highly correlated type becomes, after wavelet transforming, relatively sparse. Much theory and experiment has gone into this kind of new algorithm for efficient numerical algebra [Meyer 1992].

4) One way to effect a two-dimensional wavelet transform is to pass over all rows (of a rectangular, w-by-h signal array, both w and h being powers of 2) with a full, in-place transform; then likewise to pass over all columns. This is analogous to the two-dimensional FFT as in Project 4.2.6. But another (and faster!) way to provide a slightly different but practical transform is to perform the first T'_w stage on all the rows, then the first T'_h stage on columns, then $T'_{w/2}$ to half the rows, then $T'_{h/2}$ to half the columns, and so on. Argue that for either definition of the two-dimensional transform, the complexity is $O(wh)$. An interesting question is: for $w = h$, which one of these is equivalent to matrix multiplication $W_h S W_h^T$, where S is now an h-by-h matrix containing the signal, and W_h is the usual full wavelet transform matrix?

5) Implement two-dimensional fast wavelet transform software. Check that forward-inverse gives the identity operation, and compare performance with that for a two-dimensional FFT. Such software has many applications, such as image processing as described in Project 7.3.2.

6) Investigate the connection between wavelets and quadrature mirror filters (QMFs), the study of which in some ways predates modern wavelet theory. The basic idea is that filters exist which split signals into separate low- and high-frequency components, yet allow perfect reconstruction. The analysis of step (6), Project 5.2.2 is relevant to this analysis. See for example [Akansu et al. 1993][Daubechies 1992].

7) Investigate the connection between the discrete wavelet transform and the DFT. In this regard see [Resnikoff and Burrus 1990] who derive some interesting identites based on the relation between DFT and wavelet coefficients.

8) Project 6.2.3 involves methods for estimating fractal dimensions on the basis of scaling properties of sets and signals, which properties are often naturally exhibited under the wavelet transform, which transform, after all, is designed to reveal structure at various scales in the first place.

000	00**1**	0**1**1	01**0**	**1**10	11**1**	1**0**1	10**0**
200	2**1**0	**3**10	3**0**0	30**1**	3**1**1	**2**11	2**0**1
20**2**	2**1**2	**3**12	3**0**2	30**3**	3**1**3	**2**13	2**0**3
103	**0**03	00**2**	**1**02	1**1**2	**0**12	01**3**	**1**13
1**2**3	**0**23	02**2**	**1**22	1**3**2	**0**32	03**3**	**1**33
233	2**2**3	**3**23	3**3**3	33**2**	3**2**2	**2**22	2**3**2
23**1**	2**2**1	**3**21	3**3**1	33**0**	3**2**0	**2**20	2**3**0
130	13**1**	1**2**1	12**0**	**0**20	02**1**	0**3**1	03**0**

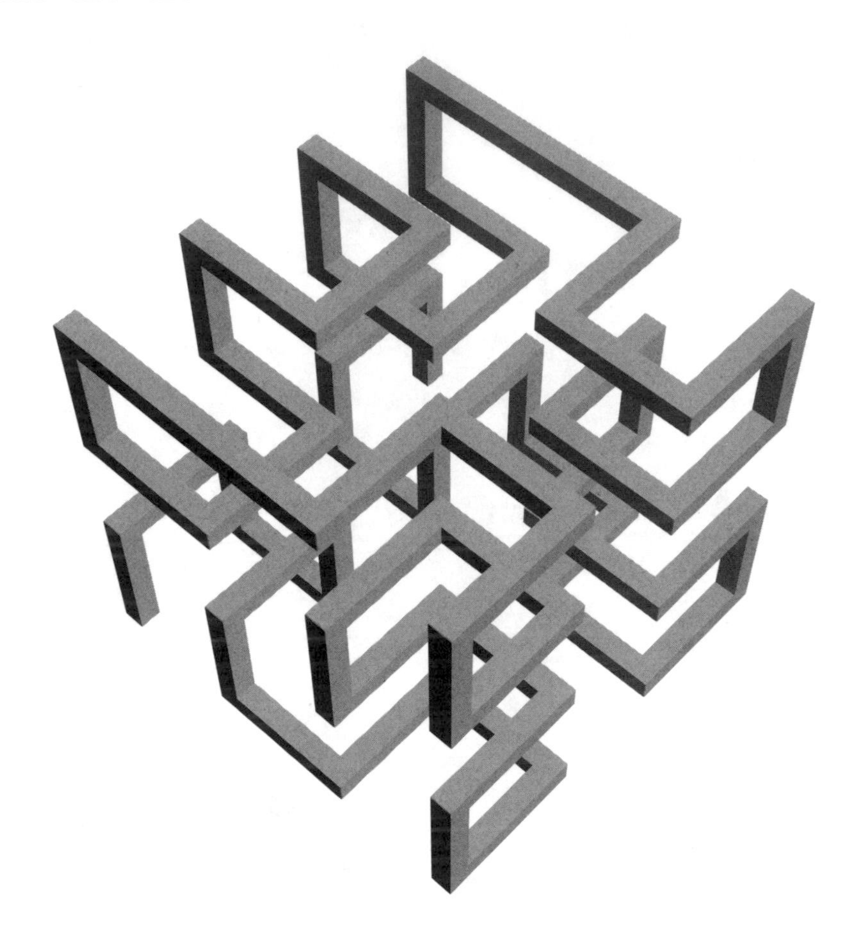

Third recursive stage in the generation of a Hilbert fractal, a space-filling curve. This stage exhibits 64 vertices, each representing a number in a base-4 Gray code: a number system shown lexicographically. As one counts, one and only one (boldfaced) digit has changed only by one. The curve manages to contact each lattice point in a cube, though always jumping just a unit distance in one of the three coordinate directions.

6 Complexity reigns

Chaos & fractals & such

The word "symmetric" *is of ancient Greek parentage and means well-proportioned, well-ordered–certainly nothing even remotely chaotic. Yet, paradoxically, self-similarity [a theme of fractal studies] alone among all the symmetries gives birth to its very antithesis:* chaos, *a state of utter confusion and disorder.*

[*Schroeder* 1991]

A meteorologist discovered the first strange attractor in an attempt to understand the unpredictability of the weather. A biologist promoted the study of the quadratic map in an attempt to understand population dynamics . . . engineers, computer scientists, and applied mathematicians gave us a wealth of problems. . . Nonlinear dynamics is interdisciplinary, and nonlinear dynamicists rely on their colleagues throughout all the sciences.

[*Tufillaro, Abbott, and Reilly* 1992]

In just these two quotes we see the words: symmetry, fractal, self-similarity, chaos, attractor, unpredictability, dynamics. It is now known that all of these concepts are intimately connected. The "chaos" in a dynamical system–which is essentially a quantifiable level of unpredictability–is exemplified in the infinitely recondite meanderings of the often-plotted Mandelbrot set, which is in turn a construct based on simple algebra and is possessed of certain symmetries. Likewise fractals can be in some sense infinitely complicated, yet just as often derivable from almost trivial rules. The author should perhaps have named this chapter "Complexity out of Simplicity."

In the Projects that follow, a balance is struck between theory and computation. It is felt important to establish at least the heuristic rules of the chaos game, so that computational problems will loom accordingly sensible.

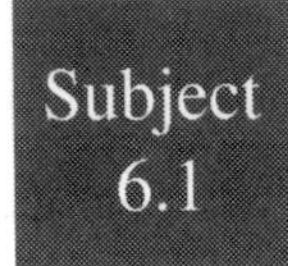

Chaos

A fundamental discrete chaotic system arises from the quadratic map:

$$x_{n+1} = ax_n(1 - x_n) \tag{6.1.1}$$

where a is a constant and the iteration starts from some chosen initial value x_0. This innocent looking iteration–which generates sequences $\{x\}$ of considerable complexity–turns out not to be "innocent" only because of discretization. In fact, the quadratic map is a discrete analog of the relatively ancient logistic equation of population biology [Verhulst 1845]:

$$\frac{dx}{dt} = \frac{r}{K}x(K - x) \tag{6.1.2}$$

Here, $x = x(t)$ is population as a function of time t, r is a (positive) growth parameter and K is a population limit. Note that some analyses of discretization of this model are also suggested in step (3), Project 2.3.1. This continuous logistic model exhibits no chaos of any kind; physically realistic solutions to (6.1.2) are always smooth and monotonic. A perturbation in the initial choice $x(0)$ causes, for a future finite time t, only a corresponding small change in the solution $x(t)$. On the other hand, the discretized logistic equation (6.1.1) exhibits chaos. Small perturbations in the initial value x_0 can give rise to wild excursions in the result x_n some fixed number n of iterates later.

Over the last decades much study has gone into the quadratic map. But the map has important relatives, some of whom are ideal for exhbiting specific aspects of chaos. The tent map is a kind of piecewise linearization of the quadratic map:

$$x_{n+1} = b - 2b\left|x_n - \frac{1}{2}\right| \tag{6.1.3}$$

and exhibits various phenomena qualitatively similar to those arising from the quadratic case. Another interesting system is the circle map:

$$\theta_{n+1} = \theta_n + \Omega - \frac{K}{2\pi} \sin 2\pi\theta_n \quad (mod\, 1) \tag{6.1.4}$$

where Ω, K are constants and (mod 1) means that every new θ_n is obtained as the fractional part of the right-hand side. This system exhibits certain "mode-locking" phenomena.

In two dimensions there are interesting discrete chaos models that often give rise to such phenomena as "fractals," "attractors," "horseshoes," "tongues," and so on. One instance is the baker's map:

$$y_{n+1} = \begin{cases} a\,y_n & ;\ 0 \le x_n < 1/2 \\ 1/2 + a y_n & ;\ 1/2 \le x_n < 1 \end{cases} \tag{6.1.5}$$

where $a \le 1/2$ is a constant. This map amounts to an x-stretching of a given region, followed by a vertical stacking of left and right halves of the stretched region. Iteration of the baker's map generates an interesting fractal. Of theoretical importance is Arnold's Cat map:

$$x_{n+1} = x_n + y_n \quad (mod\, 1) \tag{6.1.6}$$

$$y_{n+1} = x_n + 2y_n \quad (mod\, 1)$$

which has an area-preserving property typical of nonlinear Hamiltonian systems. A good general purpose two-dimensional chaos model is the Henon map:

$$x_{n+1} = 1 - a x_n^2 + y_n \tag{6.1.7}$$

$$y_{n+1} = b x_n$$

where a, b are constants. The Henon system is computationally straightforward, is possessed of attractor orbits exhibiting self-similarity, and yields experimentally measurable fractal dimensions.

Two-dimensional maps sometimes yield amusing fractal shapes, as obtains from the "gingerbread man" map:

$$x_{n+1} = 1 - y_n + |x_n| \tag{6.1.8}$$

$$y_{n+1} = x_n$$

whose simplicity denies the complexity–and one might say elegance–of the invariant set one may plot numerically.

Project 6.1.1 Quadratic map algebra

Theory, computation, graphics, difficulty level 1-3.
Support code: Appendix "TwoCycle.ma," "Invariant.ma."

1) Obtain the exact, closed-form solution to the continuous logistic differential equation (6.1.2) for given initial value $x(0) < K$. Argue that the Malthusian growth parameter r determines an exponential population growth when population itself is small. Argue that the environmental carrying capacity K is the asymptotic value of the population for large times. Optionally, create some typical plots of the population $x(t)$, on the basis of the exact solution.

Now assume that one wished to perform numerical solution for this continuous logistic model via simple Euler integration of a difference equation. Describe the relationship between one's chosen time step dt, the parameters r, K, and the quadratic map parameter a in (6.1.1), such that the latter map is the difference equation one would actually use in computations. The fact of chaotic behavior of the quadratic map, for certain a, is a valuable lesson on the potential dangers of numerical iteration as a scheme for modeling continuous systems.

2) Here we investigate some algebraic properties of the quadratic map (6.1.1). We shall assume a and all x values to be real. Note that the value $x = 0$ is a fixed point; i.e., if x_n is zero then all future iterates will be zero. What is another fixed point, this time depending on parameter $a > 1$?

Next, take the special case $a = 2$. Argue that under the iteration, *any* point in (0,1) will converge monotonically to the fixed point $x = 1/2$. This shows that there may not exist a pair of distinct points that map into each other.

Next we determine for what parameters a there do in fact exist distinct period-2 cycles. Assume the existence of points x, y such that $x = ay(1-y)$ and $y = ax(1-x)$. Show that (x, y) is one of the four solutions: $(0,0)$, $(1-1/a, 1-1/a)$, the pair:

$$\left(\frac{1+a+\sqrt{(a-3)(a+1)}}{2a}, \frac{1+a-\sqrt{(a-3)(a+1)}}{2a} \right)$$

or this last pair with the signs of the radicals swapped. Conclude that for $a > 3$ there is a genuine period-2 cycle; i.e., a distinct pair of real points that map into each other. It is tempting to try to find algebraic higher-order cycles, but for arbitrary a the algebra becomes intractable quickly. A workable special case for a is covered in the next step.

3) Here we find explicit period-n cycles for the special case $a = 4$. Show first that for this a parameter, a point $x = \sin^2 2\pi\omega$ maps into the point $\sin^2 4\pi\omega$. Describe how one can obtain successive iterates by inspecting the fractional binary expansion of ω_0 in $x_0 = \sin^2 2\pi\omega_0$. For arbitrary n, exhibit explicit starting values ω_0 that give rise to period-n cycles.

Now, still for parameter $a = 4$, we apply three different techniques for finding the so-called invariant distribution of the quadratic map. This will be a distribution–one could think of it as a population density $p(x)$ on (0,1) of a vast number of particles–that maps into itself. First, describe the probability density of a random variable $\sin^2\beta$, where β is randomly equidistributed (mod 2π). From the ideas of the previous paragraph, infer that this probability density is in fact the invariant distribution. Second, by finding what points map into an interval $(y, y+dy)$, establish the identity:

$$p(y) = \frac{1}{2}\left(p\left(\frac{1-\sqrt{1-y}}{2}\right) + p\left(\frac{1+\sqrt{1-y}}{2}\right)\right)$$

and solve this to obtain the invariant p whose integral over (0,1) is unity. Third, start with an array of points representing an equidistribution over (0,1), and numerically iterate the quadratic map with $a = 4$, to exhibit convergence to the invariant distribution. What works well is to have at least an order of magnitude more initial points than histogram bins. Iterate the quadratic map on all points, and at various times display bin populations as in Figure 6.1.1. Interesting questions are: Does the trend of *increasing* apparent noise from strips 3 through 7, say, in Figure 6.1.1 continue for future iterations? Is this noise fundamental (to an initial finite equispaced point set) or is it a computational artifact?

The $a = 4$ invariant distribution and that for $a = 3.7$ are discussed in [Holden 1986].

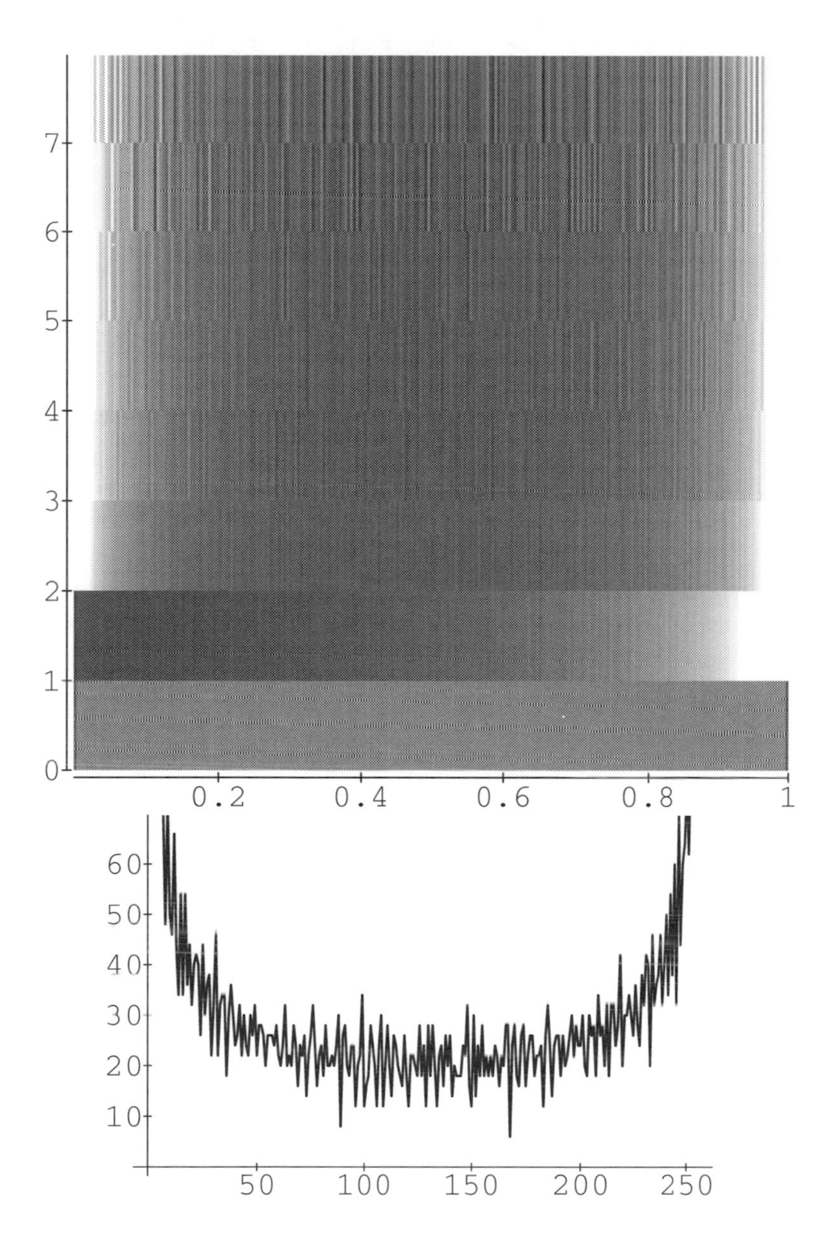

Figure 6.1.1: Iteration of the quadratic map with special parameter $a = 4$. In the top density plot, each horizontal strip is a density plot of the distribution on (0,1) at a specific iteration. Starting with the bottom strip as the equidistributed initial data, one obtains the top row (iteration 7) as an approximation to the theoretical invariant distribution. The bottom figure is another view of this 7th iteration result.

Project 6.1.2

Bifurcation and chaos

Theory, computation, graphics, difficulty level 1-3.
Support code: Appendix "Attractor8.ma."

1) Here we investigate, in a somewhat non-standard graphical fashion, the celebrated phenomenon of period-doubling. Track and plot an equidistribution of particles, using a histogram coarser than the particle density as in the last part of step (3), Project 6.1.1, but for other parameter choices such as $a = 2.5, 3.2, 3.5, 3.56$. These four cases should show, respectively, one, two, four, and eight attractor points. In other words, unlike the $a = 4$ case of the previous Project, a smooth and stable density does not necessarily develop for arbitrary a values; rather, it can happen that virtually all iterations tend to get caught in some n-cycle, in these cases for $n = 1,2,4,8$. An example plot for $a = 3.56$, showing 8-cycle attraction, appears below as Figure 6.1.2.

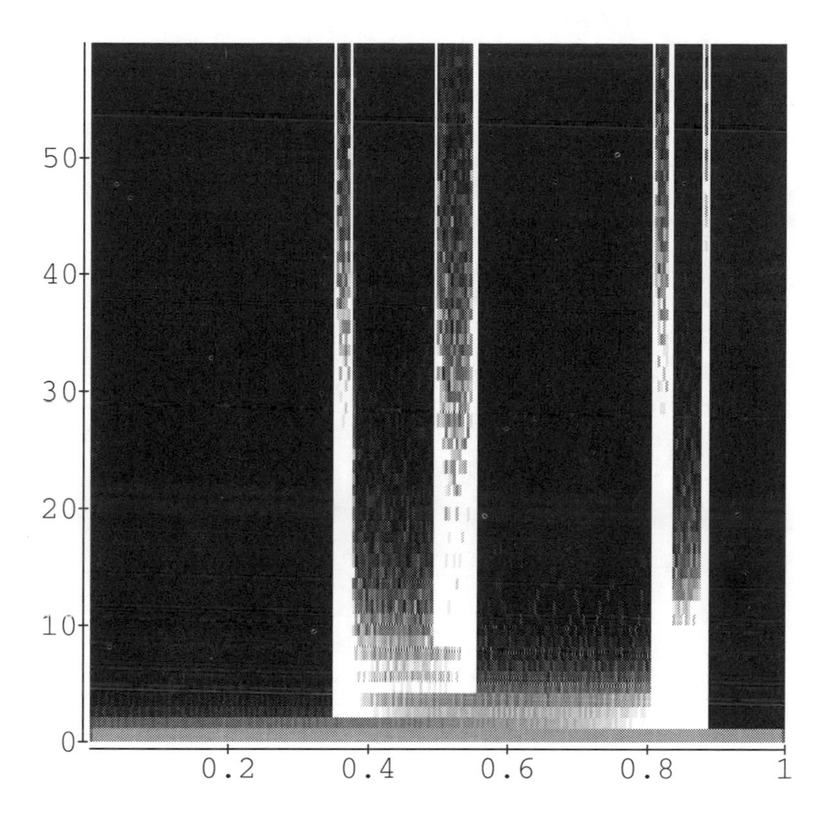

Figure 6.1.2: Density plot of the fate of an initially equidistributed point set, as in the upper part of Figure 6.1.1, except now the parameter is $a = 3.56$. The bottom horizontal row represents initial data, which evolves under iteration of the quadratic map. The first 10 or 20 iterations involve unclear dynamics, until eventually it becomes evident that some cycle of eight attractor points is the likely destiny of a starting point.

2) The attractor points of Figure 6.1.2 are, in the natural cyclic order under the quadratic map:

0.889892
0.348824
0.808639
0.550881
0.880784
0.373814

0.833314
0.49449

It is interesting that the two very closely spaced points (0.889..., 0.880...) evident in Figure 6.1.2 are four iterations apart–in fact each close pair of attractor points is likewise separated by four iterations. Using a plot of the parabola $y = 3.56\, x(1-x)$, show graphically how this 8-cycle comes about. To show such phenomena computationally, one may consider modifying software used for the some of the ensuing steps.

3) Here we investigate the classic bifurcation diagram for the quadratic map. The idea is to plot the final cyclic or chaotic states, as in the top rows of Figure 6.1.2, which is a cyclic case. The resulting plot will be the celebrated bifurcation diagram for the quadratic map. A schematic view of what should be plotted is shown in Figure 6.1.3:

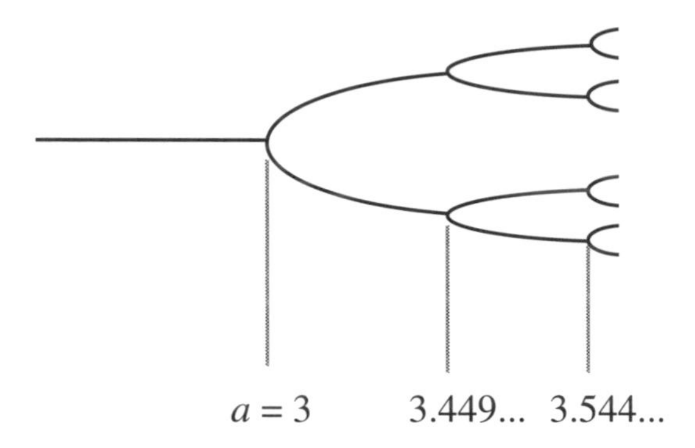

Figure 6.1.3: Schematic of the first few period doublings in a bifurcation diagram, showing approximate a values for bifurcation thresholds. This schematic is not drawn to scale, nor does it traverse a wide range such as $2 \leq a \leq 4$ as a good plot should. For one thing, just a little to the right of this schematic will be a series of regions that move from order to chaos and back.

In a good and proper plot, the horizontal axis should be the a parameter, running, say, from 2 to 4. One starts with a random x coordinate, and above each a value plots the set of points in (0,1) that are visited, but only after, say, 200 iterations *for that given a parameter* have been performed. By discarding thus the initial dithering of the map, one removes artifacts such as the bottom several rows of Figure 6.1.2. The classic bifurcation plot can be found in almost any elementary treatment of chaos [Baker and Golub 1990][Crandall 1991]. Note the consistency between Figure 6.1.2

for $a = 3.56$ and the period-8 region of the classic plot that evolves by bifurcation from the period-4 region. Note also the consistency between the invariant distribution plots of Figure 6.1.1 and the density of plotted points above $a = 4$ in the bifurcation plot.

4) Compute the universal Feigenbaum constant, which for the quadratic map is the limiting ratio of successive differences between bifurcation points. Specifically, if μ_1 is the ($a = 3$) bifurcation point where two stable attractor points appear, μ_2 is the point ($a \sim 3.45$) where the next bifurcation appears, and so on; then the Feigenbaum constant is defined as a limit [Feigenbaum 1978]:

$$F = \lim_{k \to \infty} \frac{\mu_k - \mu_{k-1}}{\mu_{k+1} - \mu_k}$$

In Figure 6.1.3, F is evident as the ratio of the distance between the first two labeled a values to the distance between the second and third values (and again we note the schematic is not drawn in linear a scale). This constant appears in many non-linear settings aside from the quadratic map. A high precision estimate of F is [Keiper 1992]:

$$F \sim 4.669201609102990671853\,20\ldots$$

which estimate enjoys the feature of having been computed directly on the basis of bifurcation points. F has been computed to extreme precision in the literature using such techniques as functional relations [Briggs 1991].

In the process of his calculation, Keiper also found that the sequence of bifurcation points starts with the following approximate values for the parameter a:

3
3.44948974278317809819728407470589139196594748065667012843154
3.54409035955192285361596598660480454058309984544457367545781
3.56440726609543259777355758652898245065773473837900855774148
3.56875941954382643129821028002531537035699383958078207940685
3.56969160980139671428826870629546660718657040829151781541733
3.56989125937812048732027120058544938979582512337441780403327

3.56993401837397640118485560188719137121928301200062290768034
3.56994317604840163635444297616231774954080666564062030611838
3.56994513734216979431472112274953091869132418785026388930119
3.56994555739124937611716994175620838011446395742974699701509
3.56994564735289979134350665376033672215937988990389898980943
3.56994566661993045219394370115113364346899845783748287442977
3.56994567074633840814496716196761457288538220359142067925408
3.56994567163008862964341529154614805624893639731770323985250
3.56994567181936086406353024354883458505367144030716584653539

The computation of Feigenbaum's F to the stated accuracy required several million CPU seconds on a workstation [Keiper 1992].

5) Choose one of the high-precision bifurcation thresholds (other than 3) of the last step and with multiprecision software take the value a few more decimals. One interesting approach is to iterate the quadratic map for some a values very slightly greater than the presumed bifurcation threshold, that is, a values slightly to the right of indicated thresholds in the schematic Figure 6.1.3. Then one may track leftward, and therefore estimate, for what projected a the relevant 2^n-cycle coalesces.

The quadratic map is the canonical, but not the only useful model of, chaos. We next investigate other one- and two-dimensional models.

Project 6.1.3 Chaos models

Theory, computation, graphics, difficulty level 1-3.
Support code: Appendix "Henon.ma," "Lorenz.ma," "Julia.ma," "NewtonTrefoil.ma."

1) For the tent map (6.1.3) perform various theoretical or computational analyses in the spirit of the quadratic map of Projects 6.1.1- 6.1.2:

- For given parameter b, is there a smooth invariant distribution?
- Is there a bifurcation behavior similar to the quadratic map's?
- Are there n-cycles for any positive integer n?

In particular, define an invariant set as a set of points which map into themselves under the tent map. For $b = 3/2$, the set of all points that *do not eventually diverge* out of (0,1) under the map is an invariant set. Show that the set in question is in fact the famous Cantor set (see the introduction of Subject 6.2). Also of interest is the case $b = 1$, for which the invariant distribution turns out to be especially simple.

2) Study the circle map (6.1.4). One interesting phenomenon associated with this kind of map is the appearance of what are known as "Arnold tongues" [Schroeder 1991], which amount to mode-locking regions in the Ω-K parameter plane. To witness mode-locking, plot via direct computation a kind of "devil's staircase" as follows. Assuming iteration of (6.1.4) starting with an initial θ_0, define an asymptotic velocity (also called a winding number):

$$\omega = \lim_{k \to \infty} \frac{\theta_k - \theta_0}{k}$$

and plot, for some fixed K such as $K = 1/2$, this velocity ω vs. Ω. The resulting plot is a staircase having plateaus *at rational values* of the parameter Ω. If in addition one plots regions of the Ω-K plane having $\omega = 1/3, 1/2$ etc. (rational values with small denominators). Attempt explanations of this mode-locking from theory alone, starting with (6.1.4).

3) Investigate the baker's map (6.1.5).

4) Investigate the gingerbread man map (6.1.8), in particular plot numerically an invariant set, to show how the map got its name.

5) Show theoretically that Arnold's Cat map (6.1.6) has the area-preserving property, by tracking what happens to an infinitesimal rectangle in the xy plane. This can be done theoretically or numerically or both.

6) Perform theoretical, computational, and graphics studies of the Henon map (6.1.7). This is an excellent map for generating numerical data that will ultimately be fed to a fractal dimension analyzer, as in Subject 6.2. Henon attractor plots should appear as in Figure 6.1.4 (for $a = 1.4$, $b = 0.3$).

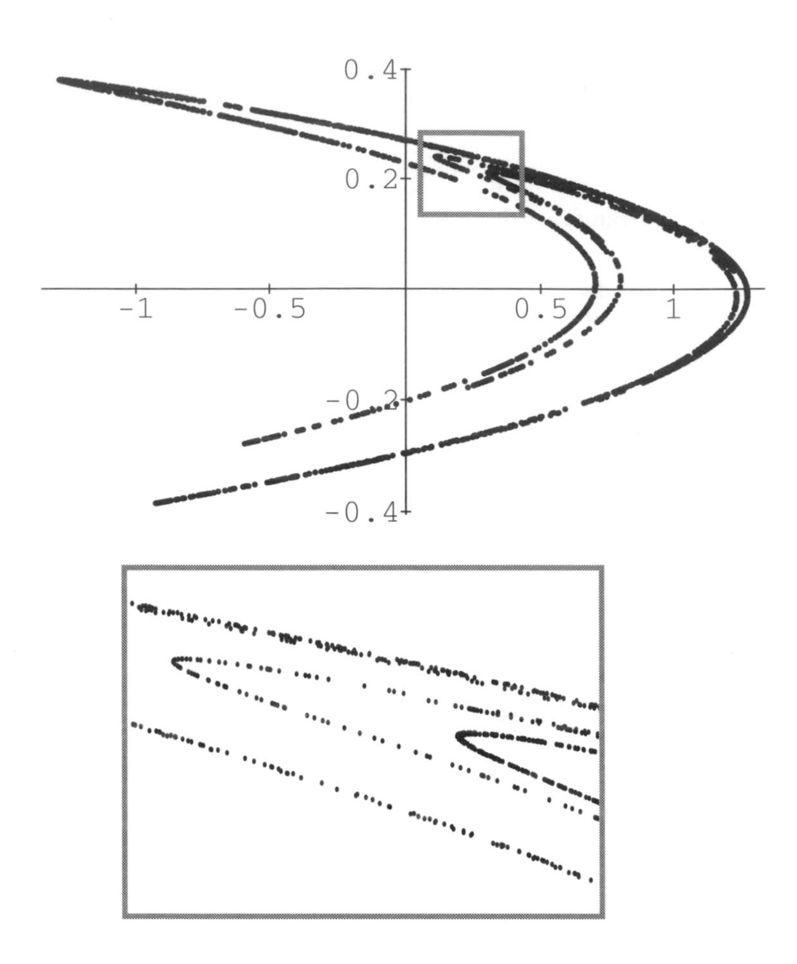

Figure 6.1.4: Henon attractor obtained by numerical iteration of the Henon system (6.1.7). The attractor has fractal properties, such as the evident self-similarity of the blown-up rectangle (bottom) to the whole main figure (top); not to mention the "dusty" appearance of the orbit. Henon data are easy to generate and prove excellent for testing of fractal and stability/chaos measurement software.

One theoretical exercise of interest is to show that an initial elliptical region centered at the origin will, under the Henon map, bend and rotate into a "boomerang" shape, essentially the shape in Figure 6.1.4.

7) Investigate the Lorenz attractor. Note that this is our very first *continuous* model of the current Subject. The Lorenz system, initially a

weather model, is taken to be:

$$\frac{dx}{dt} = \sigma(y-x) \tag{6.1.9}$$

$$\frac{dy}{dt} = rx - y - xz$$

$$\frac{dz}{dt} = xy - bz$$

Even though continuous (as opposed to the discrete iteration formulae we have been studying), this system generally gives rise to attractors having more then one basin of stability. Fixing the original parameters of [Lorenz 1963], namely $\sigma = 10$, $b = 8/3$, one may study the behavior of the system for $1 < r < \infty$. The qualitative results of the numerical solution to the differential equations should be [Sparrow, in [Holden 1986]]:

- For1 $< r <$ 13.926, numerical orbits spiral into one of two stable points;
- *For* $r \sim$ 13.96, a kind of bifurcation threshold at which long-standing orbits begin to appear;
- For 13.96 $< r <$ 24.06, orbits still spiral in but many orbits spend long times near an invariant set;
- For 24.06 $< r$, some orbits wander apparently forever near an invariant "strange attractor" set.

A plot of the trajectory $\{x(t), z(t)\}$, which is a y-projection of the 3-space trajectory, should appear as in Figure 6.1.5.

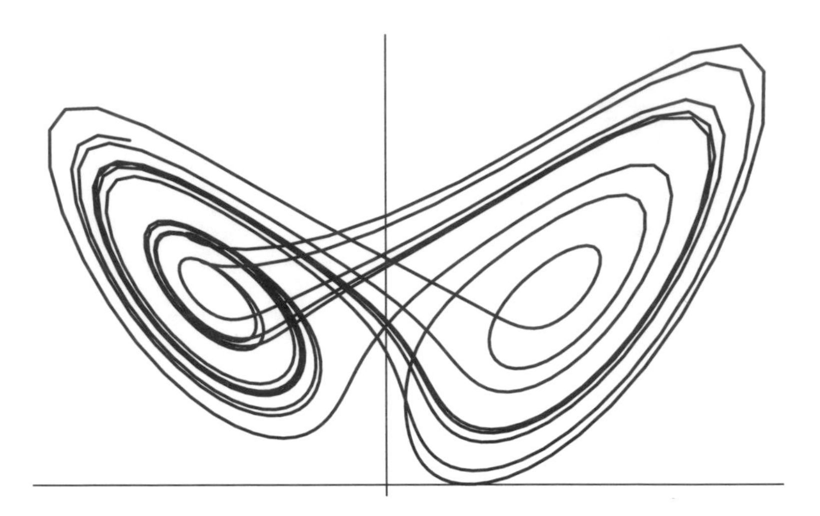

Figure 6.1.5: Lorenz attractor for parameter choices $\sigma = 10$, $b = 8/3$, $r = 28.0$. The plot axes are x (horizontal) and z. The initial point was actually at the origin of this graph, said point quickly becoming attracted to the double-basin figure shown.

Good exercises leading beyond two-dimensional plots such as Figure 6.1.5 include improving the code "Lorenz.ma" by invoking Runge-Kutta solving, and going to three dimsnions by plotting also the $y(t)$ component or various two-component planar projections of the orbit.

8) Investigate another continuous model of chaos: the forced true pendulum, whose harmonic "textbook approximation" does *not* produce chaos. The true pendulum, however (with sines of angles not approximated by the angles themselves), can exhibit dynamical properties reminiscent of the quadratic map. The equations of motion can be put in Hamiltonian first-order form:

$$\frac{d\omega}{dt} = -\frac{\omega}{q} - \sin\theta + g\cos\omega_D t \qquad (6.1.10)$$

$$\frac{d\theta}{dt} = \omega$$

where (ω, θ) are phase-space variables and q, g, ω_D are constant system parameters, the latter parameter being the angular driving frequency.

Interesting and picturesque bifurcation plots can be obtained from numerical assignments: $q = 2.0$, $\omega_D = 0.667$, $0.6 < g < 1.6$. One may proceed to graph bifurcation as follows. Choose a g (on the horizontal plotting axis) and numerically integrate, say, 30 drive cycles of (6.1.10). Then plot the instantaneous value of ω (vertical coordinate) at the *beginning* of each of the next, say, 30 drive cycles. This procedure is further explained in [Baker and Golub 1990].

9) Investigate the continuous Duffing system, which amounts to another non-linear forced oscillator, similar to driven pendulum of the previous step:

$$\frac{dy}{dt} = -ay - x^3 + b\cos t \qquad (6.1.11)$$
$$\frac{dx}{dt} = y$$

In particular, work out some theory for small-displacement trigonometric series solutions. One can find "best coefficients" of approximate solutions such as:

$$x(t) = \alpha + \beta\cos t + \gamma\sin t$$

by symbolic manipulation.

Numerical plots for the Duffing system reveal–not surprisingly–period doubling routes to chaos. For example, fix $a = 0.3$, and plot orbits for b ranging, say, from 20 to 40 [Holden and Muhamad, in [Holden 1986]].

An interesting discussion of symmetry breaking and chaos for the Duffing system can be found in [Olson and Olsson 1991].

10) Just as with the last step we moved to a continuous chaos model, we now transition to a complex-plane model. We investigate thus the celebrated Julia and Mandelbrot sets, for which graphics can hardly be avoided; there is, however, a vast treasure of theoretical opportunity here.

The key mapping is again quadratic:

$$z_{n+1} = z_n^2 + c \tag{6.1.12}$$

but is assumed to involve complex points $\{z_n\}$ and complex parameter c.

First, find a transformation on (6.1.1) that formally yields (6.1.12). The existence of such a transformation means, of course, that the behavior of the complex model–restricted to the real axis of the complex plane–has already been studied. When the full complex plane is used, marvelous manifestations of order and chaos emerge.

Second, show that if $|z_n| > 2$, further iteration of the quadratic map is unbounded. Thus, when one is looking for invariant sets, boundedness, and so on, a test of the magnitude of $|z|$ can be used to speed up computation.

Next, a Julia set is defined, for given c in (6.1.12), as the border of a "filled Julia" set, which in turn is the set of all complex z for which the map iteration remains bounded forever. It is perhaps easiest to plot filled Julia sets; for one thing more colored pixels will result. Filling can be done with elementary programs such as "Julia.ma." The ambitious investigator might attempt to plot actual Julia sets, i.e., the borders of filled Julia sets. One method for doing such is to *avoid* tracking the boundedness of points z; to use instead the known property that all the pre-images of a known point are in the set: in fact all the recursive pre-images can be expected to cover the set nicely. Say that, for the given c, z is known to be in or "near" the set. Then one inverts (6.1.12), obtaining in general two pre-images as solutions of a quadratic equation. One may choose *a random one* of these, plot it, and continue to obtain two pairs of pre-pre-images, choosing one of these at random, and so on. A filled-in Julia set is shown in Figure 6.1.6.

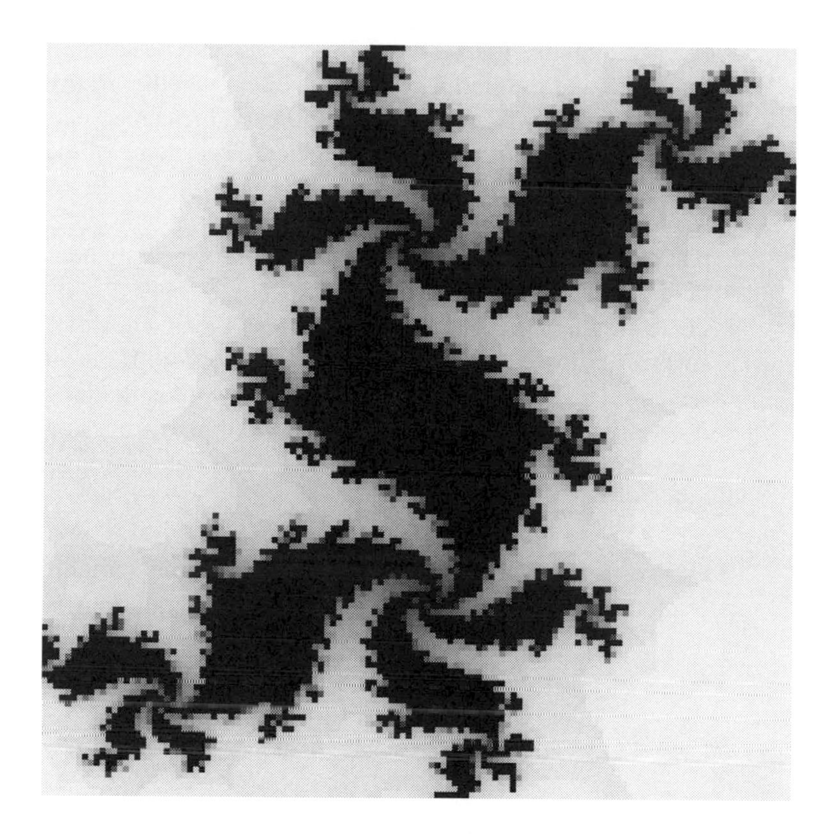

Figure 6.1.6: Representation of the filled Julia set for $c = 0.3 - 0.5i$. Gray level corresponds to the numer of map iterations required for z_n to have "escaped." This plot is a 128-by 128 pixel region

Finally, plot regions of the celebrated Mandelbrot set. Whereas for Julia sets one fixes c and searches for an invariant set of z values for which iteration is forever bounded, the Mandelbrot set is the set of all c such that, starting with z_0 = 0+0i, the iteration is likewise bounded. Theoretically, the Mandelbrot set is the set of all c such that c's Julia set is topologically connected. Note the following points that may be relevant to Mandelbrot plotting:

- The straightforward way to plot regions of the Mandelbrot set is to pick a rectangular region of the complex plane, then assign a c value to every pixel on a display, then iterate (6.1.12) starting with $z = 0+0i$, until $|z| > 2$, at which point the pixel is colored according to the number of iterations performed.

- The criterion discussed above, that $|z| > 2$, can be sped up for some machines and programs, by first checking whether $Re(z) > 2$ and possibly $Im(z) > 2$. Assuming complex values are stored as floating-point pairs, these preliminary checks remove some of the checks as to whether $Re(z)^2 + Im(z)^2 > 4$.
- A useful Mandelbrot generation program would allow "zoom," meaning that chosen complex numbers z and their immediate neighborhoods can be inspected at arbitrary scale.
- There are theoretical expedients to greater efficiency of Mandelbrot software. Investigate the following proposition: If the sides of a rectangular region lie completely in the set, then the interior of region does also. If this is true, large numbers of pixels interior to rectangular borders may immediately be colored. What about convex regions in general?

11) Investigate, via theory, computation, and graphics the Newton attractors. Such study starts with an algebraic equation such as the cubic:

$$z^3 - 1 = 0$$

which is to be solved with the classical Newton iteration:

$$z_{n+1} = z_n - \frac{z_n^3 - 1}{3z_n^2} = \frac{2}{3} z_n + \frac{1}{3z_n^2}$$

starting with some initial complex point z_0. At first it might seem like there should be three "pie piece" regions of initial complex points, in each of which regions all interior points should converge to one of the three roots of the cubic equation. But these pie pieces are "nibbled at" by the three basins of attraction–the three complex cubic roots [Schroeder 1991]. Figure 6.1.7 shows a low-resolution plot, obtained simply by coloring pixels according to which of the three attractor points $\{1, -1/2 - i\sqrt{3}/2, -1/2 + i\sqrt{3}/2\}$ is the destination of the initial point.

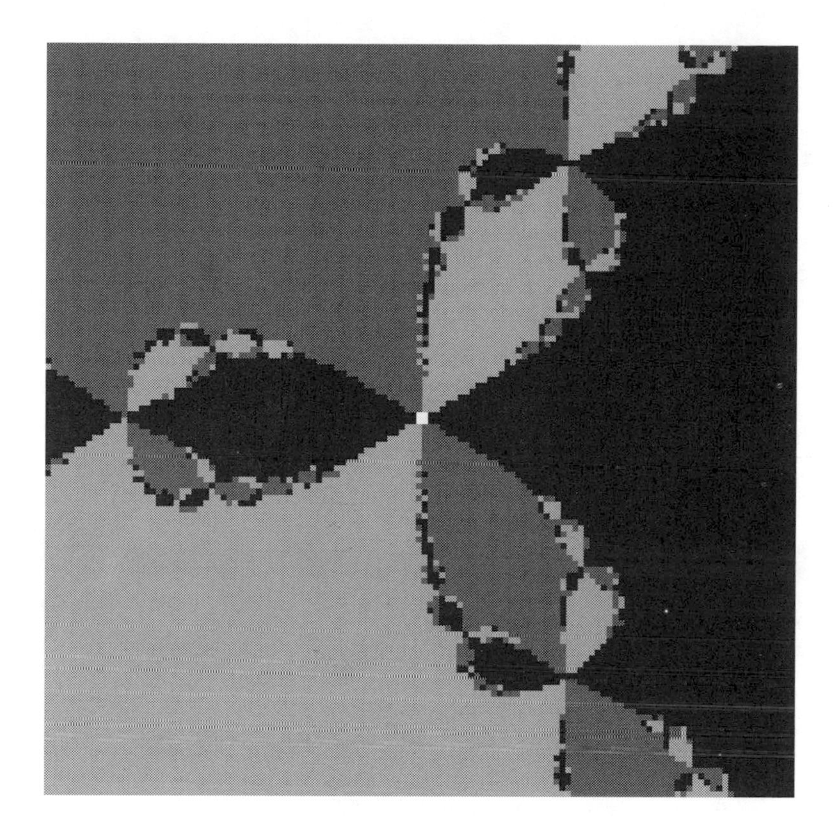

Figure 6.1.7: Newton trefoil, obtained by iterative solution of the cubic $z^3 - 1$. The resolution is 128-by-128 pixels. Initial complex points are iterated until it is clear to what attractor root (of the three cubic roots) the result is closest. Then the initial point is colored according to that destination. The three possible colorations entwine with each other in an infinite, self-similar cascade at apparent "boundaries."

A good exercise is to attempt more efficient plotting mechanisms for such figures as the Newton trefoil. One may, for example, contemplate using successive inverses, starting with a point and back-tracking the Newton iteration, selecting a root at random on each backward iteration; in this way ending up with a plot of the dusty boundaries of the regions of Figure 6.1.7. Such a boundary is analogous to the Julia set of step (10) previous, which is technically the boundary of its filled counterpart set.

Another investigative direction is to generalize to algebraic equations like $z^n - 1 = 0$, which for $n > 2$ will generally involve n "pie pieces" that intertwine in a chaotic manner.

We next turn to the general question: Precisely when is a system chaotic? One important practical indicator of the presence of chaos is the Lyapunov exponent. This exponent can be thought of in two equivalent ways:

- As a measure of how radically a small change in initial condition(s) affects the state at some finite time later;
- As a measure of how rapidly grows the information required to predict the current state (from an initial state).

In this regard, when the measure (exponent) in question is large enough, we suspect chaos. For many systems, chaos can thus be detected via the following general approach. Think of one dimension for the moment and consider an iterative map for which two neighboring initial points x_0 and $y_0 = x_0 + \varepsilon$, where ε is small, are independently propagated, so that after n iterations we have results x_n and y_n. The first of the two measures described above can often be obtained via the *ansatz*:

$$|y_n - x_n| \sim 2^{\lambda n} |y_0 - x_0| \sim 2^{\lambda n} \varepsilon \tag{6.1.13}$$

which is a statement that a deviation in initial conditions propagates in an *exponential* fashion to perturb the state at iteration n. The multiplier λ is the Lyapunov exponent. Note that some treatments will use $e^{\lambda n}$ rather than base-2 exponentiation; but using "2" has the advantage that the information-based measure above should use the very formula (6.1.13). This is because, on the basis of the *ansatz*, one must know the initial point *to a precision of about λn bits* in order to predict the state at iteration n. Thus, in such a system, we may well assume chaos is in force if the amount of information required of the initial point grows linearly with time, the constant of proportionality being the Lyapunov exponent λ.

For the reasons stated, chaotic regions are generally characterized by exponent $\lambda > 0$, with orderly or stable behavior generally having $\lambda \le 0$. In continuous dynamical systems of higher dimensions, what often occurs is that an infinitesimal, ellipsoidal initial neighborhood of principal radii $p_i(0)$, where i counts through the dimensions, has these radii growing exponentially in time. Thus, the initial ellipsoid expands exponentially, but with the exponents for the principal axes generally distinct. The collection of Lyapunov exponents in such a case is denoted $\{\lambda_i\}$, where

$$\lambda_i = \lim_{t \to \infty} \frac{1}{t} \log_2 \left| \frac{p_i(t)}{p_0(t)} \right| \tag{6.1.14}$$

There is a beautiful connection between such Lyapunov exponents and fractal dimension, as discussed in Subject 6.2.

Project 6.1.4 Chaos, stability, and Lyapunov exponents

Theory, computation, graphics, difficulty level 2-3.

1) For a one-dimensional iterative map, say $x_{n+1} = f(x_n)$, derive the approximation:

$$\frac{|y_n - x_n|}{|y_0 - x_0|} \sim \prod_{m=0}^{n-1} |f'(x_m)| \tag{6.1.15}$$

where f' is the derivative of f. This formula provides a method for detecting probable chaos: one evaluates the derivative at various points along an orbit. To take this idea further, argue from the assumption (6.1.13) that:

$$\lambda = \lim_{n \to \infty} \frac{1}{n} \sum_{m=0}^{n-1} \log_2 |f'(x_m)| \tag{6.1.16}$$

Now argue, at least heuristically (the precise argument requires a so-called "ergodic" assumption which is intuitively reasonable) that if $p(x)$ is the invariant distribution of the map, as in step (3), Project 6.1.1, the

Lyapunov exponent is:

$$\lambda = \int p(x) \log_2 \left| f'(x) \right| dx \qquad (6.1.17)$$

Thus, if the invariant distribution is known, the exponent can be computed via numerical integration.

For the special case $a = 4$ of the quadratic map, derive the remarkable result that $\lambda = 1$. In information-theoretic terms, reconcile this fact with the observation of step (3), Project 6.1.1 that iterates of the map for this $a = 4$ case can be obtained essentially by bit-shifting.

2) Plot the Lyapunov spectrum (λ vs. a) for the quadratic map (6.1.1). One possible approach is to numerically evaluate (6.1.16) for random orbits. Another is to compute directly the numerical ratio

$$\frac{|y_n - x_n|}{|y_0 - x_0|}$$

where initial points y_0, x_0 differ slightly.

3) Evaluate, numerically or perhaps theoretically, Lyapunov exponents for the Lorenz attractor of Figure 6.1.5, where two closely neighboring initial points are assumed either:

- to lie on the same immediate orbit ($\lambda \sim 0$);
- to lie on separate orbits ($\lambda > 0$);
- to lie well outside of the attractor, and to be quickly attracted into virtually the same point, to continue along the attractor ($\lambda < 0$).

For this Lorenz model the Lyapunov ellipsoid has three principal axes that behave somewhat independently, according to the precise placement of an initial, infinitesimal ellipsoidal neighborhood.

4) Evaluate Lyapunov exponents for the forced true pendulum (6.1.10). For example, the two exponents for the (ω, θ) phase space, for parameter choices $q = 4.0$, $g = 1.4954$, are reported to be $\lambda_1 = 0.16$, $\lambda_2 = -0.42$ [Baker and Golub 1990]. Consider applying such work to the connection

between the exponents and fractal dimension, as enunciated in Project 6.2.2.

5) Evaluate Lyapunov exponents for a discrete system such as the Henon attractor or another higher-dimensional system. The three basic possibilities for a pair of neighboring initial points, as in step (3) previous, should still be relevant.

We close this Subject with some research problems that connect chaos theory with other mathematical and physical studies.

Project 6.1.5 Applications of chaos theory

Theory, computation, graphics, difficulty level 3-4.

1) Design an encryption system using, say, a tent map (6.1.3) or the original quadratic map (6.1.1). The idea is to exploit the fact that a point may have two pre-images under the map, four pre-pre-images, and so on; thus, extra information (some number of binary bits, a "key") is needed to find a specific pre-image from the distant iterative past.

2) Consider the circle map as a "$1/f$" noise generator. This kind of noise is discussed in the next Subject. One way to proceed is to assess via computation the frequency spectrum of $\{\theta_n\}$, taking long power-of-two length sequences, and windowing them prior to an FFT.

3) Is there any sense in which the quadratic map can be used reliably in pseudorandom number generation? In particular, are there parameters a such that an iterated sequence can be generated and then transformed to model an equidistributed random variable?

4) Design software to handle a wide variety of chaotic models. Such general software should have a very fast map iteration core. Previous efforts in this direction include pre-compilation methods that optimize chaos and fractal calculation [Wegner et. al. 1992]. A good implementation should also have optimized graphics.

5) Investigate the notion that if a Mandelbrot set is given electrostatic charge, the lines of equipotential correspond to the "escape value" contours as discussed in step (10), Project 6.1.3. A related investigation is found in step (5), Project 6.2.1, when a fractal set is similarly charged.

6) Study the fascinating subject of "chaos plus noise." The idea is to replace the quadratic map (6.1.1) with:

$$x_{n+1} = ax_n(1 - x_n) + s_n$$

where s_n is a Gaussian random perturbation of some (usually small) standard deviation. The effect on bifurcation diagrams and Lyapunov exponents is especially interesting [Geisel, in [Buchler et. al. 1985]].

7) Apply chaos concepts to the theory of continued fractions discussed in Project 1.1.1. It turns out that continued fractions can be thought of as a dynamical consequence of the "Gauss map" $x := 1/x \pmod 1$, and even Lyapunov exponents as stability measures can be calculated for certain continued fractions. In fact, the Lyapunov exponent, when properly defined, has the attractive value:

$$\lambda = \frac{\pi^2}{6 \log 2}$$

for *almost all* continued fractions of numbers in (0,1) [Corless 1992]. This is a restatement of Levy's earlier discovery of growth criteria for the denominator corresponding to the first n fraction elements. Recent work on the dynamics of the continued fraction map can be found in [Bedford et. al. 1991].

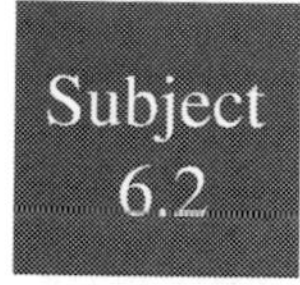

Fractals

A fundamental fractal is the standard ("middle-thirds") Cantor set. Let us take the opportunity to dovetail from the previous Subject of chaos, into the Subject of fractals by deriving this set from a chaotic map. The tent map (6.1.3) will be taken with parameter $b = 3/2$. As initimated in step (1), Project 6.1.3, the Cantor set is an invariant set of this map. To inspect this set of points whose iterates never leave (0,1), we apply successively the tent map as in Figure 6.2.1.

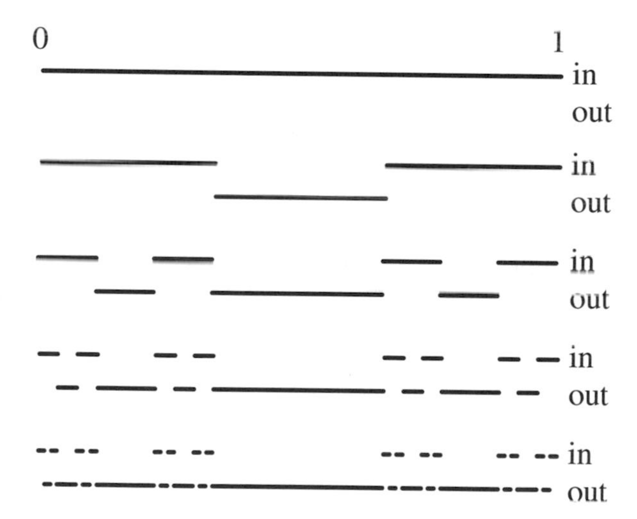

Figure 6.2.1: Construction of a Cantor set via chaotic map. Points in (0,1) are tracked under the tent map (6.1.3) for four iterations. On the first iteration the middle third of the points leave the interval (0,1); then the middle thirds of the two "in" segments leave the interval, and so on. The bottom "in" row of points is of rather small measure, suggesting that the Cantor set is somehow "sparse."

It turns out that this Cantor set has some stunning properties. For one thing, it (the limit of the "in" rows of Figure 6.2.1) has zero measure. But the set is far from empty; in fact, it has an uncountable number of points, as we see in the next Project.

Fractal dimension of the Cantor set–not to mention many other sets, curves, solids, etc.–can be estimated in an elementary way through a powerful fundamental rule. The idea is to express the effort required to cover the set in question, as a function of the size of a small covering template. The idea, of which the concept of Euclidean dimension can be thought of as "an historical special case," is that the number # of templates required should behave as:

$$\#(L) \sim \frac{1}{L^D} \tag{6.2.1}$$

where L is a (shrinking) characteristic length of the covering template and the power D is the (possibly fractional) dimension. Before we proceed with examples of fractional D, two cautionary remarks are in order. First, there are various definitions of dimension, and a case can be made for infinitely many definitions. The dimension embodied in the relation (6.2.1), which is called box dimension or capacity dimension depending subtly on where templates are placed, is just one of the many possible, some of which we discuss prior to Project 6.2.4. Second, rigorous treatments, such as the often difficult analyses of the theoretically fundamental Hausdorff dimension, are the correct way to find a set's dimension. For many well-known sets, the Hausdorff and box/capacity dimensions are known and agree; but they do not have to agree, and in some cases the Hausdorff dimension is not even known when the box dimension is known [Falconer 1990].

For the Cantor set we compute D using the relation (6.2.1). At the n-th stage of tent map iteration in Figure 6.2.1 (the top row of all "in" points being $n = 0$), we may ponder templates as line segments of characteristic length $L = 1/3$, in which case 2 templates are required. In general, $L = 1/3^n$ and we require 2^n templates to cover the "in" set. We solve

$$2^n \sim \frac{1}{(3^n)^D}$$

to get the Cantor set's dimension as $D = \log 2/\log 3 = 0.63093...$, certainly an interesting number. Note $0 < D < 1$, so the set certainly has some presence, but is sparser than a dimension-1 Euclidean line.
A second example of fractal construction and estimate of D is the Koch curve, as in Figure 6.2.2.

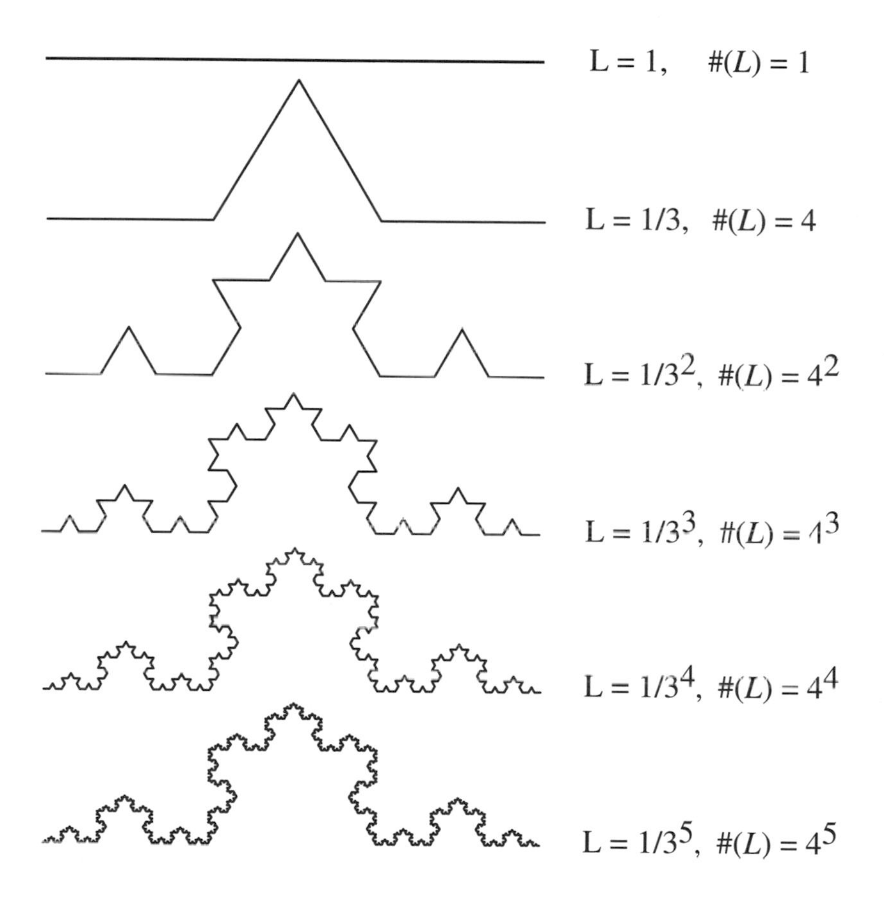

Figure 6.2.2: Development of the Koch fractal. The number $\#(L)$ of length-L segments needed to cover the curve behaves as $\# \sim L^{-D}$, where D is the dimension log 4/log 3.

The Koch curve is constructed by recursion exhibited in the top two entries of Figure 6.2.2; namely, at each step the middle-third of each segment is replaced with a "V" shaped bulge. This curve turns out to have infinite length, while enclosing (together with its natural base: the original line segment) a finite area; and, furthermore, its dimension is given by log 4/ log 3, which is greater than 1. So this curve is somehow "more than" a line but not something like a two-dimensional plane.

In this Subject we shall meet a fair variety of fractals, including statistical fractals and fractals that are kinky enough to "fill space." We shall start with a mainly theoretical Project, moving on later to computation-oriented Projects.

Project 6.2.1

Theory of fractals

Theory, a little computation, difficulty level 1-2.
Support code: Appendix "CantorGen.ma"

1) Study the properties enunciated for the Cantor set. First, show that the point 1/4 is in the set, by virtue of never leaving (0,1) under successive iteration of the tent map. Next, argue the fact of zero measure in the following amusing way. Show that the points of the Cantor set are those points of (0,1) having ternary expansions devoid of 1's. Thus a number such as:

0.02220020200022220... (base 3)

(and having nary a "1") is in the set. Now imagine all possible games involving infinite successive tosses of a three-sided die, and conclude that the measure of the Cantor set is the probability of having an infinite dice game without ever rolling "1." Finally, argue the fact of uncountability of the points of the set, by exhibiting an explicit but simple mapping from every point of the set onto the real interval (0,1).

2) Consider for base $b \geq m$, the set of all numbers in (0,1) whose base-b expansions are devoid of each of the digits $0,1,2,...,m-1$. What is the dimension D of this set, in terms of m and b?

3) For given dimension $0 < D < 1$, describe a Cantor-like set having that

dimension.

4) What is the dimension of the invariant set (if it exists) for given b in the tent map (6.1.3)?

5) Consider the dilation equation relevant to the Cantor set, as follows. Assume a population density $C_0(x)$ defined as the constant 1 for x in (0,1). Now we iterate this density under the tent map (6.1.3) for $b = 3/2$ (which iteration gave the Cantor set). After each iteration, we obtain a *normalized* density by defining $C_n(x) = 0$ outside of (0,1), and multiplying the density on (0,1) by that constant factor that renders the integral of C_n to be 1. Show that this procedure is governed by the dilation equation:

$$C_{n+1}(x) = \frac{3}{2}(C_n(3x) + C_n(3x-2))$$

Note that because of the normalization rule, the dilation equation, when integrated on each side, just gives $1 = 1$.

One can perhaps imagine dropping subscripts in the dilation equation and defining thus a stable $C(x)$. This function (actually not a function; more like a delta-function train and difficult to handle rigorously) would be the "charge density at x" if the Cantor set were given a unit net electrostatic charge. Use this notion to forge an heuristic argument leading to the relation:

$$V(y) \sim \frac{constant}{y^{1-D}}$$

for the potential, on the y-axis at $(0,y)$, when y is small. One might even contemplate finding this curious power law computationally. What is especially amusing about this scaling is that a point charge at (0,0) would give just $1/y$, and of course the dimension of a point charge is 0 (!). There exist in the literature theoretical treatments of electrostatics for fractal charge distributions [Dorren and Tip 1991].

Next, moving ahead on the heuristic notion of the existence of the C function, show that the Fourier transform $C^{\sim}(\omega)$, defined as (5.1.1) satisfies a scaling law:

$$C^{\sim}(\omega) = f\left(\frac{\omega}{3}\right) C^{\sim}\left(\frac{\omega}{3}\right)$$

and give accordingly an infinite product representation for $C^{\sim}$. Finally, deduce that, in some crude sense of local ω-average:

$$|C^{\sim}(\omega)|^2 \sim \frac{1}{2} \left|C^{\sim}\left(\frac{\omega}{3}\right)\right|^2$$

and from this scaling law deduce that the spectral power associated with the C function is, again in a crude sense of average:

$$|C^{\sim}(\omega)|^2 \propto \frac{1}{\omega^D} = \frac{1}{\omega^{\frac{\log 2}{\log 3}}}$$

Thus, we have one of many examples of fractal properties going into spectral power-law behavior. In order to test this power relation experimentally, one may proceed as follows. Choose signal length N and create a signal x consisting of elements defined: $x_j = 1$ if j has no 1's in its ternary expansion, $x_j = 0$ otherwise. The Appendix code "CantorGen.ma" has an appropriately simple Cantor signal generating function. This signal is a representation of some iterate of the dilation equation. Now window this signal (resulting in a smoother log-log plot) by creating signal y:

$$y_j = x_j \sin^2 \frac{\pi j}{N-1}$$

Take an FFT of y and plot its absolute squared values in log-log fashion, so that the slope of the plot will reveal a spectral power law. The plot will be quite noisy, requiring perhaps further local averaging as expected from theory, but at least the qualitative drop ω^{-D} should be evident.

6) Show that the dimension D for the Koch curve of Figure 6.2.2 is indeed $\log 4/ \log 3$.

7) Determine the highest dimension one can obtain from a generalized Koch curve, constructed using a recursion whereby each segment is

turned into:

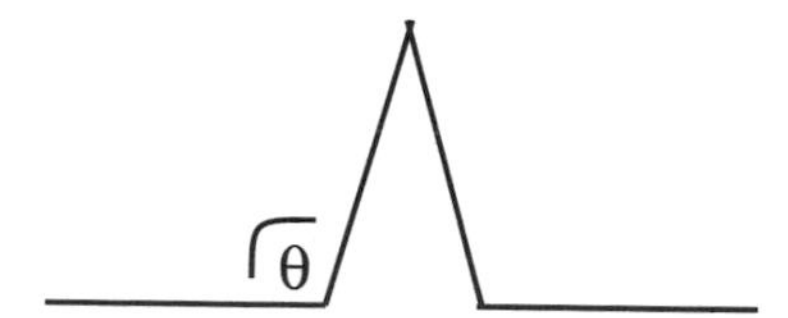

where θ is a constant angle, while the four segments shown are all equal.

8) Find the exact area under the Koch curve (and bounded by its (unit length) original base.

9) Find the dimension of the "Sierpinski arrowhead," shown here at a particular recursive level, together with a blown-up fundamental "knob:"

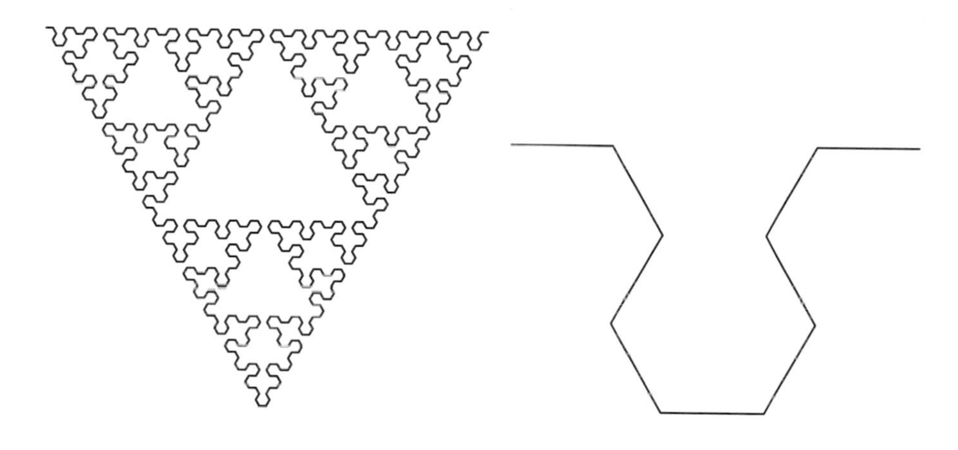

We next turn toward primarily graphics exercises pertaining to fractals, after which some statistical fractal studies and pure-computation exercises follow as the final Projects of this Chapter.

Project 6.2.2

Visualization of fractals

Graphics, a little theory, difficulty level 1-2.
Support code: Appendix "Dragon.ma," "Dragon Recursion.ps," "Hilbert.ma," "LSystem.ma," "IFS.ma."

1) Draw a "Dragon curve" using such software as in "Dragon.ma." In this implementation, one starts with a symbol "0" which means "one segment drawn to the east." Generally, "1,2,3" additionally and respectively will mean to draw "north, west, south." The substition rules:

0 –> {0,1}
1 –> {2,1}
2 –> {2,3}
3 –> {0,3}

are used, the first of which means "replace an east motion with an east followed by a north motion," and so on. The result of twelve iterations is shown in Figure 6.2.3.

Beyond this exercise there are some interesting studies. First, assessment of the fractal dimension is an interesting problem. Second, there is a physical way to build the dragon fractal, as follows. Take a long, thin ribbon of paper. Fold it in half, along its long dimension. Then fold it in half again, and again, always folding in the same sense. Then after some number of folds, unfold the paper, and–here is the key operation–manually force each fold to be a right angle, 90 degrees. The paper when viewed edge-on should be the dragon curve.

Interesting theoretical problems include proving, from the basic substitution principle, that the curve is not self-crossing. There is also the option of plotting a dragon curve in the style of L-Systems, as in step (3) below.

2) Draw a Hilbert space-filling curve, a task exemplified in "Hilbert.ma." The idea is to start with a fundamental square horseshoe shape and

correctly copy and connect rotations of that shape:

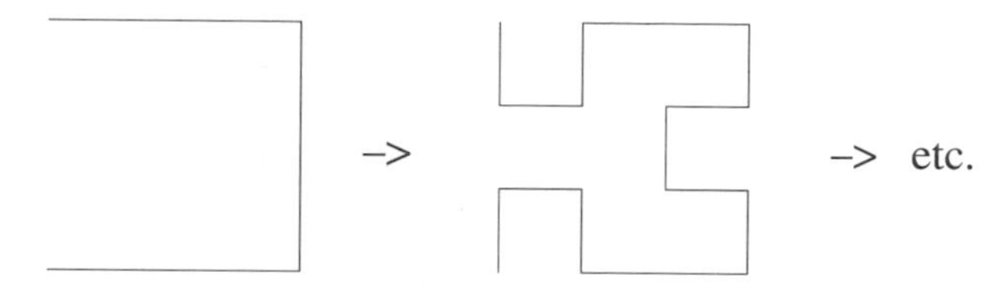

The result of several iterations is shown in Figure 6.2.3. This curve is Hilbert's example of a mapping from the unit interval onto the unit square; said mapping ultimately being everywhere continuous but non-differentiable.

Interesting exercises include construction of three-dimensional versions, also exemplified in "Hilbert.ma." Such an implementation was used to form the frontispiece to this chapter. A good theoretical problem is to prove the fractal has dimension $D = 2$. The Hilbert fractal has possible applications in image processing: the two-dimensional version always spends a "great deal of time" in one locale, thus suggesting, for example, a way to scan images, other than the traditional lexicographic scan. Yet another interesting phenomenon is that an n-dimensional Hilbert curve iterate can completely cover an n-dimensional integer lattice cube, yet every line segment undergoes just one increment/decrement in just one coordinate direction at a time. In this sense the Hilbert curve establishes the existence of general n-dimensional Gray codes.

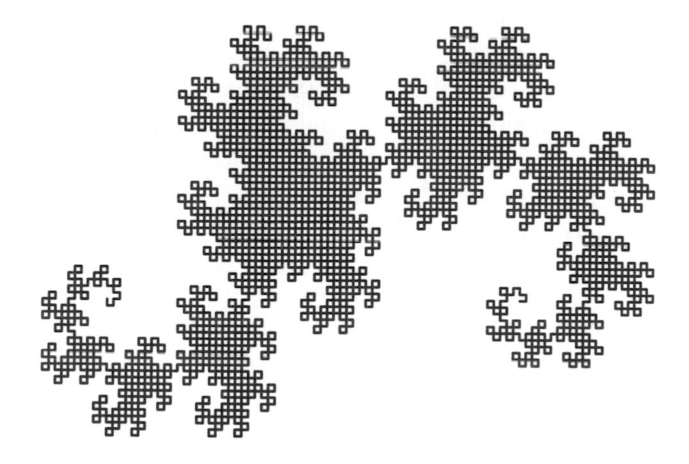

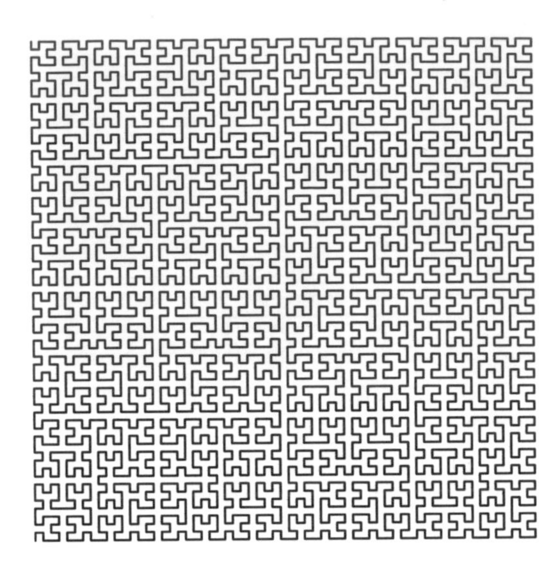

Figure 6.2.3: Dragon curve (previous page) and Hilbert space-filling curve. The dragon curve is drawn using substitution rules for line directions. The Hilbert curve is formed by recursively collecting and connecting rotates of a fundamental square horseshoe. Note that the dragon curve only touches itself precisely on certain corners, so it never passes through itself.

3) Invoke Lindemayer (L-System) algorithms, by which one directs a graphics "turtle" to substitute line segments with higher-resolution sequences. At any time there is a well-defined turtle state, consisting of position, orientation, and possibly some key parameters. One starts with an instruction set–usually a small one–which the turtle executes. Then each part of the instruction set is substituted according to fixed rules, and the turtle may be asked to draw the next level, and so on. Note that Hilbert (2-dimensional version) and dragon curves can be done with the L-System. The "Peano curve" and "tablecloth" constructions of Figure 6.2.4 and 6.2.4 are typical. Substitution rules for specific fractals are given in "LSystem.ma," and are described in the literature [Prusinkiewicz and Hanan 1989].

One excellent improvement on the Appendix code would be to invoke stack push-pop, with which the current turtle state can be saved (pushed) and later popped, at which time the turtle's orientation and position are forced to what was last put on the stack. In this way, as shown by

[Prusinkiewicz and Hanan 1989] in regard to substitution rules containing stack operations, one can model natural plants in remarkably realistic ways.

Figure 6.2.4: Peano curve, a space-filling dimension-2 fractal, generated from a general L-System algorithm, with self-similarity rule as given in step (3), Project 6.2.2.

Figure 6.2.5: "Tablecloth" fractal, created via L-System software using substitution rule given in step (3). An interesting question is: what is the dimension D of this fractal?

4) Investigate Iterated Function System (IFS) fractals [Barnsley 1988]. One manner of generating such fractals is to choose dilation and translations at random. Say, for example, that we adopt an iterative dilation/translation map:

$$\begin{bmatrix} x_{n+1} \\ y_{n+1} \end{bmatrix} = \begin{bmatrix} a_n & b_n \\ c_n & d_n \end{bmatrix} \begin{bmatrix} x_n \\ y_n \end{bmatrix} + \begin{bmatrix} e_n \\ f_n \end{bmatrix}$$

where the $\{a,b,c,d\}$ matrix and $\{e,f\}$ translation vector are chosen stochastically. Figure 6.2.6 shows some IFS ferns. The left-hand fern was generated according to matrix choices:

a	b	c	d	e	f	prob
0.0	0.0	0.0	0.16	0.0	0.0	0.01
0.85	0.04	-0.04	0.85	0.0	1.6	0.85
0.2	-0.26	0.23	0.22	0.0	1.6	0.07
-0.15	0.28	0.26	0.24	0.0	0.44	0.07

As exemplified in "IFS.ma," one simply starts with a random point in the plane, and applies one of these matrices with the random probabilities indicated, continuing to iterate on the current point *ad infinitum*. The fern, or whatever figure arises, is expected to represent the attractor of the IFS.

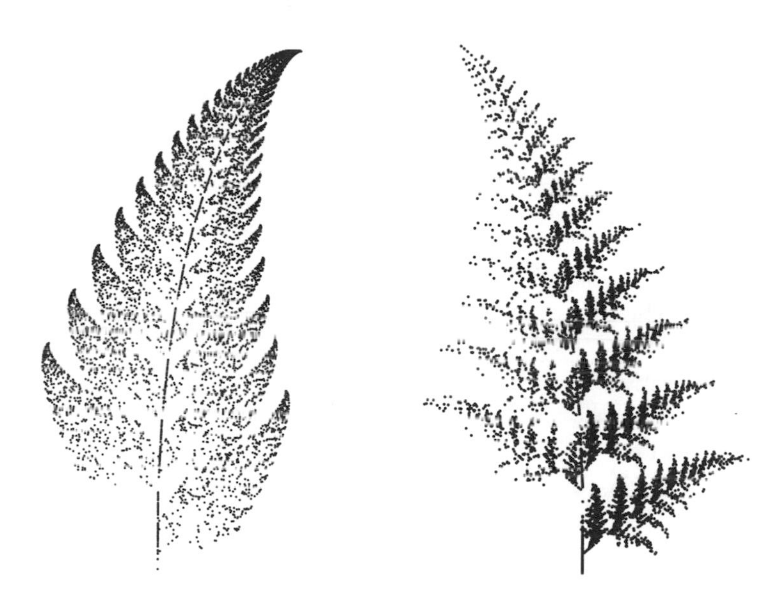

Figure 6.2.6: Iterated Function System (IFS) fractal ferns (previous page) and fractal tree. The ferns differ slightly in their numerical matrix entries. Note that each medium-sized "leaf" is essentially a copy of the whole fern, and each small frond on each leaf is again reminiscent of the whole structure. The fractal tree here, also exhibiting self-similarity at various scales, was created using a six-row IFS scheme, as exhibited in Appendix code "IFS.ma."

Interesting theory of IFS fractals can be found in [Horn, in [Crilly et. al. 1991]]. The Figure 6.2.6 tree is suggested in [Frame and Erdman 1990].

The investigator who wishes to start at a more basic level may well contemplate the elementary generating scheme of Barnsley for the Sierpinski gasket. Start with three fixed vertices of an equilateral triangle, and a random point within the triangle. Now iterate, always by taking the next point to be exactly halfway on a ray from the current point to a *random* vertex. The cloud of dots thus created will generally converge to the Sierpinski gasket as attractor [Crandall 1991]. Note that in a language such as *Mathematica* all the work can be done with vector algebra like so:

```
vertex = {{0,0},{1,0},{1/2,Sqrt[3]/2}};
Do[
 {x,y} = ({x,y} + vertex[[Random[Integer,{1,3}]]])/2;
 pointlist = Append[pointlist, {x,y}];
]
```

Of course one needs to start with some initial seed {x,y} and a list containing this seed. There are much faster and more memory-efficient ways to collect plottable points, but the Append[] approach is straightforward.

An interesting problem is this: Find some trivial deterministic pseudorandom generation on the basis of which one can actually derive rigorously the correct dimension log 4/log 3 for this random fractal.

5) Investigate diffusion-limited aggregation (DLA) and/or dielectric breakdown (DBM) fractal generation. The DLA models are especially simple to implement computationally. Start with a few central pixels colored "white" in an otherwise black sea of pixels. Let invisible particles diffuse, through random walking, from the borders of the whole region in question. When a particle manages to wander in and touch a white pixel, it sticks and is itself colored white. Thus, when a white tendril has aggregated to a large extent, it is more likely to pick up further wandering particles. A typical experiment is pictured in Figure 6.2.7.

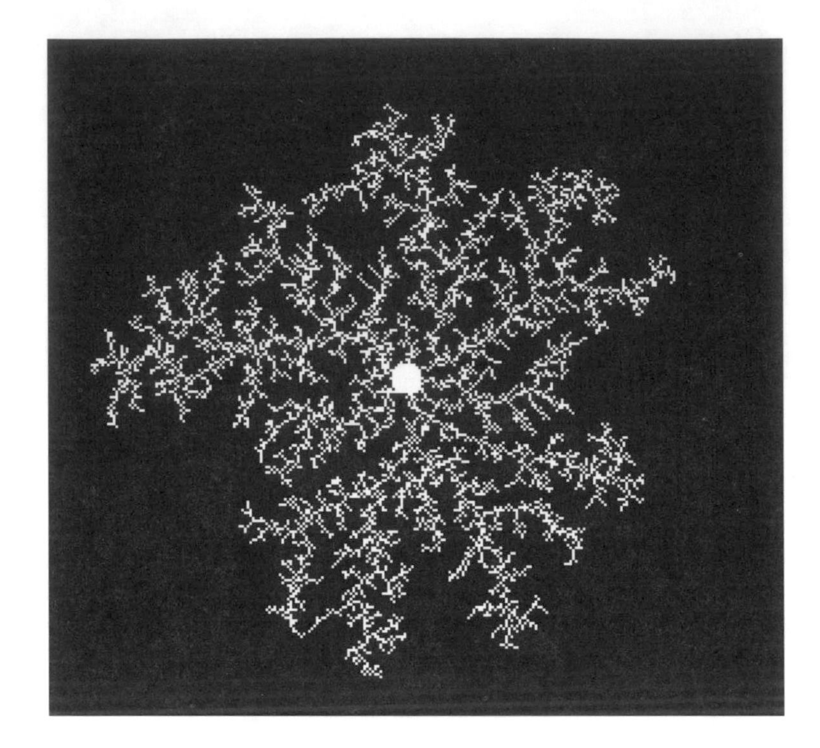

Figure 6.2.7: Diffusion-limited aggregation (DLA) fractal. Particles diffuse in and stick to any white tendrils. Such figures–for reasons not completely understood–tend to have empirical dimension $D \sim 1.7$.

There are some interesting open questions in DLA work. For one thing, the typical dimension of about 1.7 is theoretically suspect, because no aggregates of much more than 10^7 or 10^8 particles have been created computationally. It is not even clear that the approximation $D \sim 1.7$ persists for huge numbers of particles [Fleischmann et. al. 1989].

One way to assess the dimension of such DLA runs as in Figure 6.2.7 is to plot log $M(r)$ vs. log r, where $M(r)$ is the total mass of (white) aggregate pixels within a radius r. But this is not the only option; one may contemplate some of the methods intimated later in Project 6.2.4 for assessing fractal dimension. The novice observer might notice that a set of radial white tendril rays would have dimension 1, while a solid mass of white would have dimension 2.

Yet another interesting exercise is to try to create, independent of any physical model considerations, an analytic DLA tendril model that has dimension provably between 1 and 2.

We next turn to an important class of statistical fractals: fractal Brownian noise signals. Whereas the DLA experiments described in the last Project, and exemplified in Figure 6.2.7, are surely statistical fractals, and whereas statistical fractals are apparently ubiquitous in nature, the mathematical importance of the Brownian class is that analytical results are known.

Figure 6.2.8 shows typical Brownian trajectories of various dimensions, where here we refer to the dimension *of the graph* of the motion:

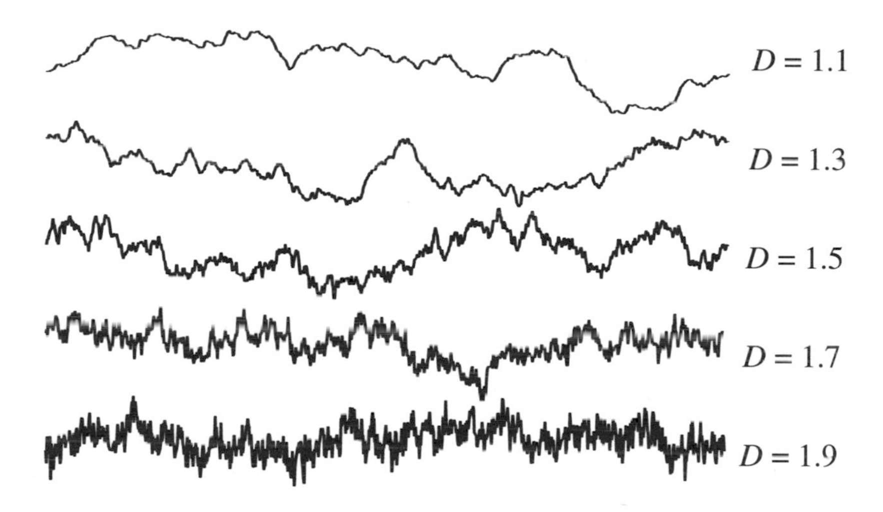

Figure 6.2.8: Fractal Brownian noise of various dimensions D, created via spectral methods as in step (2), Project 6.2.3. Note the $D = 3/2$ case, which is classical Brownian motion. The vertical of each plot is $x(t)$, the horizontal is t.

One notices finer fluctuation for higher dimensions, as expected; after all, near dimension $D = 2$ a plot should be "trying to fill the plane." The self-similarity rule that leads to such gross geometrical features is the scaling rule embodied in the very definition of fractal Brownian noise. One may define the noise by assuming a signal $x(t)$ such that the signal difference

$x(t+T)-x(t)$ is a Gaussian random variable with zero mean and a variance that scales with the time difference T as the expectation:

$$< |x(t+T)-x(t)|^2 > \quad \propto \quad |T|^{2H} \tag{6.2.2}$$

where H is the "Hurst exponent," from which the fractal dimension of the graph of $x(t)$ can be inferred. This scaling law leads to the conclusion that for positive constant r, the graphs of $x(t)$ and $x(rt)/r^H$ should be statistically similar: at least by (6.2.2) they should have the same variance. For classical Brownian motion $H = 1/2$. Figure 6.2.9 shows a typical graphics experiment in self-similarity.

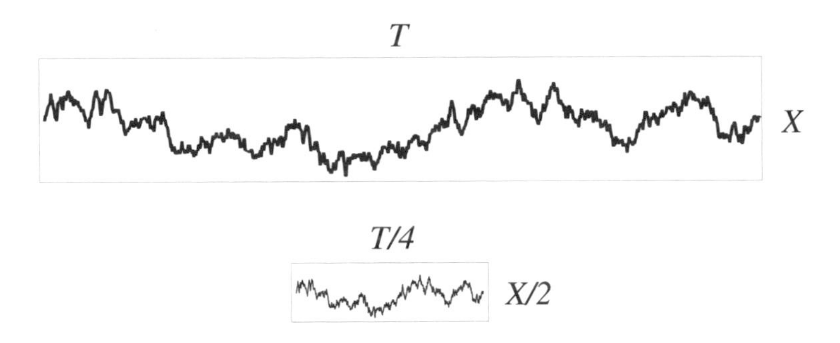

Figure 6.2.9: Self-similarity of classical Brownian noise, graph dimension $D = 3/2$. The idea idea that $x(t)$ and $x(rt)/\sqrt{r}$ are statistically similar is tested here by contracting the space and time axes by factors of two, four respectively. The resulting squashed signal is indeed qualitatively similar; in fact, on the basis of the squashed signal alone one could probably pick out $D = 3/2$ from the "lineup" of Figure 6.2.8.

It is a beautiful thing that the power spectrum can be shown to obey a scaling law for non-zero frequencies:

$$< |x^{\sim}(\omega)|^2 > \quad \propto \quad \frac{1}{\omega^{2H+1}} \tag{6.2.3}$$

where $x^{\sim}$ is the Fourier transform (5.1.1) (appropriately handled with time windows when divergent integrals are encountered), and that the graph dimension (i.e., elementary box dimension from (6.2.1) or, in rigorous studies the Hausdorff dimension [Falconer 1990]) is:

$$D = E + 1 - H \qquad (6.2.4)$$

where E is the topological dimension of the space on which the random variables live. For the one-dimensional signals $x(t)$ pictorialized above, $E = 1$, which is how we get, for classical Brownian motion with $H = 1/2$, the resulting dimension $D = 3/2$.

Project 6.2.3 Fractal Brownian noise

Theory, computation, difficulty level 1-3.
Support code: Appendix "Brownian.ma," "Mountains.ma."

1) Here we study a recursive procedure for numerical construction of classical Brownian motion ($H = 1/2$) using a random midpoint displacement method. Assume that the constant of proportionality in (6.2.2) is σ^2, and start with two fixed endpoints $x(0)$, $x(t)$. One imagines a line segment connecting these endpoints as a "zero-th approximation" to the Brownian motion signal. Then one generates a midpoint $x(t/2) = (x(0) + x(t))/2 + y$, where y is a Gaussian random variable of mean zero and variance $\sigma^2/4$. Now there are two line segments meeting at $x(t/2)$. One continues bisecting and displacing each line segment recursively, but always using a Gaussian y having variance $\sigma^2/2^{n+1}$ at the n-th approximation.

Argue that this method, with its shrinking variance assignments, properly models Brownian motion in the sense of relation (6.2.2). Then make sample plots and verify by stretching and contracting that the qualitative scaling of Figure 6.2.9 holds.

Now attempt to model Brownian noise of arbitrary dimension $1 < D < 2$, but have a care: the midpoint scheme will not work for $H \neq 1/2$ (i.e., non-classical Brownian motion) without non-trivial modification. It is not enough, for example, simply to scale the recursive variances in a certain way. There are techniques that give good approximations to general Brownian noise, even in higher dimensions [Saupe, in [Crilly et. al. 1991]].

2) Create one-dimensional numerical Brownian motion signals $\{x_j\}$ via the spectral method (this is how Figure 6.2.8 was actually created). One assumes a graph dimension D, implying that the DFT terms enjoy a power law $< |X_k|^2 > \sim k^{-5+2D}$. Create a sequence of DFT coefficients:

$$X_k = \frac{e^{2\pi i r_k}}{(k+1)^{(5-2D)/2}}$$

of some length N, where $k = 0,1,\ldots,N-1$ and the $\{r_k\}$ are equidistributed random variables (mod 1) obeying Hermiticity conditions:

$$r_{N-k} = -r_k$$

That is, one may take $r_0 = r_{N/2} = 0$, choose independent random r_k for $k = 1,2,\ldots,N/2-1$, and force the Hermitian symmetry for the rest of the k. Thus the phases of the DFT coefficients are random, while the magnitudes follow approximately the spectral law. Finally, take the inverse DFT, take its real part (there is theoretically no imaginary part, due to the Hermitian symmetry), and end up with a Brownian signal.

3) Using two independent computed Brownian signals computed as in previous steps (1) or (2) previous, create plots of Brownian walks on the *xy* plane. There is an interesting connection between such walks and–for some dimensions D–the so-called "Levy flights" of statistics, not to mention such scientific dilemmas as the distribution of galaxies in the universe (which distribution seems to be qualitatively similar to "corners" of Levy flights) [Schroeder 1991].

4) For classical Brownian motion ($H = 1/2$) with variance (6.2.2) actually equal to T, the probability of finding $x(t)$ in a small interval is:

$$Prob\{x(t) \in (y, y+dy)\} = \frac{1}{\sqrt{2\pi t}} e^{-y^2/2t}$$

Show that for positive real r the Brownian walk $x(rt)/\sqrt{r}$ has precisely the same distribution.

5) Here we develop intuition concerning the properties of white noise and

the relationship of such noise to classical ($H = 1/2$) Brownian motion. It is often noted that white noise in a kind of "derivative" of the motion. More appropriately (because Brownian paths are generally non-differentiable) the motion is an integral of white noise. Argue heuristically on the basis of the Fourier transform definition for $x^{\sim}$ (5.1.1) that the power spectrum of white noise should therefore by flat, i.e. independent of frequency.

Now assume a discrete signal $x = \{x_0, ..., x_{N-1}\}$ of independent Gaussian random variables of zero mean and unit variance. This means that

$$< x_m x_n > \;=\; \delta_{mn}$$

For the DFT vector X, argue that $< X_k > = 0$ and that $< |X_k|^2 >$ is independent of k, so that the discrete spectrum is indeed "white." Next, argue on consideration of $< |X_0|^2 >$ that the Brownian coordinate given by the sum of all the elements of x has expected square equal to N.

Now a harder but interesting question: Is the phase of X_k random? Have a care: some of these DFT elements must be real-valued. For this question it is an elegant and useful expedient to consider the process of summation for the DFT to be equivalent to a kind of random walk in the complex plane.

Finally, model a discrete Brownian signal y as partial sums of the elements of x:

$$y_0 = x_0 \;\; ; \;\; y_1 = x_0 + x_1 \;\; ; ...$$

Argue that the power $< |Y_k|^2 >$ does indeed, in some appropriate sense, behave as $O(1/k^2)$. One should assume that $k << N$, and perhaps approximate relevant sums by integrals.

6) Here we perform an heuristic analysis that leads to dimension D for the graph of fractal Brownian motion. Consider the figure:

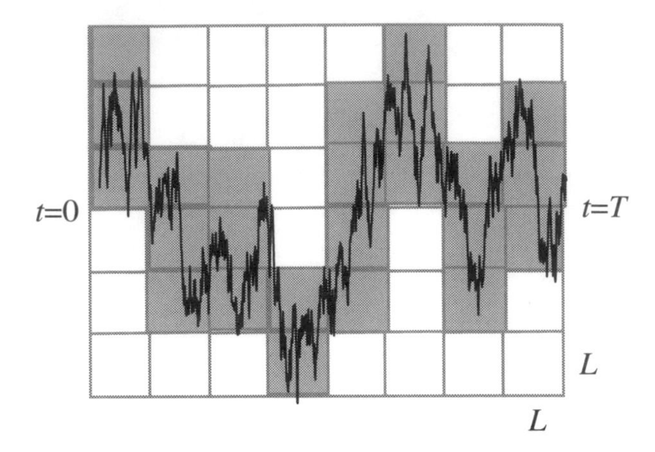

from which we are to count the number of boxes needed to cover the motion, as a function of box side L. Argue heuristically from (6.2.2) that for small L the number of boxes needed on one vertical strip of boxes is proportional to L^{H-1}. Conclude that the dimension obtained from the scaling law (6.2.1) is $D = 2 - H$.

7) Here we argue heuristically the celebrated spectral law (6.2.3) staring from the variance relation (6.2.2) for Brownian paths. Note that formally the power can be obtained by double integration of the transform definition (5.1.1) as:

$$|x^{\sim}(\omega)|^2 = \int\int x(t+s)x^*(t)\, e^{-i\omega s}\, ds\, dt$$

We say "formally" because such integrals are technically divergent–one must use time-windowing, for example, to "damp" the motion for large positive and negative times in a rigorous treatment. But proceeding forthwith, assume x is real-valued; then, performing the t integral, we obtain the expectation $<x(t+s)\ x(t)>$ for fixed s. Now argue by expanding $(x(t) - x(s))^2$ that the resulting integrand is of the form $(a + b\ |s|^{2H})$ times the exponential. Now, simply on the basis of scaling of the s-integral, deduce that the power expectation behaves as $A\delta(\omega) + B\omega^{-1-2H}$. This argument is crude, but does convey some valuable intuition in regard to scaling. As we have noted before for similar heuristics, the analysis can be made rigorous [Falconer 1990].

8) Here we establish an heuristic power-scaling relation in a somewhat different manner, using an explicit function with known fractal properties. Consider the Weierstrass function:

$$x(t) = \sum_{k=1}^{\infty} \frac{\sin a^k t}{a^{k(2-D)}}$$

where $a > 1$ and $1 < D < 2$. This function is everywhere non-differentiable and its graph has box dimension D. This is a useful function for testing fractal-based software, although evaluating x with a finite number of summands can be problematic, especially when convergence is slow.

Argue first that there is a self-similarity akin to that for graphs of fractal Brownian noise (remarks after (6.2.2)); that is, describe in what sense one might write:

$$x(t) \sim a^{D-2} x(at)$$

This at least is some evidence that perhaps the Weierstrass function has Brownian noise-like properties. Next, for the power spectrum, we would like to obtain a rough behavior $1/\omega^{5-2D}$ on the basis of (6.2.3), (6.2.4). Note that for a specific frequency $\omega = a^k$, we have a contributing power $1/\omega^{4-2D}$ simply because of the structure of the summand in the Weierstrass definition. This exponent $4-2D$ is incorrect because these frequencies a^k are not uniformly spaced, and we need some sort of power density average. Argue a sense in which we do obtain roughly the correct power law, by taking into account how these specific frequencies are spaced. Alternatively, approximate the sum with an integral over k and attempt the kind of analysis in step (7) previous.

9) Argue that the dimension of a classical ($H = 1/2$) Brownian two-dimensional walk in the plane has dimension 2. Note that this is not the dimension of a time graph, but the dimension of a permanent record of planar motion. Thus the two-dimensional random walk is 2-space-filling. Argue further that *even the* three-*dimensional* Brownian walk has dimension 2. Thus, the interesting observation obtains that a Brownian walk in 3-space is not space-filling, consistent with the famous result of

the theory of recurrence for stochastic processes that such walks recur to a specified point (such as the origin) infinitely often with probability 1 for one and two dimensions, but probability 0 for three or more dimensions.

10) Consider a Brownian path of the general type in Figure 6.2.8, and imagine the set of all zero-crossings of the horizontal t-axis. Argue, perhaps in the heuristic style of step (6) previous, that the dimension of this Cantor-like set is $1-H$, that is, one less than the dimension of the graph.

11) Plot some fractal Brownian mountains, using a spectral method in which random phases are given to two-dimensional array elements X_{jk}, with the amplitudes of these elements behaving like:

$$\frac{1}{\left(j^2 + k^2\right)^{4-D/2}}$$

which will yield, after two-dimensional inverse FFT, a height function above a two-dimensional plane, the surface thus defined having fractal dimension D. An example plot is shown in Figure 6.2.10.

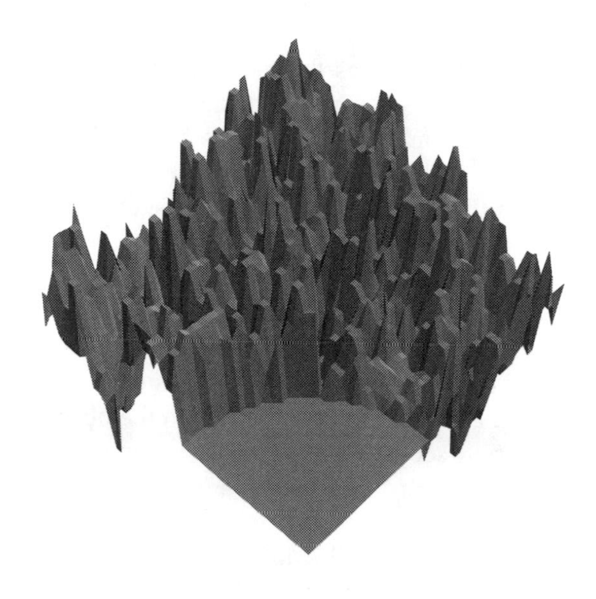

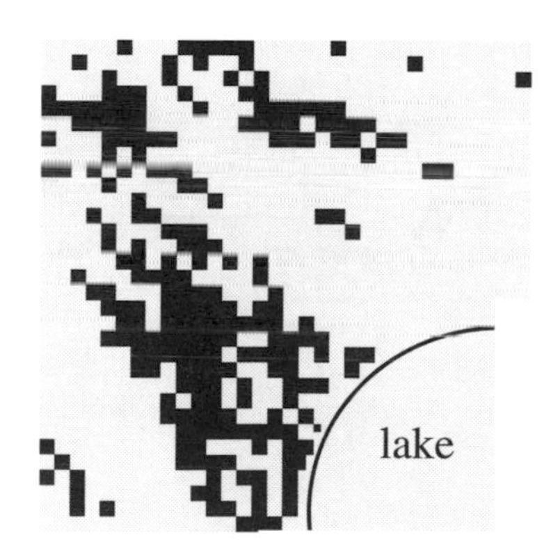

Figure 6.2.10: Fractal mountain surface (upper) of dimension $D \sim 2.85$, and its "lake" cross-section (lower), with $D \sim 1.85$. The mountains are jagged, as the surface has a dimension near 3 and therefore is "trying to be space-filling."

In the Appendix code "Mountains.ma" this procedure is carried out. Note that, to be technically pristine, we should force a Hermitian symmetry to the original two-dimensional Fourier components, but it is reasonable in practice simply to take all phases random and plot the real part of the final inverse FFT as the mountain height function. There are other ways to construct such surfaces on the basis of the Radon transform [Schroeder 1991], and on the basis of sums of self-similar statistical terms [Saupe, in Crilly et. al. 1991]. Consider software that uses generalizations of the Weierstrass function of step (8) previous, to create fractal mountain surfaces.

The bottom part of Figure 6.2.10–a plot of a cross-section of the mountain–shows "islands" whose borders should have fractal dimension 1.85, that is, one less than the surface dimension. This 1-decrement for cross-sections is a general rule, exemplified in the observation of step (10) previous that a cross-section of Brownian motion was down by one in dimension from the graph.

12) Here we investigate a possible model for the theoretically elusive–but quite naturally prevalent–"$1/f$ noise," whose power spectrum should decay as $1/\omega$. Note that the fractal Brownian motion model does not readily give such noise, because we would need the Hurst exponent to be $H = 0$. But consider an RC circuit:

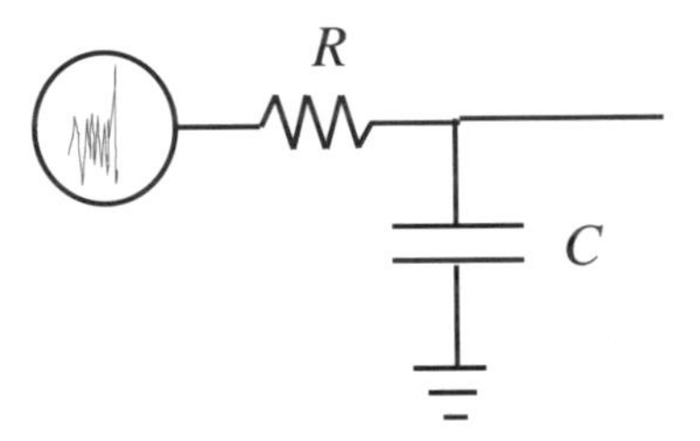

in which a white noise generator (left) is low-pass filtered. Argue first that the power spectrum of the output (right) of this circuit behaves as $1/(1 + R^2C^2\omega^2)$. Now imagine many such circuits, but with *different* R, same C, and *independent* white noise inputs on each. If we now sum all of the outputs, there will generally be a different spectrum. Explain how to choose a distribution of R values so that the resulting sum of all these output stages is has power spectrum behaving approximately like $1/\omega$.

Answer also the question: What happens if all the white noise sources are entirely dependent, i.e., they are all the same source?

This study gives rise to a possible computational method of creating white noise. One may model the RC-network with a difference equation in digital filter style, and feed to each stage a random Gaussian variable at each small time step, finally summing the results of the digital filters. It turns out that not too many filters are needed to achieve a fairly good log-log plot of power vs. frequency having slope ~ –1.

The elementary estimate of fractal dimension D in the scaling relation (6.2.1) is a proverbial tip-of-the-iceberg. It has become necessary, especially in studies of strange attractors attendant to dynamical systems, to consider other dimension measures. A convenient generalized dimension can be defined, special cases of which address themselves nicely to certain tasks of dimensional estimate. Assume a cloud of points embedded in a Euclidean space of dimension E, so that the cloud is a collection $\{y_j\}$ of E-dimensional vectors. Imagine boxes of length scale ε, of which $\#(\varepsilon)$ are non-empty of y points. Note that a box has volume ε^E. Now define a generalized dimension, based on a parameter q:

$$D_q = \frac{1}{q-1} \lim_{\varepsilon \to 0} \frac{\log \sum_{i=1}^{\#(\varepsilon)} [P_i(\varepsilon)]^q}{\log \varepsilon} \qquad (6.2.5)$$

where $q \geq 0$, $q \neq 1$. The summation is carried out over all non-empty boxes, and $P_i(\varepsilon)$ is the probability of finding a point in box i. This generalized dimension is related neatly to generalized Renyi entropy [Schroeder 1991] and signals a connection between dimension theory and information theory. The definition of D_q, while notationally forbidding, nevertheless has some useful special cases:

- $q = 0$, for the "box dimension" or closely related "capacity dimension:"

$$D_0 = -\lim_{\varepsilon \to 0} \frac{\log \#(\varepsilon)}{\log \varepsilon} \qquad (6.2.6)$$

• $q = 1$, for the "information dimension:"

$$D_1 = -\lim_{\varepsilon\to 0} \frac{\sum_{i=1}^{\#(\varepsilon)} P_i(\varepsilon)\log P_i(\varepsilon)}{\log \varepsilon} \tag{6.2.7}$$

• $q = 2$, for the "correlation dimension:"

$$D_2 = \lim_{\varepsilon\to 0} \frac{\log \sum_{i=1}^{\#(\varepsilon)} [P_i(\varepsilon)]^2}{\log \varepsilon} \tag{6.2.8}$$

Note that the D_0 dimension is essentially the elementary prescription (6.2.1), although there is a subtle theoretical distinction when one assumes freedom in choosing box locations [Bingham 1992]. The box dimension assumes natural juxtaposition of boxes of side ε, while the capacity dimension is obtained when # is the minimal number of boxes of arbitrary placement. Thus box dimension is technically an upper bound on capacity dimension, even though these two variants for D_0 are very often the same. Hereafter we shall assume unless otherwise specified that when we use explicit boxes of side ε they are always aligned in a natural, gridlike fashion. There are interesting information-theoretic aspects of these various definitions of dimension [Grassberger, in [Buchler 1985]]. The correlation dimension has reached a definite computational vogue due to the observation [Grassberger and Procaccia 1983, 1984] that, if the embedded E-dimensional point vectors are denoted $\{y_i : i = 0,...,M-1\}$, a correlation sum:

$$C(\varepsilon) = \lim_{M\to\infty} \frac{\sum_{i\neq j} \theta(\varepsilon - |y_i - y_j|)}{M^2} \tag{6.2.9}$$

where θ is the Heaviside unit step function, can be used to estimate the correlation dimension D_2. Note that the sum is simply the number of pairwise occurrences that two distinct points of the cloud are within ε of each other. It is suspected, and known in cases where the set $\{x_i\}$ is sufficiently self-similar, that the correlation dimension is:

$$D_2 = \lim_{\varepsilon \to 0} \frac{\log C(\varepsilon)}{\log \varepsilon} \tag{6.2.10}$$

In dimensional assessment work, the problem of Euclidean embedding dimension is paramount. When analyzing dynamical systems, it is customary to create, from a map iteration sequence $\{x_i\}$ of real-valued m-dimensional vectors, clouds of $2m$-dimensional points:

$$\begin{aligned} y_0 &= \{x_0, x_1\} \\ y_1 &= \{x_1, x_2\} \end{aligned}$$

... etc.

or $3m$-dimensional points:

$$\begin{aligned} y_0 &= \{x_0, x_1, x_2\} \\ y_1 &= \{x_1, x_2, x_3\} \end{aligned}$$

... etc.

and so on. The Euclidean embedding dimension is generally thus $E = km$ for the dimension definitions above. It turns out that, in general, a strange attractor dimension will "saturate" at some value when the embedding dimension exceeds the actual dimension of the attractor, so one technique is to use log-log plots for successively higher embedding dimensions, and to seek a limiting value for the dimension slope. There are other options, such as lagging the phase between successive y collections of x's by other than one iteration step, but most dimensional assessments are relatively insensitive to the precise choice of lag. When analyzing the dimension of a *graph* of a signal, the points in question would be:

$$y_i = \{x_i, i\}$$

comprising a cloud embedded in $(m+1)$ dimensions.

Project 6.2.4 Measurement of fractal dimension

Computation, difficulty level 3-4.
Support code: Appendix "corrdim.c," "wavdim.c."

1) Create software to estimate, in the style of Project 6.2.3, step (6), the box dimension of graphs of time series using the box/capacity dimension D_0. Think of the t axis as horizontal, and literally count boxes in vertical strips that contain at least one point of a given discretized signal. Test this on cases of known D values, such as graphs of fractional Brownian motion or Weierstrass functions of Project 6.2.3. Note that many enhancements over the brute force counting method are possible, as intimated in step (2) next.

2) Implement the precision-numeral-sorting-box-counting algorithm of [Bingham 1992] for D_0, and/or the correlation sum method of [Grassberger and Procaccia, 1983, 1984] for D_2. The Appendix code "corrdim.c" implements the latter algorithm–although in an unoptimized way. This code could be thought of as methodologically similar to direct DFT code; that is, one suspects that a fast algorithm is possible. Consider optional enhancements of the Appendix code, along the lines of minimum complexity discussed in step (3) next. Another code enhancement would be never to fill an auxiliary array of dimension E vectors as is done in embed() of the code example, working instead with the raw data stream. Note that the Appendix code already has one enhancement arising by departure from the precise statement of (6.2.9); namely, two points are given a "1" count in the mutual comparison if and only if both points lie within an E-dimensional cube of side ε. This small-cube check is done quickly then by comparing absolute coordinate differences. Even if the literal meaning of the Heaviside numerator of (6.2.9) is demanded, a cube pre-check still provides a speedup because of the large number of point pairs which will fail to be ε–close in any sense.

To test the Appendix code "corrdim.c" one may, for example, generate an input file of iterated (one-dimensional) x-coordinates of a chaotic map and perform trial runs of $E = 2,3,4,5$ embedding dimensions. The output data list of correlation pairs {ε, C(ε)} can then be log-log plotted in *Mathematica* style, as follows:

```
lis =
{{0.006510193199,0.332357646122},
{0.005533664219,0.310220802829},
...etc.
}}
ListPlot[Log[lis], PlotRange->{{-12,-4},{-9,-1}},
    AspectRatio->1]
```

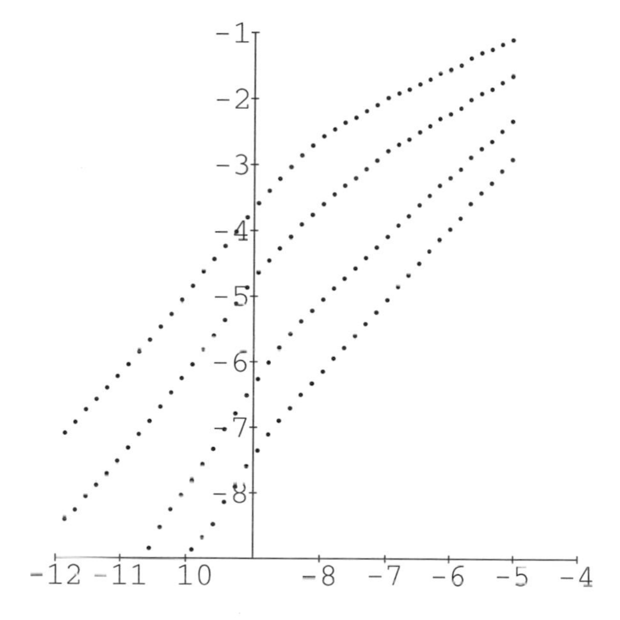

Figure 6.2.11: Measurements of fractal correlation dimension using code "corrdim.c." The Henon map was used to generate input iterations, which are embedded above in E = 2,3,4,5 dimensions respectively (top to bottom). The dimension is the common slope D_2 ~ 1.236 according to this run.

Figure 6.2.11 shows a run with only 2048 iterates of a Henon map with the parameters a, b immediately below. One can read off of the Figure the known experimental correlation dimension D_2 ~ 1.236, in good agreement with established data.

Apply whatever code is finalized to data from a Henon system (6.1.7), with parameters a = 1.4, b = 0.3, to obtain established experimental values for D_q [Ling and Schmidt 1992]:

Embedding dimension d	D_0	D_2
2	1.286	1.231
3	1.244	1.231

and for the quadratic map (6.1.1) with $a = 3.56699456$:

1	0.538	0.501

and for the Lorenz system (6.1.9) with $r = 28$, $\sigma = 10$, $b = 8/3$, and time increment $dt = 0.2$:

3	1.778	2.069
4	1.821	2.120
5	1.802	2.156

3) Here we consider the computational complexity of dimension assessment, as follows. Note that for a direct method as implemented in Appendix code "corrdim.c," the overwhelming computational price paid is the fact of $O(M^2)$ checks on mutual distances in (6.2.9). This price can be brought down by pre-partioning the cloud of points into compartments, in which case, for sufficiently small ε, the primary work is comparison only of pairs mutually within each compartment. This procedure is naturally reminiscent of algorithms for the N-body problem discussed in Chapter 2. Consider the following heuristic argument for enhancing thus the efficiency of correlation dimension computation. Assume a cloud of M points in $E = 2$ embedding dimensions. We break a rectangular bounding region of this cloud into K cells of roughly equal size, so that each cell has, roughly speaking, M/K points. Now for small ε, much of the work in computing $C(\varepsilon)$ involves mutual-distance checks within cells. Argue that the total number of within-cell mutual checks is $O(M^2/K)$. There are "border checks" that should also be performed, but these involve checking only over interstitial strips of width ε. Work out in this way a theory of complexity for this cell-covering method, ending up with an optimal value $K(\varepsilon)$, and thus an optimal cell size for each ε-based software run. It is clear that even without a variational theory, the Appendix code "corrdim.c" would benefit radically just from a choice like $K = 16^E$ or some such fixed decimation value. A box-counting approach

and other optimized approaches to estimation of correlation dimension are discussed in [Ling and Schmidt 1992].

Next, notice that breaking the fractal data cloud into cells could be done by sorting as soon as the data is input. Then the within-cell checks just mentioned would occur over consecutive sub-blocks of sorted data. But of course we mean E-dimensional sorting of some kind. This thinking is one way to lead into the $O(M)$ box-counting method of [Bingham 1992] to which we have referred. In this latter method, the sorting within cells essentially goes deeper, sorting by coordinates in the style of Cantor renumeration of space coordinates.

Whatever theoretical approach is used, and whatever the precise definition of dimension, consider making all of this rigorous by asking the question: what is the computational complexity of finding the final value D_q? This is a hard question. All we have discussed to this point is the difficulty of doing one run for given ε. Perhaps the question should be stated this way: For a fractal data cloud of M points, normalized to fit inside the unit E-cube, how long $T(M,a,b)$ does it take to compute $C(\varepsilon)$ for each value of ε:

$$\varepsilon = 2^c \sqrt{M} \qquad ; \qquad a \leq c \leq b$$

based on integers c over the stated range?

The answer would resolve the implicit question for final D_2 value, because best slope (of a resulting curve such as seen in Figure 6.2.11) can be obtained in $O(b-a)$ operations.

One more observation is that a "slope-fitting" automaton can be conceived, which looks at curves as in Figure 6.2.11 and does the following. Integrate $m*\rho$, where ρ is the radius of curvature and m is the slope, always computed discretely along known curve points, and divide the answer by the integral of ρ itself. In this way the straighter portions of a curve are given tremendously more weight. In practice, ρ would be clipped above at a large value to avoid singularity, and one might contemplate using powers of ρ to taste.

4) Implement software that assesses fractal dimension in cases where a

well-behaved power spectrum is expected to follow a scaling law. Consider the following interesting method. If spectral power density behaves as $\omega^{-\beta}$, show that β can in principle be recovered from a ratio of two areas of the spectral graph:

$$\frac{\int_u^v \omega^{-\beta} d\omega}{\int_{u/2}^{v/2} \omega^{-\beta}} = 2^{1-\beta}$$

This approach has the appeal that, by integrating thus over scaled regions, one is essentially averaging a possibly noisy spectrum. Using connection between the spectral power exponent and fractal dimension as in the introduction to Project 6.2.3, one may thus derive a dimension estimate.

5) Implement software to assess fractal dimension via the continuous wavelet transforms of Subject 5.1. Theory and practice are discussed in [Arneodo et. al., in [Meyer 1991]]. This technique has been applied to the Cantor measure, in a manner similar to the Fourier approach described in step (5), Project 6.2.1. For this measure $C(x)$, show that the continuous wavelet transform (5.1.7) satisfies a scaling relation:

$$F_{ab} = \frac{\sqrt{3}}{2}\left(F_{3a,3b} + F_{3a,3b-2}\right)$$

Now define I_a to be the integral of $F^2{}_{ab}$ over all b. Argue that if the two F terms on the right-hand side are uncorrelated, then $2I_a = I_{3a}$. Note that this last scaling relation suggests a power law of the form $I_a \sim a^D$, where D is the Cantor dimension log 2/log 3. Thus, make log-log plots of I_a vs. a to deduce a numerical value for the dimension of the Cantor set.

Now start with the Weierstrass function of Project 6.2.3, step (8), taking the special case $a = 2$. In this case there is an approximate symmetry:

$$x(t) \sim 2^{D-2} x(2t)$$

Show that the continuous wavelet transform (5.1.7) satisfies in turn an approximate relation:

$$F_{ab} \sim \frac{1}{2^{5/2-D}} F_{2a,2b}$$

As before, define I_a as an integral of the square of F and conclude:

$$I_a \sim \frac{1}{2^{6-2D}} I_{2a}$$

Therefore, on the notion that $I_a \sim a^{2D-6}$, deduce via log-log plots the dimension D for some Weierstrass signals.

6) Implement software to assess fractal dimension via the fast wavelet transforms of Subject 5.3. At least two basic scenarios are relevant, just as in step (5) previous; namely, one may attempt measures for fractal noise or for fractal measures.

In the first case of fractal noise, as discussed prior to Project 6.2.3, one may assume a power spectrum of exponent $-1-2H$ and expect a graph dimension $D = E + 1 - H$, where E is the topological dimension of the space of the relevant random variable. The interesting idea is to compute, say for an $E = 1$ discrete Brownian fractal signal of length N, the variance of the first $N/2$ "d" terms obtained via application of the T'_N matrix of Subject 5.3, and also the variance of the next $N/4$ "d" terms obtained from the $T'_{N/2}$ matrix. Call these variances σ^2_N and $\sigma^2_{N/2}$ respectively. Then according to [Mallat 1989], and for a particular normalization of the wavelet transform,

$$H = \log_2 \frac{\sigma_{N/2}}{\sigma_N}$$

or, for this case $T = 1$,

$$D = 2 - \log_2 \frac{\sigma_{N/2}}{\sigma_N}$$

These relations are reminiscent of the scaling laws for continuous transforms of step (4) previous, especially when we identify the integrals of F^2 with variances of the d-channels. The Appendix code "wavdim.c" is an attempt to implement these ideas for Weierstrass graphs.

The method appears to have good accuracy (a few per cent error in D) when the Weierstrass graph at least has $3/2 < D < 2$. The author has noticed that experimental results are somewhat better for higher-order wavelets. There are three good further exercises here. One is to find various ways to derive Mallat's discrete variance relations above, perhaps in the style that one used if one has derived the continuous scaling cases of step (5). Another exercise is to attempt to find graph dimensions for actual images of fractal noise, via higher-dimensional wavelet transforms of Project 5.3.2. Third, it would be quite interesting to perform statistical analyses of the wavelet transforms of geometrical fractals such as Hilbert and dragon curves, Sierpinski gaskets, and so on–perhaps even dynamical attractor data. The author guesses that techniques of image processing can be used in some general way; e.g., to wavelet transform a fractal curve, simply treat a discrete version of this curve as white pixels (on the curve) with black pixels elsewhere, and trasnform the resulting "image."

The other scenario, that of Cantor-like measures, would involve a discrete fast transform analog to the work of [Arneodo et. al., in [Meyer 1991]] mentioned in step (5) previous. In this case one would somehow inspect the rate of change of wavelet component amplitudes with respect to scale. If the first $N/2$ "d" terms created are denoted $d^{(1)}$, the second set of $N/4$ "d" terms denoted $d^{(2)}$, and so on; then the dimension of the Cantor measure should somehow be evident in the statistics of the sequence of $d^{(j)}$. The Appendix code "WavcletOnSignal.ma" can be used as a starting point, to plot fast wavelet transforms of a discretized Cantor measure. Due to the power-of-two lengths of the fast wavelet transforms we discussed in Section 5.3, it is recommended to take in fact the "middle-two-fourths" Cantor set, which means all numbers having neither "1" nor "2" in their base-4 expansions. Note that this set has the fractal dimension $D = \log 2/\log 4 = 1/2$. One may take signal length $N = 4^k$ and represent a Cantor signal element x_j by a "1" if j has neither "1" nor "2" in its base-4 integer expansion, and a "0" otherwise. After a fast wavelet transform, look for behavior like the following: the sum of squares of $d^{(1)}$ terms divided by the sum of squares of $d^{(3)}$ terms should be about 4^D. At least, that is the heuristic suggested by the analysis of step (5) previous.

7) Investigate the theoretical and computational aspects of the Kaplan-Yorke relation:

$$D_L = j + \frac{1}{|\lambda_{j+1}|} \sum_{k=1}^{j} \lambda_k$$

which relates the Lyapunov dimension D_L to the Lyapunov exponents, where it is assumed that the $\{\lambda_k\}$ are ordered:

$$\lambda_1 > \lambda_2 > ... > \lambda_j \geq 0$$

In other words, only a subset of the exponents is summed to obtain the dimension relation. The impressive conjecture of Kaplan and Yorke is that this D_L should *equal* the correlation dimension D_G when the points on the fractal are "approximately uniformly distributed" [Baker and Golub 1990][Grassberger and Procaccia 1983].

8) Gather numerical data, let us say from the real world, and analyze it with fractal dimension measurement software. A good example would be the geographical distribution of something: population, forest, mineral deposits, etc. One could count the number of boxes of side L required to cover all cities, for example. The coastline of a body of land usually exhibits fractal dimension noticeably greater than 1. But geography is not the only seat of such natural data. There exists a multitude of natural settings where fractal dimension makes sense. Note that, because of the natural finitude of such data, one can only hope that a reasonable straight-line fit on a log # vs. log L plot does occur over a few decades.

9) Here is, the author believes, a very hard problem. It can be shown experimentally–by literally crumpling various original sizes of paper sheets–that crumpled paper balls have a rough mass-radius relation $M(r) \sim r^{3-\varepsilon}$ where $\varepsilon > 0$. This in itself is an amusing laboratory experiment. But the problem at hand is to *model computationally* this scenario, hopefully to obtain a numerical estimate for the "mass dimension" $3-\varepsilon$. One needs software that understands a self-avoiding meandering of folded paper.

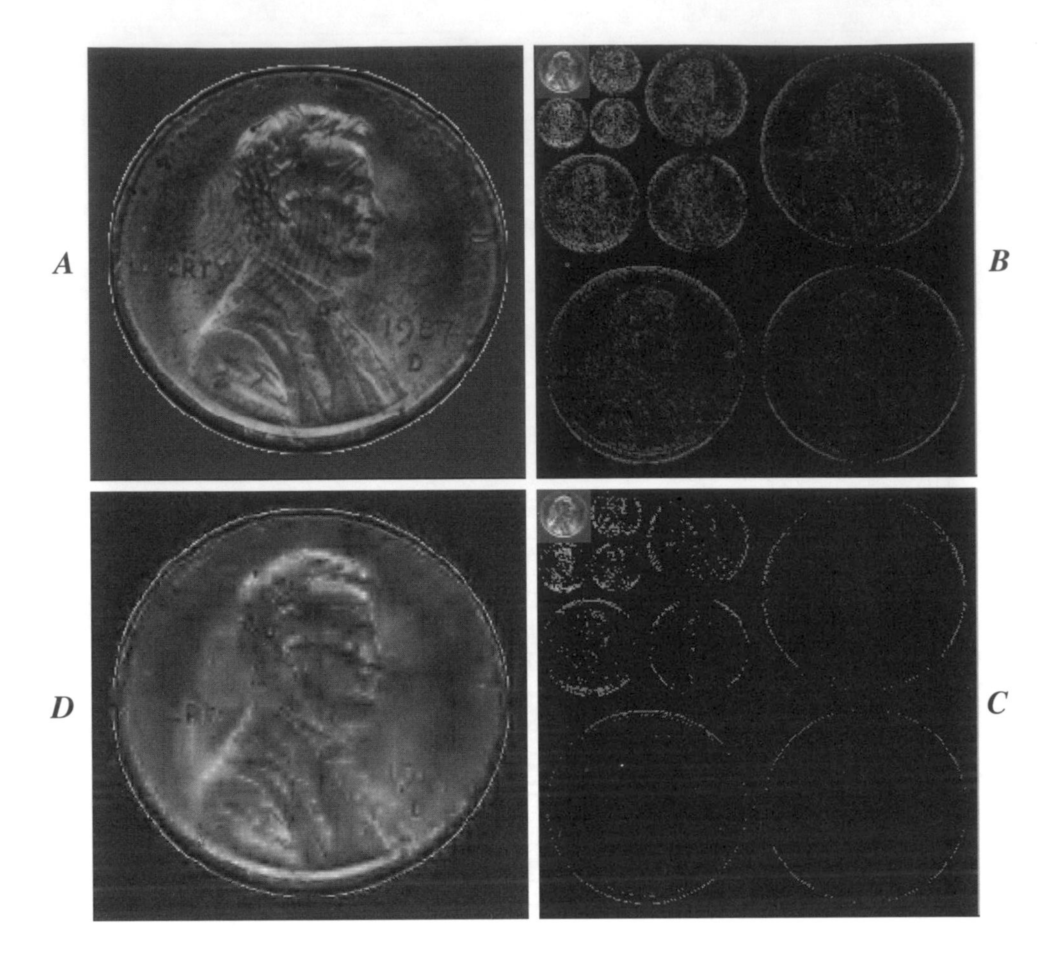

Image compression via two-dimensional transform. To an original Lincoln cent image (A) is applied a discrete, three-recursion level, two-dimensional, fourth-order wavelet transform to yield (B). Note that the three large "detail quadrants" within (B) indicate, clockwise from upper right: vertical, diagonal, and horizontal edge information respectively; and that the far upper left scalar sector is a subsampled replica of the cent. Then wavelet coefficients below a chosen threshold are dropped, yielding (C). Inverse transform yields a reconstruction (D), showing a typical 200:1 compression result.

7 Signals from the real world

Projects in signal processing

Ships that pass in the night, and speak each other in passing; Only a signal shown and a distant voice in the darkness; So on the ocean of life we pass and speak one another; Only a look and a voice; then darkness again and a silence.

[*Longfellow* 1874]

Engineers often view the eye as a kind of television camera with a certain bandwidth and dynamic range. Conversely, it is more natural for a biologist to see the visual system as an active synthesizer of the world we see . . . The visual system is dedicated to synthesizing a world from inchoate patterns of light that reach the retina. Its ability to find order in a multitude of perceptual dimensions does, indeed, sometimes seem magical. It is this magic that we must artfully exploit when we create visualizations.

[*Friedhoff* 1990]

Signals are transient by nature and by definition. Those signals emanating from the ship that "passes in the night" had better be recorded lest they be lost. We do not investigate here the recording notion; rather we explore computational avenues open to one who has recorded the data.

The elegant aspect of signal processing, then, is the notion that the time and space axes can be captured, thus becoming something as mundane as array indices or matrix locations. It is this "freezing" of transients, and subsequent analysis of them, that renders signal processing such a satisfying endeavor. It is as if one can manipulate time and space at leisure and at will.

Subject 7.1 Data compression

Luckily, most data we accumulate have some, if you will, *meaning*. We attempt to quantify the word "meaning" by invoking the concept of entropy. One could protest that we are not in the philosophy business and so should avoid notions of meaning, but it seems fair to quantify on the idea that "more meaning is generally more compressibile." It is possible to read this as "more order is more compressible." A thermodynamicist might say, for example, that random noise cannot be significantly compressed; and we shall be saying that random noise has maximum entropy (minimum meaning, minimum order).

Imagine a generator that produces a symbol sequence where symbol k occurs with respective probability p_k. Then the so-called first-order entropy of such a source is:

$$H_1 = -\sum_k p_k \log_2 p_k \tag{7.1.1}$$

measured in bits per symbol. If only one symbol occurs with probability 1, the sequence is all one symbol and we interpret $H_1 = 0$. This is the limit of complete ordering, and the entropy is zero. On the other hand, if there are M total symbols and each occurs with probability $1/M$, then we have $H_1 = \log_2 M$. But this is reasonable: one may encode such a symbol sequence by allocating $\log_2 M$ bits per symbol, simply to indicate which symbol is being sent. A more complicated example would be the following. Say that there are three symbols 0, 1, 2, having respective probabilities 1/2, 1/4, 1/4. Then the entropy per symbol is:

$$H_1 = -\frac{1}{2}\log_2\frac{1}{2} - \frac{1}{4}\log_2\frac{1}{4} - \frac{1}{4}\log_2\frac{1}{4} = \frac{3}{2}$$

How may such a sequence be encoded? One way is to allocate codes:

symbol	code
0	0
1	10
2	11

For example, the sequence 112022110000 would be encoded:

1010110111110100000

In this case the number of bits per symbol is 19/12, not far from 3/2. A little thought reveals that, asymptotically, if each symbol occurs with its proper probability, the number of bits per symbol for the encoding is indeed (1/2)*1 + (1/4)*2 + (1/4)*2 = 3/2.

The idea of entropy is exploited by virtually any compression scheme. A basic scheme is Huffman encoding, for the study of which many excellent introductions exist [Rabbani and Jones 1991][Aho et al. 1993]. The coding algorithm is, given independent symbol probabilities, to construct a Huffman tree which results in a binary code for each possible symbol. Huffman codes are close to optimal when the symbol probabilities are roughly in geometric progression by factors of 2. In our example above where the entropy was 3/2 bits per symbol, it was possible to match the exact entropy with the code. Figure 7.1.1 shows a typical tree construction pertinent for step (1), Project 7.1.1:

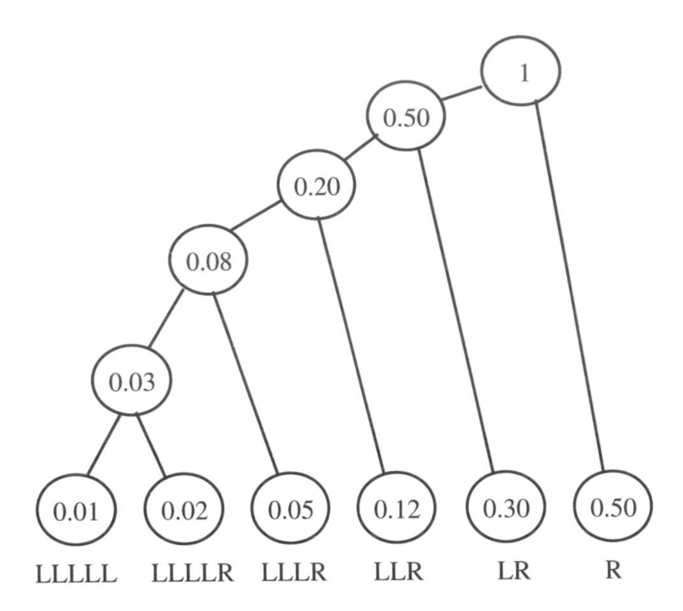

Figure 7.1.1: Example of Huffman code tree construction, with bottom row comprised of eight symbol probabilities. At every tree level, the smallest two probabilities p,q are combined into a parent leaf having a probability $p+q$. When the tree is complete (with a solitary "1" at the top) the Huffman codes are given by the sequences of left-right tree moves required to locate the original probabilities. These L-R sequences, when turned into binary 0-1 codes, yield the complete Huffman encoding.

More sophisticated compressors include arithmetic encoders, which segment the interval (0,1) into probability segments; Lempel-Ziv-Welch (LZW) compressors; and the BSTW-class of compressors that encode data according to recent occurrences, giving so-called "move-to-front" schemes. One can usually find a scenario in which one of these compressors will outperform a Huffman coder; although when the Huffman scheme is used at higher order (combining chains of symbols and using composite, multi-symbol probabilities) the differences are less dramatic. An excellent review of various modern schemes is [Fiala and Greene 1989]. For comparison of fixed-code Huffman-like schemes to BSTW methods, see [Lelewer and Hirschberg 1987].

The following Project is a brief journey through the world of lossless compressors. Later, part way into Subject 7.2, we take up the subject of lossy compressors.

Project 7.1.1 Tour of lossless data compressors

Theory, computation, difficulty level 2-4.
Support code: Appendix "huffman.c," "Compressors.ma," "eliascoder.[ch]," "eliastest.c."

1) Implement a Huffman code generator. A straightforward example is given in the Appendix as "huffman.c." A more symbolic solution is found in "Compressors.ma." In the C program, if one enters, for example:

```
% a.out 0.01 0.02 0.05 0.12 0.3 0.5
```

the resulting code scheme is:

```
0.01: 00000
0.02: 00001
0.05: 0001
0.12: 001
 0.3: 01
 0.5: 1
```

For this example, calculate the theoretical first-order entropy from (7.1.1), and compare with the expected bits/symbol using the output encoding scheme. It is a good idea to modify the program itself to supply this information in general.

2) Establish the theorem that if first-order and second-order entropies are defined:

$$H_1 = -\sum_i p_i \log p_i \quad ; \quad H_2 = -\sum_{ij} p_{ij} \log p_{ij}$$

where p_{ij} is generally the probability of occurrence of the symbol *pair* (i,j); then $H_2 \le 2H_1$. This result indicates that it may well be better to, say, Huffman encode pairs of symbols, depending on how close a Huffman scheme is to optimal for the paired case. In proving this entropy theorem, it should be evident along the way that when correlations are high between symbols, the second-order entropy is correspondingly lower. Conversely, when there is no correlation, equality should be saturated.

3) Implement an arithmetic encoder, an example of which is found in "Compressors.ma." Arithmetic encoders often use floating-point arithmetic to slice up the interval (0,1) into appropriate source probability subintervals; which probabilities are determined from the symbol stream itself. Unlike many other coders, arithemetic encoding always meets the first-order entropy bound asymptotically. Drawbacks include difficulty of implementation and usually slow execution.

4) Implement a universal Elias encoder, which encodes positive integers according to the following table:

0	1
1	010
2	011
3	00100
4	00101
5	00110
6	00111
7	0001000
...	

The rule is, to encode integer n, use a prefix of $[\log_2(n+1)]$ zeros and append to the right of this the binary representation of $n+1$. Derive a formula for the precise number of bits to encode an arbitrary n.

Then work out formulae for the average number of Elias bits per symbol given the symbol probabilities p_i, $i = 0,1,2,...$ Is there some contrived set of probabilities for occurrence of non-negative integer symbols such that this code is optimal, or near-optimal (near H_1)?

There is a second-order Elias code and a recursive procedure for shortening the asymptotic bit-lengths, and these higher-order Elias codes are asymptotically optimal (converge upon first-order entropy), but the practical benefits are dubious when small integers are to be encoded [Lelewer and Hirschberg 1987].

Though this simple Elias encoder is not asymptotically optimal, the author has found such encoders very useful, because in many complicated signal processing experiments, one has to encode integers and it is convenient during research to call on a reliable Elias function; perhaps later to optimize the transmission of the integer stream. Also it is clear that any signal stream can be put in integer form, so an Elias code library is versatile. Another reason Elias codes prove useful is that often in signal processing one has integers {...–2,–1,0,1,2,...} whose symbol probabilities are symmetrically arranged; i.e., $prob_i$ depends only upon $|i|$. But the first Elias code has one bit, while every other code length (3, 5, 7... bits) occurs an even number of times. Thus, in signal processing situations with the symmetrical symbol probabilities about the origin symbol, one may try a table:

0	1
–1	010
+1	011
–2	00100
+2	00101
–3	00110
+3	00111
–4	0001000
...	

This scheme, by being unbiased with respect to sign, proves even more effective and is often close to optimal (e.g., close to a static Huffman code), especially for small bipolar ranges of the integer symbols.

The Appendix code "eliascoder.[ch]," "eliastest.c" shows how a library can be constructed and tested for the Elias codes. In the Appendix scheme, one may encode as shown in main() any integer from 0 through 254 (= NUM_CODES-1). The typical run is:

```
% a.out 8
Stream bytes 43692; compression factor:0.666687
Encode/decode worked.
%
```

This output means the following: 65536 Poisson-distributed integers (mean value 8, exponentially decreasing $prob_i$ as $i = 0,1,2,...,254$) were encoded into 43692 bytes (compression factor about 2/3). Thus, in this particular run, about five bits per symbol were required on average. It would have taken eight bits per symbol if each integer 0 through 254 had been sent as an 8-bit byte. In this way, the Elias codes, while not optimal, are a good universal choice for general integer transmission, particularly when the probability decay, as one runs through symbols 0,1,2..., is very steep.

5) Implement a general bitstream transmit/receive system. Programmers who have worked with variable-length codes like Huffman codes know that one of the more painful stages of a compression project is the bitstream encoding.

A fast way to handle bitstreams is actually buried within the source "eliascoder.c." The key to speed is to do the decoding by lookups. The library is set up so that the steps for bitstream encoding are relatively simple:

- Allocate encode and decode tables (call them enc_new, dec_new, respective types enc_table, dec_table) as is done in the function init_elias_codes(), and jam these tables with your own custom bit codes, noting that the inverse table dec_new contains exactly 2^m entries where m is the maximum number of bits in an encode.
- To put out a coded bitstream to pointer "buff" for an array "dat" having "numdat" integers, do:

```
inb = 0;
buff += put_codes(buff, enc_new, dat, numdat, &inb);
buff += terminate_put(buff, inb);
```

The particular syntax here was designed carefully to allow *chaining* of different tables in midstream. The terminate() function rounds out the current long integer that is being filled with code, and puts it out on the stream, with "inb" always being the bit offset into the next integer. But prior to the terminate() function one has the option of:

```
buff += put_codes(buff, enc_new2, dat2, numdat2, &inb);
```

which would be the placement of an encoding of new "dat2" using a new "enc_new2" table. Note also that the number of encoded bytes is exactly given by the integer difference between the current value of "buff" and its starting origin value.

- To decode:

```
buff = &buf[0];
inb = 0;
buff += get_codes(buff, dec_elias, dat, numdat, &inb);
buff += terminate_get(buff, inb);
```

and the data array "dat" will contain a total of "numdat" decoded data. As before, extra calls of get_codes() prior to terminate_put() are allowed, perhaps with different tables, as long as the scheme is consistent with what was originally encoded.

Note that in the "eliascoder.c" library the maximum number of Elias codes is 255, meaning data arrays must contain values 0,1,...,254 and no other values. This is not a real limitation, since one can combine data values 254, 255 into "254 followed by 0 or 1," using logic completely outside of put_codes() and get_codes(). Thus, the binary encoding value for 254 could be used as a so-called "escape code," since when this was detected in the bitstream, it would unfold into one of several values (in this case 254 or 255) depending on the next value.

The previous remarks suggest ways to make Elias coding more efficient; i.e., use escape codes to avoid many long bit codes. This approach is effective in cases where larger integers are so common that there are many 11-, 13- and 15-bit Elias codes. Recall that the Elias codes start out:

```
0    1
1    010
2    011
...
```

One way to effect escapes is to change the definition of "0" encoding so that the list becomes:

0	10
ESC	11
1	010
2	011

Then any integer above 30 is encoded "ESC + byte" where the 8-bit byte is a literal encoding of the integer value 16-255 inclusive. Then this whole scheme never involves more than ten bits per symbol. Of course, what we sacrificed for this 10-bit cap on the code size was the number of bits for the "0" symbol. Clearly there are many variants on this escape scheme, and as always the goal of such a static encoding is to try to match entropy.

A good exercise is to modify and/or add to "eliascoder.c" to effect other codes, such as Fibonacci codes [Lelewer and Hirschberg 1987] and higher-order Elias codes. The Fibonacci codes are better than Elias codes for many data sets, but like the simple Elias codes presented here, are not asymptotically optimal.

Another good exercise is to combine the program "huffman.c" with this library to effect a true, stream-oriented Huffman encoder/decoder. One elementary approach is to look at the data to be compressed, create a Huffman tree from the apparent probabilities, then send the Huffman bit table itself, then send the variable-bit bitstream.

6) Implement a BSTW "move-to-front" encoder, which works according to the following example. Say a symbol sequence runs ABCADEABFD. One encodes like so:

1 A 2 B 3 C 3 4 D 5 E 3 5 6 F 5

with, for example, the integer symbols encoded via an Elias code as in the previous step. The encoded sequence here is interpreted by "moving-to-front" a symbol that has occurred in the past. One may think of a stack, on top of which one continually places the most recent symbol:

"**1**, for new symbol **A**,"
"**2**, for new symbol **B**," (now think of AB)
"**3**, for new symbol **C**,"(now think of ABC)
"**3**, to recover A from 3 deep in the past," (now think of BCA)
"**4**, for new symbol **D**,"
"**5**, for new symbol **E**,"
"**3**, to recover A again from 3 deep," (now think of BCDEA)
"**5**, to recover B from 5 deep," (now think of CDEAB)
"**6**, for new symbol **F**," (now think of CDEABF)
"**5**, to recover D from 5 deep."

Note that each literal symbol only appears a maximum of once no matter how long the stream (for this example, in other words, each symbol is boldfaced precisely once). The BSTW scheme is especially efficient when, in a long symbol stream, highly correlated bursts of symbols appear momentarily; in such cases BSTW evidently outperforms static Huffman encoders [Lelewer and Hirschberg 1987].

Investigate variants of this method, one such being that one does not move the most recent symbol all the way to the top of the stack; rather one moves it perhaps not at all (in which case the integer code sent is trivially just a fixed stack-depth), or one moves it halfway toward the top, or some such scheme. It is interesting to contemplate precisely what BSTW variant is optimal if one knows the underlying symbol probabilities *including* multi-symbol probabilities or correlations.

7) Assuming one has implemented a Huffman encoder, and also a BSTW-class encoder, or perhaps other encoders (such as LZW, one manifestation of which is Unix "compress"), a good avenue of research is to compare these on text files, sound files, image files, and compiled machine code files.

8) Here is a fascinating research problem that perennially arises in signal compression, and to which the author has never seen a completely satisfactory answer. One is given a long stream of binary data bits, split into consecutive blocks of n bits each (for example, into consecutive n = 32 bit words). One also knows the set of numbers p_m, each denoting the "probability that a block has exactly m 1's," for $m = 0,1,2,...,n$. Given such p_m what is an optimal, or near-optimal, but also fast static encoding?

Elementary Huffman coding is not recommended if the expectation $<m/n>$ is small. If $<m/n>$ is, say, 1/64, it would be better to use run-length encoding, in which a stream bit "1" means "data is a '1'" while "0" means "the next six bits give the number of 0's following." So the problem comes down to this: Determine how to transition between encoding schemes depending on the set $\{p_m\}$. Or is there a universal encoding scheme for this general signal model, for which the probabilities amount to input parameters?

Let us do some combinatorial analysis. If a block has m 1's at randomly distributed positions, the number of bits required to encode *where* these 1's lie is:

$$b(n, m) = \log_2 \binom{n}{m}$$

because the combinatorial bracket is the number of possible alternatives. Thus the whole bitstream should be encodable at a rate of about

$$\sum_{m=0}^{n} p_m \log_2 \binom{n}{m}$$

bits per block. Show that for $m << n$ the leading asymptotic term for the $\log_2$ is:

$$\log_2 \binom{n}{m} \sim m \log_2 \frac{n}{m}$$

Now if the expectation of m is denoted $M = <m>$, then it turns out that a good practical approximation to the encoding rate is:

$$\frac{b(m, n)}{n} \sim \frac{M}{n}\left(1 + \log_2 \frac{n}{M}\right)$$

bits per symbol, meaning encoded bits per original data bit. Note that this approximate formula is perfect when $M = n/2$, as would result from a completely random "0-1" coin toss data set, for in that case the formula gives $(1/2)(1+1) = 1$ bit per original bit, which is correct. Thus, the combinatorial approximation is an approximation of the entropy.

Now one may apply the combinatorial notion to encode as follows:

- For each block of n bits, determine m, the number of 1's in the block;
- Send an encoding of the integer m;
- Send an encoding of the positions of the 1's, as an integer from 0 through (n combine m) inclusive.

It may be possible to forge in this way a universal encoding scheme, one that effectively switches qualitatively between the cases where M is near $n/2$ and the cases where M is very small (in which case the combinatorial scheme should act something like run-length encoding). The hard part is to effect a rapid encoding of the precise location map of the positions of 1's, as one integer from 0 through (n combine m).

9) Here is an important, but again not completely solved research problem. The idea is to encode bits "almost losslessly," i.e., with a tight constraint on error. As in step (8) previous, let a block of n bits have m 1's and $(n–m)$ 0's, where $b(n,m)$ denotes the number of bits required to specify the bit sequence precisely. The asymptotic approximation to $b(n,m)$ given in the last step corresponds to lossless encoding of all symbols. But now consider the function $b(n,m,E)$, defined as the minimum number of bits required to encode a facsimile of the n-bit block such that the number of incorrect bits (Hamming error of the facsimile) does not exceed E. We are thus asking for the "best" way to encode a block of n bits with respect to some error constraint on the reconstructed facsimile. For example, a sequence of $n = 5$ bits, with $m = 2$ 1's might be:

0 0 1 0 1

which could be encoded as if it were instead

0 0 1 1 1

which is within the constraint $E = 1$, i.e. no more than 1 wrong bit. As we have seen, lossless encoding of a random $m = 2$, $n = 5$ sequence requires $\log_2 10 \sim 3.3$ bits. But for $E = 1$, one can show that any such sequence can be covered–up to one wrong bit–by one of:

1 1 1 0 0
1 0 0 1 1
1 1 0 0 1
0 0 1 1 1

Therefore under the error constraint, only two bits (instead of 3.3) are required to encode the block. So the problem is to generalize this situation and perhaps arrive at an asymptotic formula for $b(n,m,E)$.

A valuable analysis would be to solve the general variational problem in which error E and effective entropy b are combined. The classical method for minimizing encoding rate b/n analytically, under the constraint of fixed error rate $e = E/n$, is to minimize:

$$B = b(n,m,E)/n + \lambda e$$

where λ is a Lagrange multiplier, kept arbitrary at first. One may seek, for example, the optimal block size n. Here is an example of how to work with the Lagrange multiplier.

A solution to this problem would give rise to an encoding scheme, admittedly lossy, but which would be tunable through an error rate parameter e down to a lossless mode. Blocks of optimal length n would generally possess errors under strict error control. The experienced reader will recognize this scheme as yet another form of vector quantization, in the sense that n-dimensional vectors are given a codebook with each codebook entry "covering" multiple bit patterns. In the $m = 2$, $n = 5$, $E = 1$ example above, the codebook would consist of (11100, 10011, 11001, 00111) with 2-bit codes for every block of five bits. The whole scheme requires that for long runs of bits, the 1's density m/n be relatively stable. But such is often the case in signal processing applications.

Subject 7.2 Sound

We start next with some selected topics in sound processing. The brief tour of processing tasks is intended to prepare the investigator for the sound compression Project following.

Project 7.2.1 Examples of sound processing

Theory, a little computation, possible graphics, difficulty level 1.
Support code: Appendix "SoundEntropy.ma," "SoundFileReader.ma," "upsampler.c," "fractalsound.c."

1) Work out software possibly to create, but at least to access, sound files for signal processing of the sound data. One option is to use NEXTSTEP built-in sound functions and SoundKit. Another option is to access NEXTSTEP sound files from *Mathematica*, using such means as found in Appendix code "SoundFileReader.ma."

2) If software to generate sounds (such as "fractalsound.c" running under NEXTSTEP) is available, create a curious and entertaining "fractal sound" signal $x(t)$ calculated according to:

$$x(t) = \sum_{j=M}^{N} \sin\left(\left[(2\pi f)^{1/N}\right]^{j} t\right)$$

where N is a suitably small integer such as 32 or 64, f is the maximum frequency component in Hz. that will appear in signal x, and the minimum frequency g in x determines the lower summation limit as:

$$\frac{M}{N} = \frac{\log 2\pi g}{\log 2\pi f}$$

A good practical choice is $M/N = 1/16$, corresponding to four octaves of spectral range. If the signal is created within a loop, then t should be the integer loop counter times the sampling time of the sound output device. Evident trivially from its construction, this fractal sound has an approximate self-similarity property:

$$x(t) \sim x(at)$$

where $a = (2\pi f)^{1/N}$. The graph of a typical $x(t)$ is shown in Figure 7.2.1. Such plots can also be found in references such as [Schroeder 1991]. The signal x is, of course, reminiscent of the Weierstrass function of step (8), Project 6.2.3, involving the same basic self-similarity notion: geometrical progression of component frequencies.

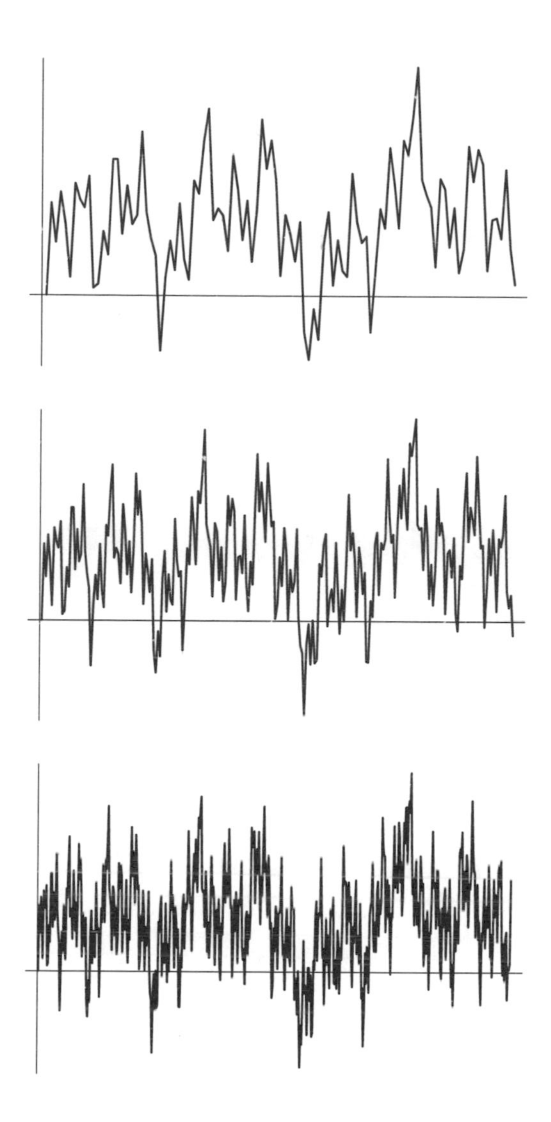

Figure 7.2.1: Fractal sound exhibiting evident self-similarity between a particular time scale (top), that scale dilated by scale factor a (middle), then dilated again by a (bottom). Because perceived pitch components are unchanged as one glides from one scale resolution to the next, there are certain sampling rates, lower than the original rate, at which the perceived "pitch" can seem *higher*.

Because of the self-similarity property, a change of time scale by a factor of a renders the signal qualitatively similar, so there should be scale changes at various geometric factors between 1 and a where such qualitative, summary impressions as "pitch" might seem lower or higher than the original. [Schroeder 1991] points out the case where a is taken to be $2^{13/12}$, whence doubling of sampling rate gives rise to a perceived drop of one semitone of the tempered musical scale. Observations of this kind suggest that the perceived audible nature of signal x is sensitive to the precise value of a, and indeed this is the case in practice.

The Appendix code "fractalsound.c" creates a NEXTSTEP formatted fractal sound file along these lines, with $a = 1.40254$. Thus (and the author, having been initially skeptical, has tried the following before placing the fact in print) a tape recording of the NEXTSTEP sound, physically slowed down by a factor of ~1.4 yields an amazingly good aural replica of the original. One can of course apply the sampling rate conversion ideas of Project 2.1.1 to mimic tape speed control, but the physical experiment is especially compelling.

Next, implement a wavelet fractal sound by creating sums of scaled and perhaps also translated wavelet functions W (from Chapter 6) in such a way that the signal obeys $x(t) \sim x(2t)$. One possibility is to attempt to sum terms $W(2^k t \pmod 1))$ over a finite set of integers k. This summation has the effect of repeating all of the smaller wavelets periodically, so that the ear can make something of such brief spikes. Note that such signals can be created with a fast inverse discrete wavelet transform, just as FFT techniques can be used to generate the sine summation fractal sound we mentioned first. In fact, it may have occurred to the reader that, formally speaking, any sound signal $s(t)$ could be used to form sums such as:

$$x(t) = \sum s\left(a^j t\right)$$

which should enjoy, in its discrete finite-sum version, an approximate self-similarity relation. It would be of interest to start with a complex recording of voices or symphony–something complex to begin with–and create such self-similar "fractal" sounds.

Another entertaining field of inquiry is to map certain fractal curves onto the musical scale. For example, the Hilbert, Peano, and Dragon curves of Chapter 6 all involve compass direction motion "NEWS." One may map each of these four directions onto tone changes of the musical scale; for example E, W could be one-semitone up/down shifts respectively, while N, S could be quarter- and half-note sustains. One thus obtains music possessed of some form of self-similarity.

3) Investigate sample rate conversion for sound data. This is a nontrivial field of inquiry, in which a central issue is the fact that the human ear is so unforgiving. For example, if one has sound array x recorded at say 8000 Hz sample rate, and one wishes to upsample this to 22000 Hz., one might be tempted simply to assign a new signal array y by:

```
y[n] = x[(8000*n)/22000];
```

With this arithmetic, the y array has on the average about three (actually 22/8) equal consecutive data values. When this y sound is played, though it will possess the correct pitch, it will be distorted due to the fact of many flat regions in the signal–those regions of about 22/8 constant data values. Another way to view this distortion crisis is to state that frequency aliases have been introduced via the (false) edges of the y signal. These need to be low-passed out of the sound.

An elegant, flexible sampling rate method due to [Smith and Gossett 1984] can easily be implemented, as exemplified in Appendix code "upsampler.c." This is actually the sampling technique used to resample in *Mathematica* itself. The Smith-Gossett formula, which can be derived in terms of the optimal, $2H$-tap antialiasing filter, is:

$$y[j] = \sum_{m=0}^{2H-1} x[m + floor(j * R)]\, sinc(\pi * (H - 1 - m + fract(j * R)))$$

Here, R is the ratio of (down frequency)/(up frequency), $2H$ is the number of filter taps, $fract(z)$ = (fractional part of z) = z (mod 1), and:

$$sinc\, z = \frac{\sin z}{z}$$

As the original authors suggest, the *sinc* values can be put into a small table, because for rational down/up ratio R there are only finitely many possible values of $fract(j*R)$.

Describe this sophisticated upsampler in the simple case of frequency doubling, i.e., $R = 1/2$. In such a case the code can be especially simple.

Implement downsampling. As Smith and Gossett point out, the formula above should have *x*[] evaluated with R = (up/down), but *sinc* should be evaluated always at R = min(up/down, down/up).

There also exist rate-conversion techniques in which a two-step process is involved: upsampling by some integer factor N and downsampling by some integer factor D to yield an overall rational resampling factor of N/D. Over the one-pass Smith-Gossett method, this N/D approach has advantages (reasonably low computational complexity and good speed) but disadvantages (impossibility of continuous rate change through sequences of rational values). In fact the latter method allows empirically smooth resampling rate changes.

4) Make histogram plots of adjacent differences between sound samples, and measure the entropy of such distributions. Figure 7.2.2 shows such a histogram:

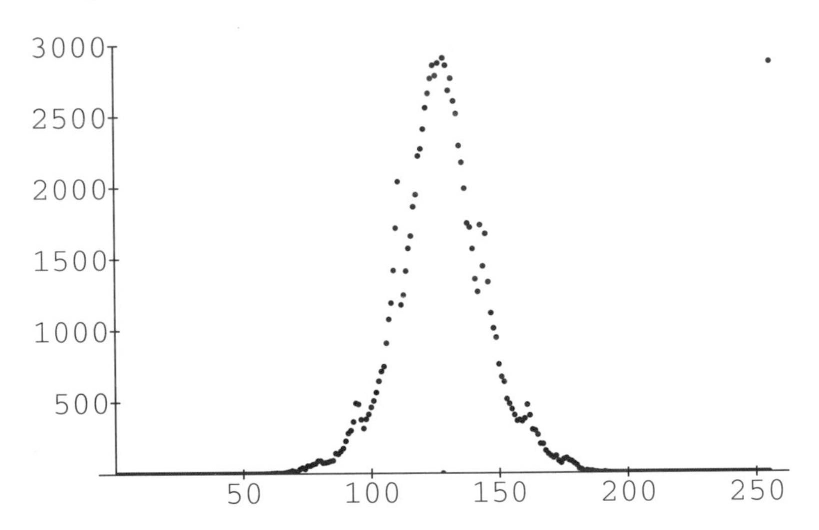

Figure 7.2.2: Sound data difference histogram. CODEC 7-bit samples were differenced by consecutive pairs and the relative probability of occurrence of each difference plotted. The entropy of this distribution calculates out to be 6.08 (as in Appendix code "SoundEntropy.ma"), which means that if simple lossless encoding of some kind is used on data differences, the data can be compressed to ~6 bits/sample.

The Appendix code "SoundEntropy.ma" shows the plotting and the calculation of the entropy at about six bits per sample. This shows that some compression can be obtained by the simple expedient of sending the first sound datum, then all subsequent differences, through a lossless compressor.

5) Make plots of sound difference correlations, meaning the horizontal axis is the difference between adjacent samples $x[n+1]-x[n]$, the *vertical* axis is the difference between the *next* pair of adjacent samples $x[n+2]-x[n+1]$, and the z-axis (or the density in such as Figure 7.2.3) is the probability of occurrence of this xy pair of differences

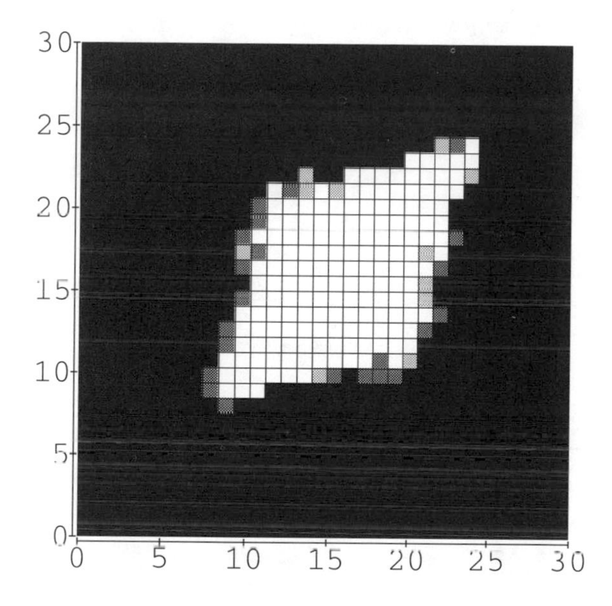

Figure 7.2.3: Second-order sound data histogram, similar to Figure 7.2.2 but this time the horizontal axis is the current difference, the vertical axis is the next difference, and correlation between successive differences can be seen as a "45 degree" tendency of the probability density. The original sound was 7 bits/sample. The second-order entropy of this distribution calculates out to be 2.6 bits/sample (as in Appendix code "SoundEntropy.ma"), which means that one can improve on the entropy assessment of Figure 7.2.2 by using higher-order entropy.

The "45 degree" effect in Figure 7.2.3 is the manifestation of the correlation between consecutive differences. The entropy (now a second-order entropy as in step (2), Project 7.1.1) turns out to be 2.6 bits/sample for this sound. This means that it should be possible to encode *paired* symbols $\{x[n+1]-x[n],\ x[n+2]-x[n+1]\}$ with Huffman or other near-entropy coding, in order to compress this particular sound to under 3 bits per sample.

In Project 7.1.1 we toured lossless compressors; but what about lossy ones? Technically speaking, the notion of lossy compressor is a vague one, especially since such a compressor can almost always be modeled as a data massage followed by a lossless compression. Thus, the subject of lossy data compression is essentially the subject of data massage, usually in the form of removal of some data components. Of course, one hopes that one has removed only negligible information.

The typical progression for lossy compression is:

- (Possibly) transform the data;
- Quantize the data;
- Put quantized data through a lossless compressor.

The typical data reconstruction steps are then:

- Decompress the encoded stream;
- Reconstruct the (possibly transformed) data from the quantized values;
- (Possibly) inverse transform to get original data.

Project 7.2.2 Examples of sound compression

Theory, a little computation, difficulty level 1.
Support code: Appendix "waveletsound.c," "WaveletOnSignal.ma," "fwt.[ch]," "walshfile.c," "walsh.c."

1) Attempt to compress sound *losslessly* via "parallel transform compression," in which one tries, on each fixed-length block of sound data, several *invertible* transforms (usually linear filters, but perhaps some non-linear ones!) and encodes which one does best. At the decode stage,

things run very quickly, because one simply enters a switch table depending on which of the finite number of decoders is indicated for the data block in question.

The author, together with M. Minnick, found that first-, second- and third-order differencing filters are the best competitors on blocks of 256 samples each. Such phenomena had been anticipated earlier by [Moorer 1979]. Sometimes, however, an exclusive-or system is superior depending on the block. This would be a non-linear system in which one stores the first sample, call it a, then stores (a^b), then (b^c), and so on. Clearly the signal can unambiguously be recovered from this data. The point is, if most of the ex-or (^) operations yield small absolute values, the runs of "0" bits can be compressed out.

Invertible filters include such as:

$$y[j] \;=\; x[j+3] \;-\; 3x[j+2] \;+\; 3x[j+1] \;-x[j]$$

which would be a third-difference filter. It is invertible in the following sense. Store $x[0]$, $x[1]$, $x[2]$ directly onto the compressed stream, then store subsequent compressed y[j] for j = 0,1,2,.... What makes this approach effective is the happy circumstance when, over a whole block of sound samples, the y values are all small, with plenty of 0's. Then in principle all the original data can be recovered from the compressed stream; for example:

$$x[3] \;=\; y[0] \;+\; 3x[2] \;-\; 3x[1] \;+\; x[0]$$

and all the terms on the right hand side live in the compressed stream. The one caution is that each possible filter in the parallel testing bank for each fixed sound block must not "overflow;" for example, the calculation for $y[j]$ above could conceivably overflow the data type range. In such a case one can separately encode the overflow to ensure invertibility, or simply move on to another filter test.

2) Implement a Walsh-Hadamard transform compressor for sound. This is a fascinating subject because the following amazing phenomenon seems to hold true. If you:

- Load in a sound file and take its Walsh transform (or Fourier, or...);
- Send the transformed data to an *alphabet-oriented* compressor such as LZW (Unix "compress");

then you almost always get noticeable compression! The theory is, since the ear understands a frequency-sorted "alphabet," it makes sense that a text-aimed (i.e., alphabet-oriented) compressor would do well on Fourier or Walsh transformed sound.

The Appendix code "walshfile.c" stores an "alphabetized," i.e., Walsh-transformed sound file, with fixed blocks of power-of-two length transformed.

Question: Does the lossless alphabet-oriented compressor work better if one stores *transposed* data; e.g., stores all first members of each block in order, then all second members of each block in order, and so on? One would thus be storing the time-dependent behavior of each sequency (or frequency) band.

The code "walshfile.c" could be modified to use FFT, DCT, Hartley, or even the new SWT of Chapter 4. The disadvantage of FFT is the fact of complex terms (which suggests using Hermitian representation as with real-signal FFTs in general) but the primary advantage is, of course, that the human ear is foremost a Fourier device.

3) Attempt wavelet compression of sound. An example of a simple–but intentionally lossy–scheme is found in Appendix code "waveletsound.c." In that example, a wavelet transform is applied to the data, then the data is massaged (thresholded and quantized), then the number of non-zero bytes corresponding to remaining transform components is reported. In such experiments an interesting question is: After the lossy thresholding and quantization, what is the best lossless compressor through which to siphon the remaining non-zero data?

4) Investigate vector quantization techniques for sound compression. Figure 7.2.3, with its "45 degree" correlation effect, shows that such a scheme makes sense. In that Figure, one would choose a "codebook" of two-dimensional vectors that "pepper" the bright density region. Then one would encode a stream of indices into the codebook. Criteria for vector quantization of speech signals are found in [Kim and Lee 1992].

5) Implement a sub-band coder for speech compression. One chooses, say, four frequency bands [Crochier and Rabiner 1988]:

0-500 Hz.
500-1000 Hz.
1000-2000 Hz.
2000-3000 Hz.

and runs through the following computational protocol:

- Use digital filtering to extract the signal bands separately;
- Quantize each band, using possibly different techniques, depending on band (usually the lower bands can be more deeply quantized);
- Compress each quantized band

Then reconstruction involves:

- Decompression of each band;
- Reconstruction of band signals from quantized data;
- Recombination of bands to form composite replica of original speech.

The parallel transform method of step (1) previous is a good choice for the sub-band compressor. Naturally, one should expect different sets of transforms to be competitive, depending on the band.

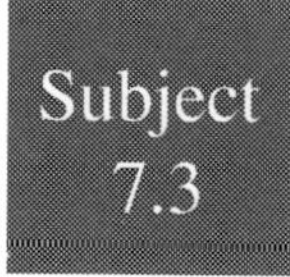

Images

As we did in the previous Subject on sound, we next consider, for images, some selected signal processing and then compression issues. Nomenclature is as follows. Denote by $p(x,y)$ the pixel ("picture element," or pel) value at location x,y. Even for color images, we shall consider p to be a one-dimensional scalar intensity. For black-white images p might be 0 through 255, i.e., representing 8-bit grayscale. For color, one could consider three separate p channels, one for each of red, green, blue (*RGB*). But generally these colors are well correlated spatially, so some investigators prefer to use *YUV* (one luminance (intensity) channel and two chroma (color) channels, respectively) or, as we see in a later Project, some simpler transformation *RGB* –> *XYZ* that gives the basic performance of a *YUV* converter.

When we speak of an "edge" in an image, we refer to a spatial derivative (of some order 1,2,...) of the $p(x,y)$ function. When we speak of a transform of an image, such as a two-dimensional FFT, we mean the transform defined on the signal vector $\{p(x,y) : 0 \leq x < M,\ 0 \leq y < N\}$, so this would be the FFT of an M-by-N rectangle of data values.

Project 7.3.1 Examples of image processing

Theory, computation, graphics, difficulty level 2-4.

1) Investigate image statistics through software. Above all, show empirically the amazing and powerful fact that the *differences* d between adjacent pixels are almost always distributed in generalized Laplacian fashion; i.e., with probability density function:

$$f(d) \ = \ C e^{-a|d|^b}$$

This general formula has profound implications for entropy assessment and therefore for compression studies.

Second, attempt to generate "synthetic paintings" by creating images for which two *successive* differences $c = p(x+2,y)-p(x+1,y)$ and $d = p(x+1,y)-p(x,y)$ have joint distribution:

$$f(c,d) \ = \ C e^{-a|c|-a|d|-b|c-d|}$$

This joint density is fit surprisingly well by many real-world images. One efficient way to model image statistics is to find empirically the a,b

parameters that fit actual plots of adjacent differences.

The author has generated "paintings" possessed of the statistics embodied in $f(c,d)$. These paintings, because of the x-y partitioning, look like plaid or woven fabrics. An interesting problem for which the author does not know a solution is to create an isotropic "painting" for which the exponential intensity correlations are a function only of *radial distance* between points. Such a painting should be devoid of x-y Cartesian artifacts.

2) Investigate methods of edge detection. An excellent theoretical summary is [Marr and Hildreth 1980], where it is argued that a good scheme for detecting edges is to find the zero values of the two-dimensional convolution:

$$p(x,y) \bullet \nabla^2 G(x,y)$$

where G is a Gaussian density. This convolution in turn may be accomplished via two-dimensional FFTs, but note that if a highly discetized approximation is desired, it may be best to perform direct convolution with small, localized sums.

Note that the Laplacian of G is a kind of two-dimensional "mexican hat" continuous wavelet as in Chapter 5. There should be theoretical relations between the Marr-Hildreth method and wavelet methods. In this regard see step (3) next.

3) Investigate a modern and intriguing approach to enhancement of image detail; which is to use non-linear diffusion for noise filtering, and reaction methods for contrast enhcancement [Cottet and Germain 1993]. One wants to reduce noise; otherwise, small image glitches, spots and so on will be brought out accidentally, as detail. Such techniques have great value, especially for medical imaging where one might want to exhibit, say, coronary arterial structure and little or nothing else. Another important application is the removal of video noise, which is the crawling, roiling effect you see on any television screen, which effect is especially pronounced when a station is off the air; but which is omnipresent to some extent regardless of picture.

4) Investigate image detail enhancement via wavelet transforms. The basic idea is to amplify the high-resolution sectors of the transform (steps (4) and (5), Project 5.3.2) to "bring out" detail. In Figure 7.3.1 the author used the image of the frontispiece to this chapter, and enhanced the three detail quadrants by a factor of four.

Figure 7.3.1: Result of enhancement of high-resolution wavelet coefficients via a resolution-dependent amplification. Compare this figure with the frontispiece to the chapter: The date (1987) and fingerprint detail are now visible.

To perform edge detection, one approach is to amplify the detail sectors, but also to *dampen* the low-resolution sectors. An example of this procedure resulted in Figure 7.3.2, where an image of dolphins is edge-detected. The printout is of a different style, showing very positive, positive, zero, negative, and very negative pixels respectively by {P, p, ., m, M}:

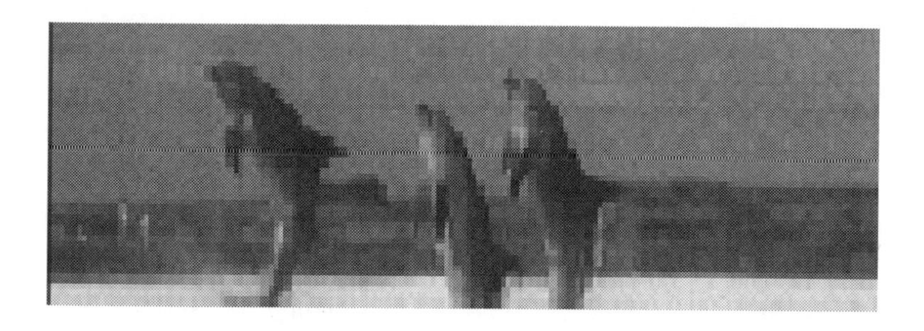

```
...M.m...mmm..mMMm....m..m.pm.p.m.m......pmp....p...........P.p.P.......p...p.
..Mmmmm...m.p.p..m...p.m.m.mm.....m.mp.p....mmmmMmm.mm.m..m.m.p...mp.pppP..mmm.p
...............P.mP.............................................
...............Mmmmmp..................Mmmp.....................
...............p....Mp.................p..m.....................
................p.m..Mp..........Mpp.......m....................
................ppP...Mp.......p..m.....mm..mp..................
.................m.....mMP......pp.m....pp..m...................
..................m......MM......p.mm.....M..m..................
..................p......pP.......p..m..m.m...mp................
.................p.p....pPpP.p....m.....P......MmPpPpppp..pPp....pp.pppppppppppp
.................ppp.....mmmmm..p..m...pp.P......mMmm.mmm.mmmmmmmmmmmmmmmmMMMMmmm
..pppppmm..p..mmmmmM.p........Mmmmm....mm.m.....mpP.m.m.p....m..p..p.pp...ppp..m
..mmmmmPPm.m.m...m..pMm....m..mmM.mm....m..p.....p..p.............mmm.....mm..pp
```

Figure 7.3.2: Edge detection via two-dimensional wavelet transform. The dolphin image was transformed, then the high-resolution components were amplified above the medium-resolution ones, with the low-resolution components damped. The resulting printout of pixel regions shows edge regions having local averages that are very positive, positive, zero, negative, and very negative as characters {P,p,.,m,M}.

The author has found that the bivariate Laplacian *ansatz* of step (1) previous often fits well the type of edge data in Fugure 7.3.2 (when actual values of detail components are used instead of threshold tokens P,p,.,m,M).

5) Investigate digital holography, in which one computes numerically the hologram of a three-dimensional data base. In practice it is feasible (if expensive) to reproduce the hologram physically via film recorders that now can exceed 10000 lines per inch.

A computer-based ray-tracing approach to digital holography is discussed in [Stein et al. 1992].

6) Investigate the fascinating and burgeoning world of optical recognition. There are filter techniques for pattern recognition [Wang Z et al. 1993], and for optical character recognition [Rosen and Shamir 1989]. The former technique evidently allows for reliable recognition of various orientations of (the same) toy model. The latter technique uses *logarithmic* radial transforms, exploiting the fact that a small character **F** and a large character **F** should both be recognized as the same character. Armed with the FFT methods of Chapter 4, and perhaps the wavelet methods of Chapter 5, the reader may find optical character recognition especially rewarding.

7) Investigate the modern technique of *morphing*, whereby one image is transformed smoothly into another. It is often not enough just to "dissolve" one image into the other by trivial weighting. For example a circle and cross, if dissolved one into the other by intensity, give the top sequence of Figure 7.3.3.

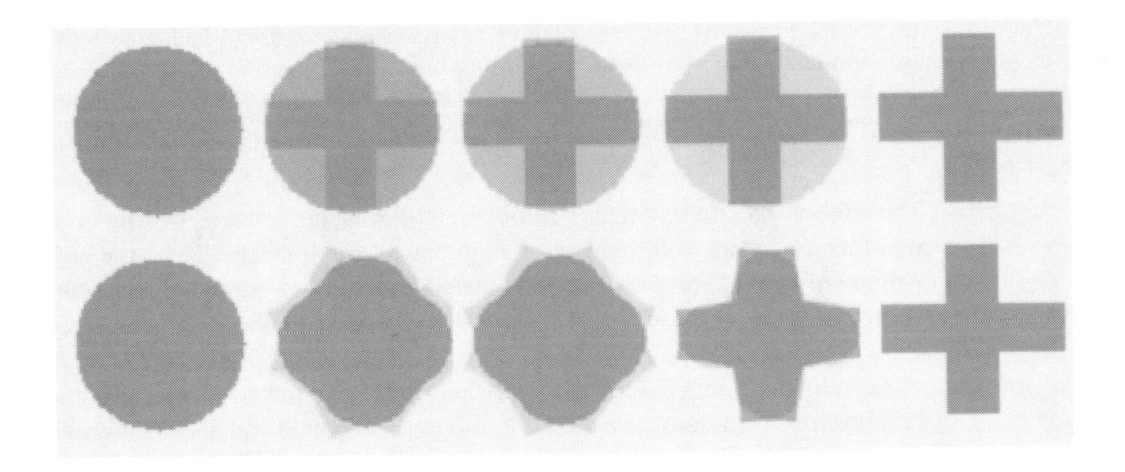

Figure 7.3.3: Comparison of trivial dissolve (top row) to a morphing technique (bottom row). The dissolve sequence simply portrays images $I(1-a) + Ja$, where I is the circle image, J is the cross image, and a is a weight parameter running from 0 to 1 (left-to-right in the top row). By contrast, the morphing sequence, though the same kind of parameter is involved, has not only intensities but coordinate systems mixing according to the a parameter.

It is evident from the figure that the morphing sequence (bottom row) is superior in modeling an actual physical warping from circle to cross. In modern movie making, morphing will transform one face into another; and if done properly, every step in the sequence will appear as if the face has "morphed" (i.e. evolved) into a new physical configuration. If trivial dissolve were used the intermediate faces would just seem like superimposed ghosts.

The circle-to-cross effect was done by using a parameter a to dissolve and to transform coordinates simultaneously. Let a run from 0 to 1 and assume a transformation T_1 takes the J image's coordinates into the I coordinates; and that another transformation S_1 takes the coordinates the other way. S might be the inverse of T for example. Then a parametrized image sequence that is more general than trivial dissolve is defined at pixel vectors r by assuming that the S, T transformations themselves can be continuously parameterized, to give an overall pixel intensity at r of:

$$(1-a)\, I(S_a(r)) \; + a\, J(T_{1-a}(r))$$

Note that if both S and T are identity mappings one has trivial dissolve. Because we have well-defined endpoint images I and J, we require that both S_0 and T_0 are identity mappings, corresponding to the limits $a = 0$ and $a = 1$. The key to morphing is to demand that some physical constraint be preserved by the coordinate transformations. For example one might demand that a line (p,q) in the I image, where p and q denote the line's vector endpoints, will map onto the line (p', q') of the J image. One reasonable and practical way to do this is described in [Beier and Neely 1992]. Set, for a point r in the I system, two scalar variables:

$$u \;=\; \frac{(r-p)\cdot(q-p)}{|q-p|^2}$$

$$v \;=\; \frac{(r-p)\cdot n}{|n|}$$

where the vector n is normal to and has the same length as $q-p$. Then define the fully transformed coordinate $S_1(r)$ to be:

$$r' = p' + u(q' - p') + v\frac{n'}{|n'|}$$

The interested reader should verify that the set of r comprising the line from p to q does indeed get carried into the line from p' to q'. A second instructive exercise is to think of the situation geometrically and come up with a parameterization for S_a with a running from 0 to 1, so that $a = 0$ gives S_a as the identity while $a = 1$ yields the transformation just described. Yet another good exercise is to generalize this transformation of Beier and Neely to constrain multiple line segments. The circle-to-cross demonstration was created, in fact, using four simultaneous line segment pair preservations, as allowed in a NEXTSTEP application written by [Snyder 1993]. Two line segments were drawn to join (diagonally, in pairs) the four inner crevices of the cross, while two other segments joined the vertical and horizontal tips of the cross. All four segments were given, as transformation destinations, the obvious corresponding segments on the circle image. Preservation of these four line segments is the means by which such a morphing application manages to cause actual shape-changing.

The author suspects that other classes of line-segment-preserving transformations would be useful. If one wishes to preserve a w-by-h image rectangle's borders, one might note that sums of terms

$$\sin\frac{\pi x}{w}\sin\frac{\pi y}{h}$$

with arbitrary coefficients always must vanish along all four image edges. An example would be the Fourier decomposition of the function $x(w-x)y(h-y)$, which manifestly vanishes along all image borders. Similar Fourier sums can in principle be constructed to vanish on line segments, and therefore could be applied to the problem of segment preservation. What would be especially interesting, if it is at all possible, would be a two-dimensional FFT-based morphing algorithm.

We have seen that, for sounds, lossless compression comes down to entropy considerations (perhaps in higher-dimensional spaces). At first sight, the basic goal of *lossy* image compression may be high compression ratio, but of course what one really wants is a high quotient of the form (compression ratio)/distortion. One distortion measure which is clearly not completely compatible with visual perception, but does have useful analytical properties, is the root-mean-square-error (RMSE). This error pertains to how poor is the reconstructed, final image with respect to the original image; and is defined by:

$$RMSE^2 = \frac{1}{MN} \sum_{x=0}^{M-1} \sum_{y=0}^{N-1} [p(x,y) - p'(x,y)]^2$$

where the image is assumed to be M-by-N and p,p' denote original and reconstructed image, respectively. One fine property of this distortion measure is that the Parseval theorems we have investigated for Fourier and wavelet transforms (generally, any unitary transform) work well with it. In fact, for unitary transforms, the RMSE can be cast precisely as a properly normalized sum of squares of *discarded* transform components. A similar statement can be made about the error due to quantization of components: one simply adds up all the squared componentwise errors due to quantization or whatever cause.

Some thoughts are in order on the subject of color images. Image processing experts often contemplate theoretical schemes, not for color but just for 8-bit grayscale, where each $p(x,y)$ runs from 0 through 255. The reason why color issues can often be avoided is that one may transform the RGB color problem into a YUV-type problem and treat the three converted channels as three independent gray images. In almost every case, one should *avoid* compressing R,G,B independently, for the simple reason that these channels are usually quite dependent. Note, for example, that if the sky is brightening in some direction, all three color components are probably increasing, so invoking three separate encoders is wasting something. Instead, a Y channel (luminosity) will indeed be increasing, but chroma channels (U,V) are probably fairly stationary (and furthermore fairly small) for the brightening sky. The customary standard NTSC video transform is:

$$\begin{bmatrix} Y \\ U \\ V \end{bmatrix} = \begin{bmatrix} 0.299 & 0.587 & 0.114 \\ 0.596 & -0.274 & -0.322 \\ 0.212 & -0.523 & 0.311 \end{bmatrix} \begin{bmatrix} R \\ G \\ B \end{bmatrix}$$

So one would take the RGB pixel values, calculate three independent YUV planes, and compress each of the latter three. Then the inverse matrix would be used to reconstruct RGB from decompressed YUV. It is a stunning fact that the time-integrated energy of commercial video is, historically, about 90 per cent Y channel. This is because pastels (whiteish tints) are so common, and these are possessed of low U,V values. It turns out in practice, for most compression projects, that simpler and faster transforms also do the job. The author has found that for transform-based compressors:

$$\begin{bmatrix} Y \\ U \\ V \end{bmatrix} = \begin{bmatrix} 1/3 & 1/3 & 1/3 \\ -1/3 & -1/3 & 2/3 \\ 2/3 & -1/3 & -1/3 \end{bmatrix} \begin{bmatrix} R \\ G \\ B \end{bmatrix}$$

is an effective choice, consistent with the remarks of other researchers who suggest that the effectiveness of the chosen color-space transform is largely independent of the precise matrix elements [Rabbani and Jones 1991].

Project 7.3.2 Image compression

Computation, a little theory, difficulty level 3-4.
Support code: Appendix "fwt.[ch]."

1) Implement a predictor-corrector (lossless or lossy; both versions are possible here) compressor as follows. Consider from Figure 7.3.3 the pixel value d, denote by "predictor" the quantity $P = b+c-a$, and denote by "corrector" the difference $C = (d-P)$:

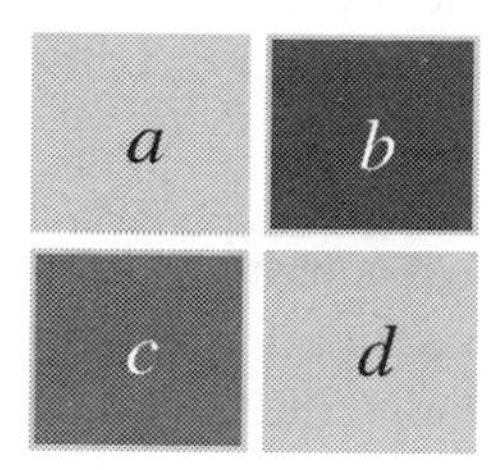

$$\begin{aligned} d &= (b+c-a) + (d-(b+c-a)) \\ &= \text{predictor} + \text{corrector} \end{aligned}$$

Figure 7.3.4: Schematic for an elementary predictor-corrector approach to image compression. Generally, as one scans an image lexicographically, when one is at pixel *d*, one already knows *a,b,c*. Thus, both the compressor and decompressor can compute the "predictor," so that all the compression bitstream need convey is the "corrector."

As one scans an image lexicographically, *a,b,c* are known for almost all pixels *d* in the Figure. So an encoding of correctors *C* comprises almost all of the compression effort for this method.

An implementation can be rendered lossy–to yield more compression–by quantizing the corrector to some limited range of values. One must take care, though, in this lossy case, to avoid runaway effects by accumulated error. To control error, quantized correctors must constantly be introduced into the feedback loop, so that new *d* pixel values are computed based on the *erroneous* values of *a,b,c* pixels.

Yet another approach is not to compute *P* in the indicated way, but to encode *which of* (*a,b,c*) is closest to *d.* Thus, the predictor *P* would be coded as an integer out of the set {0,1,2}. This method might be effective when the image is composed of colors from a limited palette.

One has the option of compressing the *C* terms losslessly, or quantizing the *C* terms to some limited set. Caution is in order for the latter, lossy scheme: since general pixels will exhibit error, the predictor for the next missing pixel must be carefully interpreted so that the decompressed image does not become unstable, accumulating errors as it is reconstructed lexicographically.

As an interesting theoretical exercise, find all possible vector directions for a linear intensity gradient (over a region containing all of a,b,c,d) such that the corrector C vanishes.

Contemplate a dynamical predictor-corrector scheme, in which one generalizes the linear algebra for the predictor, and inserts into the compressed bitstream enough information to specify the predictor algebra used over image regions. By analogy with the preceding paragraph, one could in principle detect parabolic or cubic gradients and act accordingly to alter the algebra.

2) Investigate vector quantization (VQ) compression for images [Rabbani and Jones 1991]. Attempt to create a codebook that is fixed for all images, or attempt a codebook that is taken from *the image itself*. In the latter case, for a given patch of pixels, one finds somewhere else in the plane a good match to the patch in question.

3) Investigate frequency-domain image compression; e.g., adaptive DCT (ADCT)/JPEG systems (of which there are many commercial implementations), or FFT methods.

The FFT methods all suffer from aliases as discussed in step (2), Project 4.3.2. However, there is one tremendous advantage of the traditional two-dimensional FFT that should be exploited if possible. Show that if an image is N-by-N pixels, and:

$$p(x,y) \;=\; q(x-vy)$$

where v is an integer with GCD(N, y) = 1, then the two-dimensional DFT of p can be written in a simple way in terms of the *one-dimensional* transform of q. This kind of image involves just one row of pixels "moving," with periodic boundary conditions (mod N), across the planar image area at a "spatial velocity" v. The entropy of such an image is essentially that of just one row, and the DFT theorem here reveals this. So when an image is made up of subsections in which some pattern is spatially propagating, the FFT should do quite well in collapsing the image to a well-compressible set of components. Clearly these observations have some implication with respect to time-domain compression, where the notion of rigid, rectilinear motion is commonplace, and three-dimensional FFTs should tend to "collapse" into two-dimensional ones in regions of such motion.

4) Implement a wavelet compressor for images. One follows the essential advice for two-dimensional wavelet transforms in steps (4) and (5), Project 5.3.2. Fast wavelet transform C code (which can be made faster via optimizations of various kinds) appears as Appendix code "fwt.[ch]." Then one may threshold the transform components, for example by zeroing all those whose absolute value is below some threshold. It turns out to be lucrative to make the threshold "resolution sector dependent" as intimated in [Antonini et al. 1992]. There is also the option of using vector quantization on the wavelet components (which also has a thresholding effect). The choice of a lossless compressor, either to encode non-zero remaining components or to encode VQ code indices, is always an interesting step.

The author has found that with careful application of wavelet transforms (careful especially in the *quantizer* and *compressor* design), it is possible to better the RMSE-vs.-compression curves obtained with the best ADCT/JPEG schemes, for virtually any test image.

5) Design a compressor that simply uses *dithering* as the compression mechanism. One chooses a (small) palette of colors (grays) and, as the image is scanned, continually chooses a "best" color from said palette. One good choice for the reconstruction algorithm is *error diffusion*, in which every error difference ((true color) – (palette color)) is split up into fixed fractions and literally "fed" (added to) to pixels that live further down in the scan order.

6) Investigate IFS fractal compression methods (the basic IFS notion is described in Chapter 6). The basic idea is to find affine transformations and initial seed sets so that iteration of the whole system generates the image. This approach appears to be vaguely competitive–at least in compression ratio–with the best wavelet and ADCT implementations. An implementaion of IFS compression is described in [Jacobs et al. 1992].

Appendix

Support code for the book Projects

This Appendix contains reference source code for the Projects of the book. The code is primarily *Mathematica* (.ma extension) and *C* source (.c extension). An alphabetical list of the programs follows, together with page references.

AitkenLimits.ma	**333**, 6, 16
Attractor8.ma	**333**, 236
Brownian.ma	**334**, 273
CantorGen.ma	**334**, 258
Compressors.ma	**335**, 297
ContFract.ma	**341**, 6, 16, 30, 138
corrdim.c	**342**, 284
DFTgraphs.ma	**346**, 160
DanielsonLanczos.ma	**344**, 166
Daubechies.ma	**345**, 212
Dragon.ma	**349**, 262
DragonRecursion.ps	**347**, 262
Eigenvalue.ma	**350**, 33
eliascoder.c	**350**, 297
eliascoder.h	**352**, 297
eliastest.c	**353**, 297
Equipotential.ma	**355**, 93
Erfc.ma	**355**, 30
FFTs.ma	**357**, 174
fft.c	**367**, 162, 174, 177, 178, 179

fft2D_real.c **359**, 179
fft_real.c **375**, 177
fractalsound.c **381**, 307
FSWT.ma **383**, 191
fwt.c **386**, 222, 224, 314, 326
fwt.h **394**, 222, 224, 314, 326
FWTInPlace.ma **383**, 222
FWTNotInPlace.ma **384**, 222
FermatConvolution.ma **356**, 145
Genetic.ma **395**, 77
giants.h **396**, 115, 129, 133, 135
giants.c [ON DISK ONLY] 115, 129, 133, 135
Henon.ma **399**, 240
Hilbert.ma **400**, 262
huffman.c **402**, 297
IFS.ma **404**, 262
Invariant.ma **405**, 232
inverse.c **405**, 33
Julia.ma **408**, 240
LSystem.ma **409**, 262
Lorenz.ma **409**, 240
Mountains.ma **410**, 273
MultiNewton.ma **411**, 38
Neural.ma **411**, 70
NewtonMatrixInverse.ma **412**, 33
NewtonTrefoil.ma **412**, 240
PhiHat.ma **413**, 212
Pi.ma **414**, 6
PolynomialSolve.ma **414**, 38, 79
quasi.c **415**, 47
RecursiveFHT.ma **417**, 183
RombergLimits.ma **417**, 6, 16
Sessile.ma **419**, 93
SloshingPacket.ma **419**, 102
SoundEntropy.ma **419**, 307
SoundFileReader.ma **420**, 307
sphere.c **422**, 47
strassen.c **426**, 33
strassen.h **428**, 33
Strassen.ma **430**, 33
strassentest.c **423**, 33
threebody.c **431**, 93
timing.c **432**
TwoCycle.ma **433**, 232
upsampler.c **433**, 307
walsh.c **436**, 188, 314
walshfile.c **434**, 314
wavdim.c **438**, 284
WaveletMatrix.ma **439**, 222
WaveletOnSignal.ma **440**, 224, 314
waveletsound.c **443**, 314
ZetaGamma.ma **445**, 30

AitkenLimits.ma

```
(* Aitken extrapolation example.
*)

(* Next, use Aitken extrapolation to solve x = cos x. *)
lim = 6;
x = N[1/4.0,20];
q = Table[0,{i,1,lim},{j,1,lim}];
Do[
     q[[n+1,1]] = N[Cos[q[[n,1]]]],
     {n,1,lim-1}
];
(* Next do first differences. *)
Do[
    q[[n,2]] = N[q[[n,1]] - q[[n-1,1]],20],
    {n,2,lim}
];
(* Next do second differences. *)
Do[
    q[[n,3]] = N[q[[n+1,2]] - q[[n,2]],20],
    {n,2,lim-1}
];
(* Next compute Aitken Delta^2 correction. *)
(* Optionally, restart original iteration every time a
    new Aitken extrapolated value q[[n,4]] is found. *)

Do[
    q[[n,4]] = N[q[[n,1]] - q[[n,2]]^2/q[[n-1,3]],20],
    {n,3,lim}
];
Print[MatrixForm[q]]
```

Attractor8.ma

```
(* Convergence of the quadratic map
   to 8 attractor points.
 *)

num = 256;
dil = 16;
a = 3.56;
f[x_] := a x (1-x);

particles = Table[1.0*q/(dil*num),{q,0,dil*num}];
Do[
     hist[q] = Table[0,{r,0,num-1}];
     his = hist[q];
     Do[i = Floor[particles[[r]]*(num-1)+0.5];
          his[[i+1]] += 1,
          {r,1,dil*num+1}
     ];
     hist[q] = his;
     particles = Map[f, particles];
```

```
    ,
    {q,0,127}
];

f[x_,y_] := hist[Floor[y]][[Floor[(num-1)*x]+1]];
DensityPlot[f[x,y],{x,1/(2*num),1},{y,0,59.999},

Axes->Automatic, Mesh->False,PlotPoints->num];
```

Brownian.ma

```
(* Plotting of fractional Brownian motion,
    dimensions 1.1 to 1.9 in steps of 0.2
 *)

size = 1024;
 Do[
    beta = 5-2*dim;
    phase = Table[Exp[ 1.0 I Random[Real, N[2 Pi]]],{v,1,size}];
    phase[[1]] = phase[[size/2+1]] = 1;
    Do[phase[[size-q+2]] = Conjugate[phase[[q]]], {q,2,size/2}];
    spect = phase * Table[N[1/k^(beta/2)],{k,1,size}];
    brown = Re[Fourier[spect]];
    ListPlot[brown, PlotJoined->True, Axes->None],
    {dim,1.1,1.9,0.2}
 ]
```

CantorGen.ma

```
(* Tent map invariant set search, showing convergence to a
   Cantor set.
 *)
num = 3^6;
dil = 1;
b = 1.5;
f[x_] := b - 2 b Abs[x-0.5];
char[x_] := If[x > 0, 1, 0];

particles = Table[1.0*q/(dil*num),{q,1,dil*num}];
copy = particles;
Do[
    ListPlot[Map[char,particles],
        Axes->None, PlotRange->{-0.1,10}];
    copy = Map[f, copy];
    Do[
        part = copy[[s]];
        If[(part<0) || (part>1),
            particles[[s]] = 0
        ],
    {s,1,Length[particles]}
    ]
    ,
    {q,0,4}
```

```
];

(* Cantor measure function, simply returns 1 iff
   the argument's ternary expansion has no "1"s.
 *)

cantor[m_] :=
    Block[{e,dig},


        e = m;
        While[e>0,
            dig = Mod[e,3];
            If[dig==1,Return[0]];
            e = Floor[e/3];
        ];
        Return[1];
];
ListPlot[Table[cantor[q-1],{q,1,3^6}]]
```

Compressors.ma

```
(* Huffman and Arithmetic Coding compressors.
   Adapted from [Winfree et. al. 1993].
   Mathematica v. 2.0 is expected.
   Various ways of doing the compressors are
   listed here.

   There is a test sequence at
   the end of this source.
*)

(* type plist: sorted e.g. "plist = {0.07, 0.08, 0.15, 0.15, 0.2,
0.35};" *)

HuffmanCode[ plist_ ] :=
    Last /@ Sort[ HBits[ HMerge[ #[]& /@ plist ]]]

HMerge[ {sub1_,sub2_} ] := (Head[sub1]+Head[sub2])[sub1,sub2]
HMerge[ {sub1_,sub2_,rest__} ] :=
    HMerge[ Sort[ {HMerge[{sub1,sub2}], rest} ]]

HBits[tree_] := {tree[[##,0]]&@@#, SequenceForm@@(#-1)}& /@
Position[tree,_[]]

HuffmanEncode[ plist_, message_ ] :=
    Flatten[ SequenceForm@@( HuffmanCode[ plist ] [[message]] ) ]

HuffmanTree[ plist_ ] :=
    HMerge[ MapIndexed[#1[First[#2]]&, N[plist]] ] //.
    { _[n_] -> n, _Real -> List }


(* type pairs: e.g. "pairs =
```

```
{{0.5,c},{0.2,a},{0.15,d},{0.08,b},{0.07,e}};"

For explicit codes, it is convenient to represent the code as rules
which
take tokens to bit-strings.
*)

ExplicitHuffmanCode[ pairs_ ] :=
    MapThread[ Rule, {Last /@ #, HuffmanCode[ First /@ # ]} ]& @
Sort[pairs]

ExplicitHuffmanEncode[ pairs_, message_ ] :=
    Flatten[ SequenceForm@@( message /. ExplicitHuffmanCode[ pairs ] )
]

ExplicitHuffmanTree[ pairs_ ] :=
    HMerge[ N[First[#]][Last[#]]& /@ Sort[pairs]] //.
        { _[n_] -> n, _Real -> List }

(*
HuffmanDecode[] works for both explicit and implicit token types,
since
the trees share a common form.
e.g.
intmessage = { 6,4,3,3,2,1,4,3,2 }; message = {c, a, a, d, a, e, b, e,
b, b};
HuffmanDecode[ HuffmanTree[plist], HuffmanEncode[plist, intmessage] ]
HuffmanDecode[ ExplicitHuffmanTree[pairs],
ExplicitHuffmanEncode[pairs,message]]
*)

HuffmanDecode[ tree_, bits_ ] :=
    Fold[ {##} ~Replace~ {  {{tokens_,path_List},b_} :> {tokens,
        b[path]},
        {alltokens_, b_} :> {Append@@alltokens, b[tree]} }&,
      { {}, tree }, bits /. { 0 -> First, 1 -> Last }
    ] \
    ~Replace~ { {tokens_,path_List} :> tokens, alltokens_ :>
    Append@@alltokens }

(* Let's verify that this all works...*)
ht[n_,m_] := { #1, #2,

HuffmanDecode[ HuffmanTree[#1], HuffmanEncode[#1, #2] ] === #2 } & @@



    ( { #, RandomMessage[#,n] }& @ RandomPlist[m] )

htcheck[times_:200,n_:100, m_:15] :=
    Cases[
      Table[ht[Random[Integer,{1,n}],Random[Integer,{2,m}]],{times}],
      {_,_,False}
    ]

(* Next, arithmetic coding
 *)
ArithBits[ {low_,high_} ] :=
```

```
    SequenceForm@@IntegerDigits[Round[(low+high)2^(#-1)],2,#]& @
    (-Floor[N[Log[2,high-low]]])
ArithFrac[ bits_ ] := Fold[ 2 #1 + #2 &, 0, bits ] / 2^Length[bits]

(* a few functions to check that this works...
 *)
ct[exp_] := ( {#, ArithFrac[ArithBits[#]], Length[ArithBits[#]]}& @
    (#+{0,#[[1]]}& [ Rationalize[#,0]& /@
    {.9 Random[], .1 //. (r_ /; Random[] > exp) :> r Random[] } ]) )
    //
    ( Append[#, #[[1,1]] < #[[2]] < #[[1,2]]] &)

ctcheck[times_:200,exp_:.1] := Cases[ Table[ct[exp], {times}],
{_,_,_,False} ]


(* Arithmetic Coding operates analogous to encoding a message (in k-
symbol alphabet) using a base-k real number, except that the interval
"alloted" to each symbol is proportional to its probability.  (Thus
the equiprobably distribution _is_ base-k.)  Results are read out in
base 2, of course.  To determine the end of decoding, the terminating
symbol "1" MUST be sent, and is reserved for this purpose -- thus
prob(1) is the exponential length-of-message factor; the probabilities
need not otherwise be sorted.
*)

ArithEncode = ArithEncode1; ArithDecode = ArithDecode;

ArithEncode1[ probs_, message_ ] :=
    With[ {pspan = PSpan[probs]},
        { {0,1}, message } //. {
        { {lo_,hi_}, {} } :> ArithBits[{lo,hi}],
        { {lo_,hi_}, {tok_,rest___} } :> {pspan[[tok]](hi-lo) + lo,
        {rest}} }
    ]


ArithDecode1[ probs_, bits_ ] :=
    With[ {pspan = PSpan[probs], code = ArithFrac[bits]},
        Append[ ( { {0,1}, {} } //. {
        { {lo_,hi_}, toks_ } :> { {lo,hi}, toks,
        Position[#[[1]] <= code < #[[2]]& /@ (pspan (hi-lo) +
            lo),True]},
        { {lo_,hi_}, toks_, {{i_},___} /; i>1 } :>
        { pspan[[i]] (hi-lo) + lo, Append[toks, i] }
        } ) [[2]], 1]
    ]

ArithEncode2[ probs_, message_ ] :=
    With[ {pspan = PSpan[probs]}, ArithBits @

Fold[ -pspan[[#2]] Subtract@@#1 + First[#1]&, {0,1}, message ]]


ArithDecode2[ probs_, bits_ ] :=
    Module[
        {pspan = PSpan[probs], code = ArithFrac[bits],
```

```
        i=2,lo=0,hi=1,toks={},phi},
        phi = Last /@ pspan; While[ i > 1,
        i = Position[Sort[Prepend[phi (hi-lo) + lo,
            code]],code][[1,1]];
        If[ i>1, {lo,hi} = pspan[[i]] (hi-lo) + lo; AppendTo[toks, i]
        ] ];
        Append[toks,1]
    ]

(* Convert probabilities to intervals of {0,1}, & RATIONALIZE for
   accuracy
 *)
PSpan[ probs_ ] := Thread[ { Drop[#,-1], Drop[#,1] } ]& @

FoldList[Plus,0,Rationalize[#,0]& /@ probs]

(* Let's verify that this all works...*)
at[exp_,m_] :=
    { #1, #2, ArithDecode[ #1, ArithEncode[#1, #2] ] === #2 } & @@

    ( { #, RandomEMessage[#] }& @ RandomProbs[exp,m] )

atcheck[times_:200,exp_:.1, m_:15] := Cases[

Table[at[Random[] exp,Random[Integer,{2,m}]],{times}],
    {_,_,False}
]

(* Now for the adaptive version of arithmetic coding. *)

(* Here we use the representation { lo, hi-lo } for encoding *)
   During encoding and decoding, "probabilities" are incrementally
   updated.
   The "probs" needn't sum to one.
 *)

AdaptiveEncode[ k_, message_ ] :=
    Module[ {probs = ACinit[k], lo=0, s=1 },

Scan[ ( {lo,s} =
    s { Plus@@Take[probs,#-1], probs[[#]] } / Plus@@probs + {lo,0};
    ACupdate[probs,#] )&, message ];
    ArithBits[{lo,lo+s}]   ]
    AdaptiveDecode[ k_, bits_ ] := Module[
        {probs = ACinit[k], code = ArithFrac[bits],
        i=2,lo=0,hi=1,toks={},divs},
    While[ i > 1,
        divs = lo + (hi-lo) FoldList[Plus,0,probs] / Plus@@probs;
        i = Position[Sort[Prepend[ divs, code]],code][[1,1]] - 1;
    If[ i>1,
        {lo,hi} = divs[[{i,i+1}]]; ACupdate[probs,i]; AppendTo[toks,i]
    ] ];
    Append[toks,1]
]

SetAttributes[ACupdate, HoldFirst];
```

```
(* Use these for standard adaptive coding, where probabilities just
reflect the cumulative frequencies of each symbol seen so far *)

ACinit[k_] := Table[1,{k}]
ACupdate[probs_, i_] := probs[[i]]++

(* Use these for another adaptive coding, where knowledge of
symbols seen long ago dies off exponentially. *)

ACinit[k_] := Table[1/k,{k}]
ACupdate[probs_, i_] := (probs[[i]]+=ACconst; probs /= Plus@@probs)
ACconst = .001;

(* Using these definitions, we can simulate the non-adaptive code *)
ACinit[k_] := {0.003, 0.347, 0.2, 0.1, 0.25, 0.1}
ACupdate[probs_,i_] := Null


(* some functions for comparing actual and theoretic entropies *)

info[p_] := p Log[2, 1/p ]
Entropy[ plist_ ] := Plus@@(info /@ plist)
Entropy[ plist_, codes_ ] := Plus@@( plist (Length /@ codes) )

testHuffman[ plist_List, n_:100 ] :=
    { Entropy[plist],
      Entropy[plist,HuffmanCode[plist]],
      Length[HuffmanEncode[plist, RandomMessage[plist,n]]] / n //N }
testHuffman[ s_Integer, n_:1000 ] := testHuffman[ RandomPlist[s], n]

testPlotH[dots_] :=
    Show[Graphics[ { Thickness[.001],
        Line[{ {0,0}, {1.1,1.1} } Max[Flatten[#]] ],

{Circle[{#1,#2},Scaled[{.008,.008}]],

 Disk[{#1,#3},Scaled[{.005,.005}]]}&@@#& /@ #
    } ], Frame -> True, AspectRatio -> 1 ]& @
    Table[ testHuffman[ Random[Integer,{2,30}] ], {dots} ]


testArith[ probs_List, n_:10 ] :=
    { Entropy[probs],
      (Plus@@(Length[ ArithEncode[probs, #]]/
      (Length[#]+1)& /@ #) )/n //N,
      Length /@ #}  & @ Table[ RandomEMessage[probs],{n}]

testArith[ s_Integer, exp_:.1, n_:10 ] := testArith[
RandomProbs[exp,s], n]

testPlotA[dots_] :=
    Show[Graphics[ { Thickness[.001],
        Line[{ {0,0}, {1.1,1.1} } Max[Flatten[#]] ],

Disk[#[[{1,2}]],Scaled[{.005,.005}]]& /@ #
    } ], Frame -> True, AspectRatio -> 1 ]& @
    Table[ testArith[ Random[Integer,{2,30}] ], {dots} ]
```

```
Recurse[ init_, rules_] := FixedPoint[ # ~Replace~ rules &, init ]

(* message utilities *)

RandomProbs[exp_,s_] := #/(Plus@@#)& @ Prepend[RandomPlist[s], exp]
RandomPlist[s_] := #/(Plus@@#)& @ Sort[Table[Random[],{s}]]

RandomEMessage[probs_] := Append[ FixedPoint[ Function[ m,

If[#==0,m,Append[m,#+1]]&[Random[WeightedInteger,probs]] ], {2} ], 1]
RandomMessage[ probs_, len_ ] :=
Table[Random[WeightedInteger,probs]+1, {len}]

Begin["System'"];
Unprotect[Random];
WeightedInteger::usage =

"Random[WeightedInteger, {p0, ..., pn}] returns an Integer

between 0 and n, with probability pi for i.";
End[];

Begin["System'Private'"];
Random[WeightedInteger,probs_List] :=
    Position[Sort[Prepend[Take[FoldList[Plus,0,probs],{2,-
2}],#]],#][[1,1]]-1& @
         Random[Real,Plus @@ probs]
Protect[Random];
End[];


ArithmeticEncode[s_String]:=
    Module[{high=1, low=0, sum=0, ranges, charlist, len, data,
                coderange, decimalplaces
            },
            charlist = Characters[s];
            len = Length[charlist];
            data = {Count[charlist,#]/len,#}& /@ Union[charlist];
            Scan[ (ranges[#[[2]]] = {sum,sum+=#[[1]]})&, data];
            Scan[(coderange = N[high-low,2len];
                 high = low+coderange ranges[#][[2]];
                 low = low+coderange ranges[#][[1]]
                )&,
                charlist
            ];
            decimalplaces = - MantissaExponent[high-low][[2]] +1;
            {N[Floor[high
10^decimalplaces]/10^decimalplaces,decimalplaces],
                data, len}
    ]


ArithmeticDecode[{code_Real,freqs:{{_Rational,_}..},len_}] :=
    Module[{total = 0, data, result = Table[Null,{len}],
            num = SetPrecision[code,Precision[code]+2] (* ???? *)
            },
```

```
        data = {total,total+=#[[1]],#[[2]]}& /@ N[freqs,2len];
        For[i=1,i<=len,i++,
                Scan[If[num<#[[2]],
                        range=#[[2]]-#[[1]];
                        num-=#[[1]];
                        num/=range;
                        result[[i]]=#[[3]];
                        Return[];
                    ]&,
                    data
                ]
        ];
        StringJoin@@result
    ]

(* Begin actual tests. *)

probs = Sort[RandomProbs[.01, 5]]; Print[probs];
mess = RandomEMessage[probs]; Print[mess];
encstream = HuffmanEncode[probs,mess]; Print[encstream];
huffout = HuffmanDecode[HuffmanTree[probs], encstream];
Print[huffout];
encstream = ArithEncode1[probs,mess]; Print[encstream];
a1 = ArithDecode1[probs, encstream]; Print[a1];
encstream = ArithEncode2[probs,mess]; Print[encstream];
a2 = ArithDecode2[probs, encstream]; Print[a2];
ACinit[k_] := Table[1,{k}];
ACupdate[probs_, i_] := probs[[i]]++;
encstream = AdaptiveEncode[6,mess]; Print[encstream];
AdaptiveDecode[6, encstream];
encstream = ArithmeticEncode["This is a sample string!"];
Print[encstream];
arithout = ArithmeticDecode[encstream]; Print[arithout];
```

ContFract.ma

```
(* Continued fraction tools.
 *)

(* Next, calculate the fraction from knowledge
   of its elements.
 *)
len = 60;
(* Fraction elements are thought of like so:
   a = {2,1,2,1,1,4,1,1,6,1,1,8,1,1,10};
   b = {0,1,1,1,1,1,1,1,1,1,1,1,1,1,1};
*)
(* We'll use the elements for Pi, next. *)
a = Table[2,{r,1,len}];
b = Table[If[r==0,0,(2r-1)^2],{r,0,len}];
pp = 1; pc = a[[1]];
qp = 0; qc = 1;
Do[
```

```
    pf = a[[n]] pc + b[[n]] pp;
    qf = a[[n]] qc + b[[n]] qp;
    pp = pc; qp = qc; pc = pf; qc = qf,
    {n,2,Length[a]}
]
Print[pf/qf," ",N[4/(pf/qf -1),40]]

(* Next, obtain the fraction's elements from
   knowledge of its value.
 *)
scf[x_, len_]:=
    Block[{lis,a,q,y},
        lis = {};
        y = x;
        Do[
            a = Floor[y];
            lis = Append[lis,a];
            y = 1/(y-a);
            ,{q,1,len}
        ];
        Return[lis]
  ];
scf[N[E,40],10] (* Give fraction for E. *)
```

corrdim.c

```
/*
 * corrdim.c
 *
 * Simple program for estimating fractal correlation
 * dimension of attractor data.
 *
 * % cc -O corrdim.c -o corrdim
 * % corrdim d < datafile
 *
 * where d is embedding dimension and datafile consists of ASCII
 * floats or ints in one column.
 *
 * The output is a Mathematica List format, for optional
 * plotting and assessment of dimensional slopes.
 *
 */

#include <stdio.h>
#include <stdlib.h>
#include <math.h>

#define MAX_NUM (1<<17)
#define MAX_EMBED 5
#define SCALE_RATIO 0.85

int embed_dim, num_data, num_vects;
double max, min;
double x[MAX_NUM], y[MAX_NUM][MAX_EMBED];
```

```
void embed()
{
    int j, k;
    for (j = 0; j < num_vects; j++) {
        for (k = 0; k < embed_dim; k++) {
        y[j][k] = x[j + k];
        }
    }
}

double corr(double eps)
/* Return the number of mutual pairs within eps of each other. */
{
    int close = 0, tilt, j, k, b;
    double delta;

    for (j = 0; j < num_vects; j++) {
        for (k = j + 1; k < num_vects; k++) {
            tilt = 0;
            for (b = 0; b < embed_dim; b++) {
                delta = y[j][b] - y[k][b];
                if ((delta > eps) || (delta < -eps)) {
                    tilt = 1; break;
                }
            }
            if (tilt) continue;

/* More precise E-dimensional ball check.

            dist = 0;
            for (b = 0; b < embed_dim; b++) {
                le = y[j][b] - y[k][b];
                le *= le;
                dist += le;
            }
            if (dist < eps2) ++close;
*/
            ++close;
        }
    }
    return (2.0 * close / ((double)num_vects * (double)num_vects));
}

void main(int argc, char **argv)
{
    int j;
    double eps;

    embed_dim = atoi(argv[1]);
    num_data = 0;
    while ((fscanf(stdin, "%lf", &x[num_data++]) != EOF));
    --num_data;
    num_vects = num_data - embed_dim + 1;
    max = min = x[0];
    for (j = 1; j < num_data; j++) {
        if (max < x[j])
```

```
                max = x[j];
            else if (min > x[j]) min = x[j];
        }
        embed(num_data, embed_dim);
        eps = (max - min) * 0.1;
        while (eps / (max - min) > 0.0001) {
            eps *= SCALE_RATIO;
            fprintf(stdout, "{%2.12f,%2.12f},\n", eps, corr(eps));
            fflush(stdout);
        }
}
```

DanielsonLanczos.ma

```
(* Danielson-Lanczos FFT recursion
   This can be tested via built-in functions:
   Sqrt[Length[x]] * InverseFourier[x]
 *)

fft[x_] :=
   Block[{n, xeven, xodd, xsplit, g, w},
        If[(n=Length[x])==1,Return[x]];
        xsplit = Transpose[Partition[x,2]];
        xeven = fft[xsplit[[1]]]; xodd = fft[xsplit[[2]]];
        g = Exp[-2 Pi I/n]; w = Table[g^m,{m,0,n-1}];
        Return[Join[xeven,xeven] + w * Join[xodd,xodd]];
   ]

x = {1,1,1,1,-1,-1,-1,-1};
Print[fft[x]]

(* Next, version with explicit loop for even/odd splitting. *)

fft[x_] :=
   Block[{n, xeven, xodd, w},
        If[(n=Length[x])==1,Return[x]];
        xeven = xodd = x;
        Do[xeven = Drop[xeven,{m+1,m+1}];
            xodd  = Drop[xodd,{m,m}],{m,1,n/2}];
        xeven = fft[xeven]; xodd = fft[xodd];
        w = Table[Exp[-2 Pi I m/n],{m,0,n-1}];
        Return[Join[xeven,xeven] + w * Join[xodd,xodd]];
   ]

x = {1,1,1,1,-1,-1,-1,-1};
Print[fft[x]]

(* Next, version with passed primitive root and smaller table. *)

fft[x_, g_] :=
   Block[{n, xeven, xodd, xsplit, w},
        If[(n=Length[x])==1,Return[x]];
        xsplit = Transpose[Partition[x,2]];
        xeven = fft[xsplit[[1]],g^2]; xodd = fft[xsplit[[2]],g^2];
        w = Table[g^m,{m,0,n/2-1}];
```

```
        Return[Join[xeven+w*xodd,xeven-w*xodd]]
    ]


x = {1,1,1,1,-1,-1,-1,-1};
Print[fft[x, Exp[-2Pi I/Length[x]]]]

(* Next, a 'minimal' version. *)

fft[x_] :=
   Block[{n, xeven, xodd, g, w},
        If[(n=Length[x])==1,Return[x]];
        {xeven, xodd} = Map[fft, Transpose[Partition[x,2]]];
        g = Exp[-2 Pi I/n]; w = Table[g^m,{m,0,n-1}];
        Return[Join[xeven,xeven] + w * Join[xodd,xodd]];
    ]

x = {1,1,1,1,-1,-1,-1,-1};
Print[fft[x]]

(* Next, a combined version. *)

fft[x_, g_] :=
   Block[{n, xeven, xodd, w},
        If[(n=Length[x])==1,Return[x]];
        xeven = fft[x[[ Range[1, n, 2] ]], g^2];
        xodd  = fft[x[[ Range[2, n, 2] ]], g^2];
        w = Table[g^m,{m,0,n/2-1}];
        Return[Join[xeven+w*xodd,xeven-w*xodd]]
    ]

x = {1,1,1,1,-1,-1,-1,-1};
Print[fft[x, Exp[-2 Pi I/Length[x]]]]
```

Daubechies.ma

```
(* Plot wavelets by iteration of the mother function's
   dilation equation.
 *)

(* Define Daubechies' wavelet. *)
hh[n_] := Block[{ },
    If[n==0,Return[(1+Sqrt[3])/(4)]];
    If[n==1,Return[(3+Sqrt[3])/(4)]];
    If[n==2,Return[(3-Sqrt[3])/(4)]];
    If[n==3,Return[(1-Sqrt[3])/(4)]];

    Return[0];

]
(* Next, define rational wavelet (it's well known). *)
hh2[n_] :=
    Block[{ },
         If[n==0,Return[3/5]];
         If[n==1,Return[6/5]];
```

```
        If[n==2,Return[2/5]];
        If[n==3,Return[-1/5]];
        Return[0];
    ]

(* Next, choose a wavelet for this run. *)
h[n_] := N[hh[n]];
size = 3*128;
(* Next, make the hat function table. *)
tphi = Table[1,{p,0,size-1}];
uphi = tphi;
phi[j_] :=
    Block[{},
        If[j<0, Return[0]];
        If[j>=size,Return[0]];
        Return[tphi[[j+1]]];
    ];
(* Next, iterate the dilation equation. *)
For[q=0,q<5,q++,
    For[r=0,r<size,r++,
        uphi[[r+1]] = Sum[h[y] phi[2r - y size/3],{y,0,3}];
    ];
    tphi = uphi;
    ListPlot[tphi,PlotJoined->True];
];
(* Next, compute the wavelet from the scaling function. *)
For[r=0,r<size,r++,
    uphi[[r+1]] =
        Sum[h[3-y] (-1)^y phi[2r - y size/3],{y,0,3}];
];

(* Next, finally show the wavelet. *)
ListPlot[uphi,PlotJoined->True]
```

DFTgraphs.ma

```
(* Graphs of the Discrete Fourier Transform
   (DFT) for special tcst signals.
 *)

(* DFT amplitude plot for square wave. *)

n = 128;
x = Table[If[q<n/2,1.,-1.],{q,0,n-1}];
ListPlot[Abs[Sqrt[n] InverseFourier[x]]]

(* DFT decibels/octave plot for square wave. *)

n = 128;
x = Table[If[q<n/2,1.,-1.],{q,0,n-1}];
y = N[Abs[Sqrt[n] InverseFourier[x]]];
ymax = Max[y];
y = Table[{Log[2,q-1.0], 20 Log[y[[q]]/ymax]},{q,2,n,2}];

ListPlot[y, PlotJoined->True]
```

```
(* DFT amplitude plot for chirp wave. *)

n = 128;
x = Table[N[Exp[Pi I q^2/n]],{q,0,n-1}];
ListPlot[Abs[Sqrt[n] InverseFourier[x]], PlotRange->{0,12}]
```

DragonRecursion.ps

```
%!PS-Adobe-2.0
%%Pages: 20 1
%%EndComments

% DragonRecursion.ps
% PostScript recursive dragon curve
% [Marion Becca Sturtevant and John Kenton Newlin 1987]

/inch { 72 mul } def
/height 11 inch def       % paper (screen) dimensions
/width 8.5 inch def
/right true def           % turn constants
/left false def

/turn {            % takes direction from stack (4.06 sec for order 10)
    0 0 moveto
    {                             % push args on stack for
        -90 1 1 0 .6 .4 1 1 1     % curve right
    }{
        90 -1 1 0 .6 -.4 1 -1 1 % curve left
    } ifelse
    curveto stroke translate rotate
} bind def

/dragon {                 % takes direction, order
    dup 1 le {            % dup order, test if recursion bottoms out
        pop turn          % pop order, pass direction to turn
    }{                    % else descend recursively
        1 sub             % compute order-1 for children
        //right 1 index dragon
        exch turn         % pass direction to turn
        //left exch dragon
    } ifelse
} bind def

/dragonPage {    % takes order from stack, assumes gstate as set just
above
    gsave
    /order exch def
    /size      % for 1/2 a seg
        1 2 order 2 idiv exp div        % 1/(2^(n div 2))
        order 2 mod 1 eq {              % odd orders only
            1 2 sqrt div mul } if       % * (1 / sqrt(2))
        8 25 div height mul mul         % * 8/25 * height
        def
    1 inch -17 25 div height mul moveto % 17/25 * height below start
```

```
point
    erasepage (order ) show               % print title: order n dragon
curve
    order 20 string cvs show
    ( dragon curve) show
    order 8 mod 45 mul 180 add rotate     % rotate to start
    size dup scale                        % scale so we can make unit
steps
    0 0 moveto 0 1 lineto stroke          % 1st 1/2 seg
    0 1 translate
    right order dragon                    % dragon curve
    0 0 moveto 0 1 lineto stroke          % last 1/2 seg
    showpage
    grestore
} def
%%EndProlog

%%BeginSetup
%%PaperSize: Letter
6 17 div width mul       % x = 6/17 * width
19 25 div height mul     % y = 19/25 * height
translate                % moveto start point
0 setlinewidth
/Helvetica findfont 16 scalefont setfont
%%EndSetup

%%Page: 1 1
1 dragonPage

%%Page: 2 2
2 dragonPage

%%Page: 3 3
3 dragonPage

%%Page: 4 4
4 dragonPage

%%Page: 5 5
5 dragonPage

%%Page: 6 6
6 dragonPage

%%Page: 7 7
7 dragonPage

%%Page: 8 8
8 dragonPage

%%Page: 9 9
9 dragonPage

%%Page: 10 10
10 dragonPage

%%Page: 11 11
11 dragonPage
```

```
%%Page: 12 12
12 dragonPage

%%Page: 13 13
13 dragonPage

%%Page: 14 14
14 dragonPage

%%Page: 15 15
15 dragonPage

%%Page: 16 16
16 dragonPage

%%Page: 17 17
17 dragonPage

%%Page: 18 18
18 dragonPage

%%Page: 19 19
19 dragonPage

%%Page: 20 20
20 dragonPage
```

Dragon.ma

```
(* One style of Dragon Fractal generation:
   replace each direction with a special direction
   sequence, equivalent to the "folded paper strip"
   rule for this fractal.
 *)

depth = 13;  (* Maximum depth of Dragon Fractal. *)
drag = {0};
Do[
     len = Length[drag];
     Do[ rep = Switch[drag[[q]],0,{0,1},1,{2,1},2,{2,3},3,{0,3}];
         drag = Insert[Drop[drag,{q,q}],rep,q],
         {q,1,len}
     ];
     drag = Flatten[drag];
     pt = {0,0};
     pts = {pt};
     Do[
            pt += Switch[drag[[q]],0,{1,0},1,{0,1},2,{-1,0},3,{0,-1}];
            pts = Append[pts,pt],
            {q,1,len}
     ];
     Show[Graphics[Line[pts]], AspectRatio->Automatic],
     {r,1,depth}
];
```

Eigenvalue.ma

```
(* Power method for leading eigenvalue estimate.
   Printed out are approximations to the
   leading eigenvalue and an associated
   eigenvector.
 *)

a = {{4,3,0},
     {0,3,0},
     {1,1,2}};

     v = {3,0,0};
Do[
    m = Max[N[Abs[v]]];
    Print[m," ",v/m];
    v = a .(v/m),
    {q,1,10}
];

Eigensystem[N[a]]
```

eliascoder.c

```
/*
 * eliascoder.c
 *
 * Elias integer coding routines. Note that integers from 0 through
 * NUM_CODES-1 (that's 254) inclusive are encoded.
 *
 */

/* Input data is assumed to be big endian */

#include <stdlib.h>
#include <stdio.h>
#include <math.h>
#include "eliascoder.h"

enc_table *enc_elias = NULL;
dec_table *dec_elias = NULL;

void init_elias_codes()
{
    int maxb, j, k, lg, c, nones, mask;

    enc_elias = (enc_table *)malloc(sizeof(enc_table));
    enc_elias->entry = (enc_entry *)malloc(NUM_CODES *
sizeof(enc_entry));
    maxb = 1;
    for (j = 0; j < NUM_CODES; j++) {
        lg = (int)(log((double)(j + 1)) / log(2.0) + (1e-10));
```

```
            enc_elias->entry[j].code =
              (((1 << lg) - 1) << (lg + 1)) | (j + 1 - (1 << lg));
            enc_elias->entry[j].bits = 2 * lg + 1;
            if (maxb < 2 * lg + 1)
                maxb = 2 * lg + 1;
        }
        enc_elias->maxbits = maxb;
        dec_elias = (dec_table *)malloc(sizeof(dec_table));
        dec_elias->entry = (dec_entry *)malloc((1 << maxb) *
                                               sizeof(dec_entry));
        dec_elias->maxbits = maxb;
        for (j = 0; j < (1 << maxb); j++) {
            c = 1 << (maxb - 1);
            nones = 0;
            for (k = 0; k < maxb; k++) {
                if ((j & c) == 0)
                    break;
                c >>= 1;
                ++nones;
            }
            mask = (1 << nones) - 1;
            dec_elias->entry[j].val =
              mask + ((j >> (maxb - (2 * nones + 1))) & mask);
            dec_elias->entry[j].bits = 2 * nones + 1;
        }
}

int put_codes(unsigned char *out, enc_table *enc,
              unsigned char *data, int numdata, int *inbits)
{
    int dep, nbits, j, code, b, c;
    unsigned int *iout = (unsigned int *)out;

    if (*inbits == 0)
        dep = 0;
    else
        dep = *iout;
    nbits = *inbits;
    for (j = 0; j < numdata; j++) {
        code = enc->entry[data[j]].code;
        b = enc->entry[data[j]].bits;
        if (nbits + b < BITS_PER_INT) {
            nbits += b;
            dep = (dep << b) | code;
        } else {
            c = BITS_PER_INT - nbits;
            nbits = b - c;
            dep = (dep << c) | (code >> nbits);
            *iout++ = dep;
            dep = code & ((1 << nbits) - 1);
        }
    }
    if (nbits > 0)
        *iout = dep;
    *inbits = nbits;
    return ((int)((unsigned char *)iout - out));
}
```

```
int terminate_put(unsigned char *out, int inbits)
{
    unsigned int *iout = (unsigned int *)out;

    if (inbits > 0)
        *iout++ = *iout << (BITS_PER_INT - inbits);
    return ((int)((unsigned char *)iout - out));
}

int terminate_get(unsigned char *in, int inbits)
{
    return ((inbits > 0) ? sizeof(int) : 0);
}

int get_codes(unsigned char *in, dec_table *dec, unsigned char *data,
              int numdata, int *inbits)
{
    unsigned int *iin = (unsigned int *)in;
    unsigned char *dat = data;
    unsigned int c, d;
    int j, totbits, mbits, leftbits, bits, rot;
    dec_entry *den = dec->entry;
    dec_entry *dentry;

    c = *iin++;
    totbits = *inbits;
    c <<= totbits;
    d = 0;
    mbits = dec->maxbits;
    rot = BITS_PER_INT - mbits;
    leftbits = BITS_PER_INT - totbits;
    for (j = 0; j < numdata; j++) {
        if (leftbits < BITS_PER_INT) {
            d = *iin++;
            c |= (d >> leftbits);
            d <<= BITS_PER_INT - leftbits;
            leftbits += BITS_PER_INT;
        }
        dentry = den + (c >> rot);
        *dat++ = dentry->val;
        bits = dentry->bits;
        c <<= bits;
        c |= (d >> (BITS_PER_INT - bits));
        d <<= bits;
        totbits += bits;
        leftbits -= bits;
    }
    *inbits = totbits & (BITS_PER_INT - 1);
    return ((totbits / BITS_PER_INT) * sizeof(int));
}
```

eliascoder.h

```
/*
```

```
 * eliascoder.h
 *
 * Header file for eliascoder.c
 */

#define NUM_CODES 255  /* Symbols
                          0 through (NUM_CODES-1) may be encoded. */

#define BITS_PER_INT 32

typedef struct {
    unsigned short code;
    unsigned char bits;
} enc_entry;

typedef struct {
    unsigned char maxbits;
    enc_entry *entry;
} enc_table;

typedef struct {
    unsigned char bits;
    unsigned short val;
} dec_entry;

typedef struct {
    unsigned char maxbits;
    dec_entry *entry;
} dec_table;

extern enc_table *enc_elias;     /* The fundamental elias table:
                                          0       0
                                          1       100
                                          2       101
                                          3       11000
                                          4       11001
                                          5       11010
                                          6       11011
                                          ...
                                  */
extern dec_table *dec_elias;

extern void init_elias_codes();

extern int put_codes(unsigned char *out, enc_table *enc,
                unsigned char *data, int numdata, int *inbits);

extern int terminate_put(unsigned char *out, int inbits);

extern int terminate_get(unsigned char *in, int inbits);

extern int get_codes(unsigned char *in, dec_table *dec,
                    unsigned char *data, int numdata, int *inbits);
```

eliastest.c

```
/*
 * eliastest.c
 *
 * Test of Elias integer coding routines. Compile with:
 *
 * % cc eliastest.c eliascoder.c -o eliastest
 *
 * Run with:
 *
 * % eliastest m
 *
 * where m is a mean value for internal Poisson generation of
 * symbols.  Compression data are reported.
 *
 * Note eliascoder.c encodes integer data from 0 through
 * (NUM_CODES-1) inclusive.
 */

#include <stdlib.h>
#include <stdio.h>
#include <math.h>
#include "eliascoder.h"

#define SIZE 65536
#define DEN ((double)((1<<31)-1))
unsigned char dat[SIZE];
unsigned char *dat2;

void main(int argc, char **argv)
{
    int dev, j, inbits, streamcount;
    double poiss;
    unsigned char *out0;
    unsigned char *outbuff;
    unsigned char *inbuff;

    /*
     * Next, input a mean "dev" for the Poisson-distributed integer
     * data.
     */
    dev = atoi(argv[1]);
    init_elias_codes();
    outbuff = (unsigned char *)malloc(2 * SIZE);
    dat2 = (unsigned char *)malloc(SIZE);
    out0 = outbuff;         /* To provide a basal stream pointer. */

    /* Next, create Poisson-distributed integer data. */
    for (j = 0; j < SIZE; j++) {
        /* Next, obtain a Poisson random variable. */
        poiss = dev * (-log(random() / DEN));
        dat[j] = (poiss < NUM_CODES) ? poiss : NUM_CODES - 1;
        /* This last check prevents any code > NUM_CODES-1. */
    }

    /* Next, encode "dat" and output the encoding "outbuff." */
    inbits = 0;
```

```
    outbuff += put_codes(outbuff, enc_elias, dat, SIZE, &inbits);
    outbuff += terminate_put(outbuff, inbits);

    streamcount = (int)(outbuff - out0);
    printf("Stream bytes %d; compression factor:%f\n",
            streamcount, streamcount / ((double)SIZE));
    inbuff = out0;

    /* Next, decode "inbuff" output the decoding to "dat2." */
    inbits = 0;
    inbuff += get_codes(inbuff, dec_elias, dat2, SIZE, &inbits);
    inbuff += terminate_get(inbuff, inbits);

    /* Next, test data integrity for overall encode/decode. */
    for (j = 0; j < SIZE; j++) {
        if (dat2[j] != dat[j]) {
            printf("Encode/decode error at %d.\n", j);
            exit(0);
        }
    }
    printf("Encode/decode worked.\n");
}
```

Equipotential.ma

```
(* Plot various equipotential curves for
   given potential functions.  The Karman vortex
   sheet arises also in fluid dynamics.
 *)

(* Next, electrostatic dipole. *)
vd[x_,y_] := 1/Sqrt[(x-1/4)^2 + y^2] - 1/Sqrt[(x+1/4)^2 + y^2];
(* Next, electrostatic linear wire quadrupole. *)
vq[x_,y_] := Log[(x-1/16)^2 + y^2] + Log[(x+1/16)^2 + y^2] -
             Log[(y-1/16)^2 + x^2] - Log[(y+1/16)^2 + x^2];
(* Next, Karman vortex street. *)
a = 0.05; b = 0.1;
vk[x_,y_] := Re[I Log[Sin[Pi(x+I y)/a] - I b/a] +
                I Log[Sin[Pi(x+I y)/a] - I b/a]];

ContourPlot[vq[x,y],{x,-0.2,0.2},{y,-0.2,0.2},PlotPoints->32,
    Axes->None]
```

Erfc.ma

```
(* Asymptotic breakover method for
   computation of the error function.
 *)

k = 10.0;  (* The number of good decimal digits required. *)

myerfc[x_] :=
```

```
     Block[{m},
        If[x < 0, Return[2-myerfc[-x]]];
        Return[
            N[If[x^2 < N[k Log[10]],
                     1-2/Sqrt[Pi] Sum[(-1)^m/(2m+1) x^(2m+1)/m!
                                  ,{m,0,Floor[E^1.5 k Log[10]]}]
                    ,
                     Exp[-x^2](1/x+ Sum[(-1/2)^m (2m-1)!!/x^(2m+1),
                                  {m,1,Floor[k Log[10]]}])/Sqrt[Pi]]
              ],30
            ]
        ];
     ];

Print[Table[myerfc[q],{q,0,30}]];
Print[Table[N[1-Erf[q],30],{q,0,30}]];
```

FermatConvolution.ma

```
(* Fast multiplication modulo Fermat Numbers
   via Discrete Weighted Transform (DWT).
   For the Fermat number w^m+1, where m is a power
   of 2, there are two historical choices:

   1) Zero-pad the digits of x and y up to length 2m,
      perform cyclic convolution, and take the result
      (mod w^m+1).  Three transforms each have length 2m.
   2) Perform negacyclic
      convolution.  Three transforms each have length m,
      weight constant a = Exp[I Pi/m].

   The method here is to represent each of x, y as a
   real digit signal (the first m/2 of the digits),
   plus an imaginary signal (the last m/2 of the digits).
   Then:

   3) Zero-pad to length m/2 and perform "right-angle"
      convolution.  Three transforms each have length m/2,
      weight constant a = Exp[I Pi/m] again.

   The DWT is, instead of the Fourier
   transform of a signal {x[j]}, the Fourier transform of
   {x[j]} a^j, where a is the "weight signal".
  *)

x = 78979789771111117787877887878111l;
y = 5675656567655787787878787878787;
w = 65536;
m = 8;  (* We shall work (mod 2^128 + 1), or (mod w^8 + 1). *)
digits[x_, base_, run_] :=
    Block[{t, k, j, i},
        t = {};
        i = x;
        Do[ k = Mod[i, base];
             t = Append[t, k];
```

```
                i = (i-k)/base,
                {j,1,run}
            ];
            Return[t];
        ];
digx = digits[x, 65536, m];
digy = digits[y, 65536, m];
Print["Base w digits of x:"];
Print[digx];
Print["Base w digits of y:"];
Print[digy];
n = m/2;
ycomplex = Table[digy[[j]] + I digy[[j+n]], {j,1,n}];
xcomplex = Table[digx[[i]] + I digx[[i+n]], {i,1,n}];
Print["Complex digits of x:"];
Print[xcomplex];
Print["Complex digits of y:"];
Print[ycomplex];

(* Next, do two weighted transforms. *)
a = N[Exp[I Pi/(2n)]];
Do[ ycomplex[[k]] *= a^(k-1); xcomplex[[k]] *= a^(k-1), {k,1,n}];
xhat = N[Sqrt[n]] * InverseFourier[xcomplex];
yhat = N[Sqrt[n]] * InverseFourier[ycomplex];

(* Next, multiply in spectral space. *)
xhat *= yhat;

(* Next, take inverse transform,
   normalize, round to integer elements. *)
xhat = N[Sqrt[n]] * Fourier[xhat];
Do[ xhat[[p]] *= (1/a)^(p-1) / n, {p,1,n}];
squaredig = Join[Re[xhat], Im[xhat]];
(* squaredig is now the length-m list of negacyclic
   convolution elements (mod w^m+1).  The point is,
   we did this with transforms of length n = m/2. *)
Print["Negacyclic elements:"];
Print[N[squaredig,20]];
squaredig = Round[squaredig];
Print[squaredig];
(* At this point a complete implementation would do
   add-with-carry (mod w^m+1), but we'll just prove
   the square is correct. *)
square = Sum[squaredig[[q]] w^(q-1), {q,1,m}];
Print["Transform square: ",Mod[square, w^m+1]];
Print["Direct square   : ",Mod[x y, w^m+1]];
```

FFTs.ma

```
(* Gentleman-Sande, decimation in frequency radix-2 complex FFT.
   Scramble bits *after* this call. *)

fftfreq[x_] :=
      Block[{y,n,m,i,j,a,ii,im},
          y = x;
          n = Length[x];
```

```
        For[m = Floor[n/2] , m > 0 , m = Floor[m/2],
          For[j = 0, j<m, j++,
            For[ i = j, i < n, i += 2m,
                ii = i+1; im = i+m+1; (* List indices. *)
                a = Exp[-2 Pi I j/(2*m)];
                {y[[ii]],y[[im]]} =
                   {y[[ii]]+y[[im]], a*(y[[ii]]-y[[im]])};
            ];
          ];
        ];
        Return[y];
     ];

(* Cooley-Tukey, decimation in time radix-2 complex FFT.
   Scramble bits *before* this call. *)

ffttime[x_] :=
     Block[{y, n,m,i,j,a,ii,im},
        y = x;
        n = Length[x];
        For[m = 1, m < n , m += m,
          For[j = 0, j<m, j++,
            For[ i = j, i < n, i += 2m,
                ii = i+1; im = i+m+1; (* List indices. *)
                a = Exp[-2 Pi I j/(2*m)];
                {y[[ii]],y[[im]]} =
                   {y[[ii]]+a*y[[im]], y[[ii]]-a*y[[im]]};
            ];
          ];
        ];
        Return[y];
     ];

(* Next, bit scrambler intended to be used after a DIF FFT, but
before a DIT FFT. *)

scramble[x_] := Block[{y, i, j, k, tmp},
        y = x;
        n = Length[x];
        For[i=0;j=0,i<n-1,i++,
                If[i<j,
                  tmp = y[[j+1]];
                  y[[j+1]]=y[[i+1]];
                  y[[i+1]]=tmp;
                ];
                k = Floor[n/2];
                While[k<=j,
                        j -= k;
                        k = Floor[k/2]
                ];
                j += k;
        ];
        Return[y];
]

z = {1,3,2,4,5,6,7,9};
```

```
xfreq = scramble[fftfreq[z]];
xtime = ffttime[scramble[z]];
xtest = Sqrt[Length[z]] InverseFourier[z];

Print[xfreq]
Print[xtime]
Print[N[xfreq-xtest]]
Print[N[xtime-xtest]]

(* Next, a cyclic convolution example.
   Note that scramble/unscramble is unnecessary.
 *)

a = {1,2,3,4,5,6,7,8};
b = {3,1,4,2,7,6,5,8};

N[ffttime[Conjugate[N[fftfreq[a] * fftfreq[b]]]]/Length[a]]


(* Next, direct cyclic convolution. *)

len = Length[a];
conv = Table[Sum[ a[[1+Mod[q,len]]] * b[[1+Mod[k-q,len]]],
                  {q,0,len-1}
             ],
             {k,0,len-1}
       ]
```

fft2D_real.c

```
/*
 * fft2D_real.c
 *
 * Routines for fast, in-place, Hermitian-ordered 2-Dimensional FFT
 * for real image source (or for real image result).
 *
 * The real valued 1-Dimensional FFT routines follow a split-radix
 * scheme adapted from [Sorenson 1987]
 *
 *
 * The routines of main interest are:
 *
 *      fft2D_real_to_hermitian(x, w, h);
 *      fftinv2D_hermitian_to_real(x, w, h);
 *
 * which, acting together in the stated order, should amount to an
 * identity operation on images x (arrays of type double) of width w
 * and height h. The w,h values must both be powers of 2.
 *
 * The forward transform fft2D_real_to_hermitian() is an in-place
 * operation that leaves Hermitian components of the 2D FFT in a
 * certain order, an order that the inverse transform
 * fftinv2D_hermitian_to_real() expects.  From this ordering one can
 * extract easily the actual complex FFT, as below.
 *
```

```
 * Precise definitions are as follows.  The FFT is defined as:
 *
 *      X[m,n] = Sum{0<=j<w,0<=k<h} x[j,k] Exp[-2 Pi I (mj/w + nk/h)]
 *
 * The transform computed here assumes real-valued x[j,k] and
 * produces
 *
 *      Y[m,n] = Sum{0<=j<w,0<=k<h} x[j,k] e(j, m, w) e(k, n, h)
 *
 * where the e() function is:
 *
 *      e(a,b,c) =  Cos[2 Pi I a b/c]  ; if 0 <= b <= c/2
 *                 -Sin[2 Pi I a b/c]  ; if c/2 < b
 *
 * It requires a few add operations in general to compute the complex
 * X[m,n] from the Y values
 */


#include <stdlib.h>
#include <stdio.h>
#include <math.h>

#define TWOPI (2*3.14159265358979323846)
#define SQRTHALF 0.70710678118654752440

int cur_run = 0;
double *sincos = NULL, *aux = NULL;

void init_sincos(int n)
{
    int j;
    double e = TWOPI / n;

    if (n == cur_run)
        return;
    cur_run = n;
    if (!aux)
        aux = (double *)malloc(sizeof(double) * n);
    else
        aux = (double *)realloc(aux, sizeof(double) * n);
    if (!sincos)
        sincos = (double *)malloc(sizeof(double) * (1 + (n >> 2)));
    else
        sincos =
          (double *)realloc(sincos, sizeof(double) * (1 + (n >> 2)));
    for (j = 0; j <= (n >> 2); j++) {
        sincos[j] = sin(e * j);
    }
}

double s_sin(int n)
{
    int seg = n / (cur_run >> 2);

    switch (seg) {
        case 0: return (sincos[n]);
        case 1: return (sincos[(cur_run>>1)-n]);
```

```
        case 2: return (-sincos[n-(cur_run>>1)]);
        case 3: return (-sincos[cur_run-n]);
    }
}

double s_cos(int n)
{
    int quart = (cur_run >> 2);

    if (n < quart)
        return (s_sin(n + quart));
    return (-s_sin(n - quart));
}

void scramble_real(double *x, int n)
{
    register int i, j, k;
    double tmp;

    for (i = 0, j = 0; i < n - 1; i++) {
        if (i < j) {
            tmp = x[j];
            x[j] = x[i];
            x[i] = tmp;
        }
        k = n / 2;
        while (k <= j) {
            j -= k;
            k >>= 1;
        }
        j += k;
    }
}

void fft_real_to_hermitian(double *z, int n)
/*
 * Output is {Re(z^[0]),...,Re(z^[n/2),Im(z^[n/2-1]),...,Im(z^[1]).
 * This is a decimation-in-time, split-radix algorithm.
 */
{
    register double cc1, ss1, cc3, ss3;
    register int is, id, i0, i1, i2, i3, i4, i5, i6, i7, i8, a, a3, b,
b3, nminus = n - 1, dil, expand;
    register double *x, e;
    int nn = n >> 1;
    double t1, t2, t3, t4, t5, t6;
    register int n2, n4, n8, i, j;

    init_sincos(n);
    expand = cur_run / n;
    scramble_real(z, n);
    x = z - 1;                    /* FORTRAN compatibility. */
    is = 1;
    id = 4;
    do {
        for (i0 = is; i0 <= n; i0 += id) {
            i1 = i0 + 1;
```

```
            e = x[i0];
            x[i0] = e + x[i1];
            x[i1] = e - x[i1];
        }
        is = (id << 1) - 1;
        id <<= 2;
    } while (is < n);
    n2 = 2;
    while (nn >>= 1) {
        n2 <<= 1;
        n4 = n2 >> 2;
        n8 = n2 >> 3;
        is = 0;
        id = n2 << 1;
        do {
            for (i = is; i < n; i += id) {
                i1 = i + 1;
                i2 = i1 + n4;
                i3 = i2 + n4;
                i4 = i3 + n4;
                t1 = x[i4] + x[i3];
                x[i4] -= x[i3];
                x[i3] = x[i1] - t1;
                x[i1] += t1;
                if (n4 == 1)
                    continue;
                i1 += n8;
                i2 += n8;
                i3 += n8;
                i4 += n8;
                t1 = (x[i3] + x[i4]) * SQRTHALF;
                t2 = (x[i3] - x[i4]) * SQRTHALF;
                x[i4] = x[i2] - t1;
                x[i3] = -x[i2] - t1;
                x[i2] = x[i1] - t2;
                x[i1] += t2;
            }
            is = (id << 1) - n2;
            id <<= 2;
        } while (is < n);
        dil = n / n2;
        a = dil;
        for (j = 2; j <= n8; j++) {
            a3 = (a + (a << 1)) & nminus;
            b = a * expand;
            b3 = a3 * expand;
            cc1 = s_cos(b);
            ss1 = s_sin(b);
            cc3 = s_cos(b3);
            ss3 = s_sin(b3);
            a = (a + dil) & nminus;
            is = 0;
            id = n2 << 1;
            do {
                for (i = is; i < n; i += id) {
                    i1 = i + j;
                    i2 = i1 + n4;
```

```
                    i3 = i2 + n4;
                    i4 = i3 + n4;
                    i5 = i + n4 - j + 2;
                    i6 = i5 + n4;
                    i7 = i6 + n4;
                    i8 = i7 + n4;
                    t1 = x[i3] * cc1 + x[i7] * ss1;
                    t2 = x[i7] * cc1 - x[i3] * ss1;
                    t3 = x[i4] * cc3 + x[i8] * ss3;
                    t4 = x[i8] * cc3 - x[i4] * ss3;
                    t5 = t1 + t3;
                    t6 = t2 + t4;
                    t3 = t1 - t3;
                    t4 = t2 - t4;
                    t2 = x[i6] + t6;
                    x[i3] = t6 - x[i6];
                    x[i8] = t2;
                    t2 = x[i2] - t3;
                    x[i7] = -x[i2] - t3;
                    x[i4] = t2;
                    t1 = x[i1] + t5;
                    x[i6] = x[i1] - t5;
                    x[i1] = t1;
                    t1 = x[i5] + t4;
                    x[i5] -= t4;
                    x[i2] = t1;
                }
                is = (id << 1) - n2;
                id <<= 2;
            } while (is < n);
        }
    }
}

void fftinv_hermitian_to_real(double *z, int n)
/*
 * Input is {Re(z^[0]),...,Re(z^[n/2),Im(z^[n/2-1]),...,Im(z^[1]).
 * This is a decimation-in-frequency, split-radix algorithm.
 */
{
    register double cc1, ss1, cc3, ss3;
    register int is, id, i0, i1, i2, i3, i4, i5, i6, i7, i8, a, a3, b,
b3, nminus = n - 1, dil, expand;
    register double *x, e;
    int nn = n >> 1;
    double t1, t2, t3, t4, t5;
    int n2, n4, n8, i, j;

    init_sincos(n);
    expand = cur_run / n;
    x = z - 1;
    n2 = n << 1;
    while (nn >>= 1) {
        is = 0;
        id = n2;
        n2 >>= 1;
        n4 = n2 >> 2;
        n8 = n4 >> 1;
```

```
do {
    for (i = is; i < n; i += id) {
        i1 = i + 1;
        i2 = i1 + n4;
        i3 = i2 + n4;
        i4 = i3 + n4;
        t1 = x[i1] - x[i3];
        x[i1] += x[i3];
        x[i2] += x[i2];
        x[i3] = t1 - 2.0 * x[i4];
        x[i4] = t1 + 2.0 * x[i4];
        if (n4 == 1)
            continue;
        i1 += n8;
        i2 += n8;
        i3 += n8;
        i4 += n8;
        t1 = (x[i2] - x[i1]) * SQRTHALF;
        t2 = (x[i4] + x[i3]) * SQRTHALF;
        x[i1] += x[i2];
        x[i2] = x[i4] - x[i3];
        x[i3] = -2.0 * (t2 + t1);
        x[i4] = 2.0 * (t1 - t2);
    }
    is = (id << 1) - n2;
    id <<= 2;
} while (is < n - 1);
dil = n / n2;
a = dil;
for (j = 2; j <= n8; j++) {
    a3 = (a + (a << 1)) & nminus;
    b = a * expand;
    b3 = a3 * expand;
    cc1 = s_cos(b);
    ss1 = s_sin(b);
    cc3 = s_cos(b3);
    ss3 = s_sin(b3);
    a = (a + dil) & nminus;
    is = 0;
    id = n2 << 1;
    do {
        for (i = is; i < n; i += id) {
            i1 = i + j;
            i2 = i1 + n4;
            i3 = i2 + n4;
            i4 = i3 + n4;
            i5 = i + n4 - j + 2;
            i6 = i5 + n4;
            i7 = i6 + n4;
            i8 = i7 + n4;
            t1 = x[i1] - x[i6];
            x[i1] += x[i6];
            t2 = x[i5] - x[i2];
            x[i5] += x[i2];
            t3 = x[i8] + x[i3];
            x[i6] = x[i8] - x[i3];
            t4 = x[i4] + x[i7];
```

```
                    x[i2] = x[i4] - x[i7];
                    t5 = t1 - t4;
                    t1 += t4;
                    t4 = t2 - t3;
                    t2 += t3;
                    x[i3] = t5 * cc1 + t4 * ss1;
                    x[i7] = -t4 * cc1 + t5 * ss1;
                    x[i4] = t1 * cc3 - t2 * ss3;
                    x[i8] = t2 * cc3 + t1 * ss3;
                }
                is = (id << 1) - n2;
                id <<= 2;
            } while (is < n - 1);
        }
    }
    is = 1;
    id = 4;
    do {
        for (i0 = is; i0 <= n; i0 += id) {
            i1 = i0 + 1;
            e = x[i0];
            x[i0] = e + x[i1];
            x[i1] = e - x[i1];
        }
        is = (id << 1) - 1;
        id <<= 2;
    } while (is < n);
    scramble_real(z, n);
    e = 1 / (double)n;
    for (i = 0; i < n; i++)
        z[i] *= e;
}

void mul_hermitian(double *a, double *b, int n)
/* b becomes b*a in Hermitian representation. */
{
    int k, half = n >> 1;
    register double c, d, e, f;

    b[0] *= a[0];
    b[half] *= a[half];
    for (k = 1; k < half; k++) {
        c = a[k];
        d = b[k];
        e = a[n - k];
        f = b[n - k];
        b[n - k] = c * f + d * e;
        b[k] = c * d - e * f;
    }
}

void fft_offset_real_to_hermitian(double *z, int off, int n)
{
    int j, k;

    for (j = 0, k = 0; j < n; j++, k += off)
        aux[j] = z[k];
```

```
    fft_real_to_hermitian(aux, n);
    for (j = 0, k = 0; j < n; j++, k += off)
        z[k] = aux[j];
}

void fftinv_offset_hermitian_to_real(double *z, int off, int n)
{
    int j, k;

    for (j = 0, k = 0; j < n; j++, k += off)
        aux[j] = z[k];
    fftinv_hermitian_to_real(aux, n);
    for (j = 0, k = 0; j < n; j++, k += off)
        z[k] = aux[j];
}

void fft2D_real_to_hermitian(double *x, int w, int h)
{
    int row, col;

    for (row = 0; row < h; row++) {
        fft_real_to_hermitian(x + w * row, w);
    }
    for (col = 0; col < w; col++) {
        fft_offset_real_to_hermitian(x + col, w, h);
    }
}

void fftinv2D_hermitian_to_real(double *x, int w, int h)
{
    int row, col;

    for (col = 0; col < w; col++) {
        fftinv_offset_hermitian_to_real(x + col, w, h);
    }
    for (row = 0; row < h; row++) {
        fftinv_hermitian_to_real(x + w * row, w);
    }
}

#define MAX_WH (512*512)

void main(int argc, char *argv[])
{
    double x[MAX_WH];
    int w = atoi(argv[1]), h = atoi(argv[2]), j, k;

    for (j = 0; j < w * h; j++)
        x[j] = j;

    fft2D_real_to_hermitian(x, w, h);

    /* Next, print out the Hermitian spectrum. */
    if (argc > 3)
        for (j = 0; j < h; j++) {
            for (k = 0; k < w; k++) {
                printf("%4.2f ", x[k + w * j]);
            }
```

```
            printf("\n");
        }

    fftinv2D_hermitian_to_real(x, w, h);

    /*
     * Next, print out the final data, which should coincide with
     * original data.
     */
    if (argc > 3) {
        printf("\n\n");
        for (j = 0; j < h; j++) {
            for (k = 0; k < w; k++) {
                printf("%4.2f ", x[k + w * j]);
            }
            printf("\n");
        }
    }
}
```

fft.c

```
/*
 * fft.c, Fast Fourier Transform C library
 */

/*
 * Routines include:
 *      FFT2raw : Radix-2 FFT, in-place, with yet-scrambled result.
 *      FFT2    : Radix-2 FFT, in-place and in-order.
 *      FFTreal : Radix-2 FFT, with real data assumed, in-place and
 *                in-order (this routine is the fastest of the lot).
 *      FFTarb  : Arbitrary-radix FFT, in-order but not in-place.
 *      FFT2dim : Image FFT, for real data, easily modified for
 *                other purposes.
 *
 * To call an FFT, one must first assign the complex exponential
 * factors. A call such as
 *      e = AssignBasis(NULL, size)
 * will set up the complex *e to be the array of cos, sin pairs
 * corresponding to total number of complex data = size.  This call
 * allocates the (cos, sin) array memory for you.  If you already
 * have such memory allocated, pass the allocated pointer instead
 * of NULL.
 */

#include <stdlib.h>
#include <stdio.h>
#include <math.h>

#define PI 3.14159265358979323846

typedef struct {
    double re, im;
```

```
} complex;

complex *expn;

void conjugate(x, n)
    complex x[];
    int n;
{
    register int j;

    for (j = 0; j < n; j++) {
        x[j].im = -x[j].im;
    }
}

int scancomplexdata(x)
    complex x[];
{
    int n = 0;
    while (scanf("%lf %lf\n", &x[n].re, &x[n].im) != -1)
        ++n;
    return (n);
}

void putcomplexdata(x, n)
    complex x[];
    int n;
{
    int j;

    for (j = 0; j < n; j++)
        printf("%f %f\n", x[j].re, x[j].im);
}

void putdatawithamps(x, n)
    complex x[];
    int n;
{
    int j;

    for (j = 0; j < n; j++)
        printf("%f %f %f\n", x[j].re, x[j].im,
                  sqrt(x[j].re * x[j].re + x[j].im * x[j].im));
}


/* Query whether the data is pure real. */
int PureReal(x, n)
    complex x[];
    int n;
{
    register int m;

    for (m = 0; m < n; m++) {
        if (x[m].im != 0.0)
            return (0);
    }
```

```
    return (1);
}

/* Query whether n is a pure power of 2 */
int IsPowerOfTwo(n)
    int n;
{
    while (n > 1) {
        if (n % 2)
            break;
        n >>= 1;
    }
    return (n == 1);
}

complex *AssignBasis(ex, n)
    complex ex[];
    int n;
{
    register int j;
    register double a = 0, inc = 2 * PI / n;

    if (ex != NULL)
        expn = ex;
    else {
        expn = (complex *)malloc(n * sizeof(complex));
        if (expn == NULL)
            return (NULL);
    }
    for (j = 0; j < n; j++) {
        expn[j].re = cos(a);
        expn[j].im = -sin(a);
        a += inc;
    }
    return (expn);
}

void scramble(x, n, skip)
    complex x[];
    int n, skip;
{
    register int i, j, k, jj, ii;
    complex tmp;

    for (i = 0, j = 0; i < n - 1; i++) {
        if (i < j) {
            jj = j * skip;
            ii = i * skip;
            tmp = x[jj];
            x[jj] = x[ii];
            x[ii] = tmp;
        }
        k = n / 2;
        while (k <= j) {
            j -= k;
            k >>= 1;
        }
        j += k;
```

```
    }
}

void FFT2dim(x, w, h)
/*
 * Perform 2D FFT on image of width w, height h (both powers of 2).
 * IMAGE IS ASSUMED PURE-REAL.  If not, the FFT2real call should be
 * just FFT2.
 */
    complex *x;
    int w, h;
{
    register int j;

    for (j = 0; j < h; j++)
        FFT2real(x + j * w, w, 1);
    for (j = 0; j < w; j++)
        FFT2(x + j, h, w);
}

void FFT2real(x, n, skip)
/*
 * Perform real FFT, data arranged as {re, 0, re, 0, re, 0...};
 * leaving full complex result in x.
 */
    complex x[];
    int n, skip;
{
    register int half = n >> 1, quarter = half >> 1, m, mm;
    register double tmp;
    complex a, b, s, t;

    if (!quarter)
        return;
    for (mm = 0; mm < half; mm++) {
        m = mm * skip;
        x[m].re = 0.5 * x[m << 1].re;
        x[m].im = 0.5 * x[(m << 1) + skip].re;
    }
    FFT2raw(x, half, 2, skip);
    scramble(x, half, skip);
    half *= skip;
    n *= skip;
    x[half] = x[0];
    for (mm = 0; mm <= quarter; mm++) {
        m = mm * skip;
        s.re = 1 + expn[mm].im;
        s.im = -expn[mm].re;
        t.re = 2 - s.re;
        t.im = -s.im;
        a = x[m];
        b = x[half - m];
        b.im = -b.im;

        tmp = a.re;
        a.re = a.re * s.re - a.im * s.im;
        a.im = tmp * s.im + a.im * s.re;
```

```
        tmp = b.re;
        b.re = b.re * t.re - b.im * t.im;
        b.im = tmp * t.im + b.im * t.re;

        b.re += a.re;
        b.im += a.im;

        a = x[m];
        a.im = -a.im;
        x[m] = b;
        if (m) {
            b.im = -b.im;
            x[n - m] = b;
        }
        b = x[half - m];

        tmp = a.re;
        a.re = a.re * t.re + a.im * t.im;
        a.im = -tmp * t.im + a.im * t.re;

        tmp = b.re;
        b.re = b.re * s.re + b.im * s.im;
        b.im = -tmp * s.im + b.im * s.re;

        b.re += a.re;
        b.im += a.im;

        x[half   m] = b;
        if (m) {
            b.im = -b.im;
            x[half + m] = b;
        }
    }
}

void FFT2(x, n, skip)
/*
 * Perform FFT for n = a power of 2. The relevant data are the
 * complex numbers x[0], x[skip], x[2*skip], ...
 */
    complex x[];
    int n, skip;
{
    FFT2raw(x, n, 1, skip);
    scramble(x, n, skip);
}

void FFT2split(z, n)
    complex z[];
    int n;
{
    FFT2splitraw(z, n);
    scramble(z, n, 1);
}

void FFT2splitraw(z, n)
```

```
    complex z[];
    int n;
{
    int m, n2, j, is, id;
    register int i0, i1, i2, i3, n4;
    double r1, r2, s1, s2, s3, cc1, ss1, cc3, ss3;
    int a, a3, ndec = n - 1;
    complex *x;

    x = z - 1;
    n2 = n << 1;
    m = 1;
    while (m < n / 2) {
        n2 >>= 1;
        n4 = n2 >> 2;
        a = 0;
        for (j = 1; j <= n4; j++) {
            a3 = (a + (a << 1)) & ndec;
            cc1 = expn[a].re;
            ss1 = -expn[a].im;
            cc3 = expn[a3].re;
            ss3 = -expn[a3].im;
            a = (a + m) & ndec;
            is = j;
            id = n2 << 1;
            do {
                for (i0 = is; i0 <= n - 1; i0 += id) {
                    i1 = i0 + n4;
                    i2 = i1 + n4;
                    i3 = i2 + n4;
                    r1 = x[i0].re - x[i2].re;
                    x[i0].re += x[i2].re;
                    r2 = x[i1].re - x[i3].re;
                    x[i1].re += x[i3].re;
                    s1 = x[i0].im - x[i2].im;
                    x[i0].im += x[i2].im;
                    s2 = x[i1].im - x[i3].im;
                    x[i1].im += x[i3].im;
                    s3 = r1 - s2;
                    r1 += s2;
                    s2 = r2 - s1;
                    r2 += s1;
                    x[i2].re = r1 * cc1 - s2 * ss1;
                    x[i2].im = -s2 * cc1 - r1 * ss1;
                    x[i3].re = s3 * cc3 + r2 * ss3;
                    x[i3].im = r2 * cc3 - s3 * ss3;
                }
                is = (id << 1) - n2 + j;
                id <<= 2;
            } while (is < n);
        }
        m <<= 1;
    }
    is = 1;
    id = 4;
    do {
        for (i0 = is; i0 <= n; i0 += id) {
            i1 = i0 + 1;
```

```
            r1 = x[i0].re;
            x[i0].re = r1 + x[i1].re;
            x[i1].re = r1 - x[i1].re;
            r1 = x[i0].im;
            x[i0].im = r1 + x[i1].im;
            x[i1].im = r1 - x[i1].im;
        }
        is = (id << 1) - 1;
        id <<= 2;
    } while (is < n);
}

void FFT2raw(x, n, dilate, skip)
/*
 * Data is x, data size is n, dilate means: library global expn is
 * the (cos, -j sin) array, EXCEPT for effective data size n/dilate,
 * skip is the offset of each successive data term, as in "FFT2"
 * above.
 */
    complex x[];
    int n, dilate, skip;
{
    register int j, m = 1, p, q, i, k, n2 = n, n1;
    register double c, s, rtmp, itmp;

    while (m < n) {
        n1 = n2;
        n2 >>= 1;
        for (j = 0, q = 0; j < n2; j++) {
            c = expn[q].re;
            s = expn[q].im;
            q += m * dilate;
            for (k = j; k < n; k += n1) {
                p = (k + n2) * skip;
                i = k * skip;
                rtmp = x[i].re - x[p].re;
                x[i].re += x[p].re;
                itmp = x[i].im - x[p].im;
                x[i].im += x[p].im;
                x[p].re = c * rtmp - s * itmp;
                x[p].im = c * itmp + s * rtmp;
            }
        }
        m <<= 1;
    }
}

#define MAX_FACTORS 20

void FFTarb(data, result, n)
/*
 * Compute FFT for arbitrary n, with limitation  n <= 2^MAX_FACTORS.
 */
    complex data[], result[];
    int n;
{
    int p0, i, j, a, b, c, d, e, v, k;
```

```
register int p, q, arg;
register double x, y;
register double e0, e1, r0, r1;
int sum, car;
int aa[MAX_FACTORS], pr[MAX_FACTORS], cc[MAX_FACTORS];

/* Next, get the prime factors of n */
q = n;
v = 0;
j = 2;
while (q != 1) {
    while (q % j == 0) {
        q /= j;
        pr[v++] = j;
    }
    j += 2;
    if (j == 4)
        j = 3;
}
/*
 * pr[] is now the array of prime factors of n, with v relevant
 * elements
 */

/* Next, re-order the array in reverse-complement binary */
cc[0] = 1;
for (i = 0; i < v - 1; i++) {
    cc[i + 1] = cc[i] * pr[i];
    aa[i] = 0;
}
aa[v - 1] = 0;
for (i = 1; i < n; i++) {
    j = v - 1;
    car = 1;
    while (car) {
        aa[j] += car;
        car = aa[j] / pr[j];
        aa[j] %= pr[j];
        --j;
    }
    sum = 0;
    for (q = 0; q < v; q++)
        sum += aa[q] * cc[q];
    result[i] = data[sum];
}
c = v;
a = 1;
b = 1;
d = n;
while (c--) {
    a *= pr[c];
    d /= pr[c];
    e = -1;
    for (k = 0; k < n; k++) {
        if (k % a == 0)
            ++e;
        arg = a * e + k % b;
```

```
            p0 = (k * d) % n;
            p = 0;
            x = y = 0;
            for (q = 0; q < pr[c]; q++, arg += b) {
                e0 = expn[p].re;
                e1 = expn[p].im;
                r0 = result[arg].re;
                r1 = result[arg].im;
                x += r0 * e0 - r1 * e1;
                y += r0 * e1 + r1 * e0;
                p += p0;
                if (p >= n)
                    p -= n;
            }
            data[k].re = x;
            data[k].im = y;
        }
        bcopy((char *)data, (char *)result, n * sizeof(complex));
        b = a;
    }
}

void DFT(data, result, n)
/* Perform direct Discrete Fourier Transform. */
    complex data[], result[];
    int n;
{
    int j, k, m;
    double s, c;

    for (j = 0; j < n; j++) {
        result[j].re = 0;
        result[j].im = 0;
        for (k = 0; k < n; k++) {
            m = (j * k) % n;
            c = expn[m].re;
            s = expn[m].im;
            result[j].re += data[k].re * c - data[k].im * s;
            result[j].im += data[k].re * s + data[k].im * c;
        }
    }
}
```

fft_real.c

```
/*
 * fft_real.c
 *
 * Routines for split-radix, real-only transforms.
 *
 * These routines are adapted from [Sorenson 1987]
 *
 * When all x[j] are real the standard DFT of (x[0],x[1],...,x[N-1]),
 * call it x^, has the property of Hermitian symmetry: x^[j] =
 * x^[N-j]*.
 * Thus we only need to find the set
```

```
 *  (x^[0].re, x^[1].re,..., x^[N/2].re, x^[N/2-1].im, ..., x^[1].im)
 * which, like the original signal x, has N elements.
 * The two key routines perform forward (real-to-Hermitian) FFT, and
 * backward (Hermitian-to-real) FFT, respectively. For example, the
 * sequence:
 *
 *      fft_real_to_hermitian(x, N);
 *      fftinv_hermitian_to_real(x, N);
 *
 * is an identity operation on the signal x. To convolve two
 * pure-real signals x and y, one does:
 *
 *      fft_real_to_hermitian(x, N);
 *      fft_real_to_hermitian(y, N);
 *      mul_hermitian(y, x, N);
 *      fftinv_hermitian_to_real(x, N);
 *
 * and x is the pure-real cyclic convolution of x and y.
 */

#include <stdlib.h>
#include <math.h>

#define TWOPI (2*3.14159265358979323846)
#define SQRTHALF 0.70710678118654752440
int cur_run = 0;
double *sincos = NULL;

void init_sincos(int n)
{
    int j;
    double e = TWOPI / n;

    if (n <= cur_run)
        return;
    cur_run = n;
    if (sincos)
        free(sincos);
    sincos = (double *)malloc(sizeof(double) * (1 + (n >> 2)));
    for (j = 0; j <= (n >> 2); j++) {
        sincos[j] = sin(e * j);
    }
}

double s_sin(int n)
{
    int seg = n / (cur_run >> 2);

    switch (seg) {
    case 0:
        return (sincos[n]);
    case 1:
        return (sincos[(cur_run >> 1) - n]);
    case 2:
        return (-sincos[n - (cur_run >> 1)]);
    case 3:
        return (-sincos[cur_run - n]);
```

```
    }
}

double s_cos(int n)
{
    int quart = (cur_run >> 2);

    if (n < quart)
        return (s_sin(n + quart));
    return (-s_sin(n - quart));
}

void scramble_real(double *x, int n)
{
    register int i, j, k;
    double tmp;

    for (i = 0, j = 0; i < n - 1; i++) {
        if (i < j) {
            tmp = x[j];
            x[j] = x[i];
            x[i] = tmp;
        }
        k = n / 2;
        while (k <= j) {
            j -= k;
            k >>= 1;
        }
        j += k;
    }
}

void fft_real_to_hermitian(double *z, int n)
/*
 * Output is {Re(z^[0]),...,Re(z^[n/2),Im(z^[n/2-1]),...,Im(z^[1]).
 * This is a decimation-in-time, split-radix algorithm.
 */
{
    register double cc1, ss1, cc3, ss3,
    register int is, id, i0, i1, i2, i3, i4, i5, i6, i7, i8, a, a3, b,
b3, nminus = n - 1, dil, expand;
    register double *x, e;
    int nn = n >> 1;
    double t1, t2, t3, t4, t5, t6;
    register int n2, n4, n8, i, j;

    init_sincos(n);
    expand = cur_run / n;
    scramble_real(z, n);
    x = z - 1;                    /* FORTRAN compatibility. */
    is = 1;
    id = 4;
    do {
        for (i0 = is; i0 <= n; i0 += id) {
            i1 = i0 + 1;
            e = x[i0];
            x[i0] = e + x[i1];
```

```
            x[i1] = e - x[i1];
        }
        is = (id << 1) - 1;
        id <<= 2;
    } while (is < n);
    n2 = 2;
    while (nn >>= 1) {
        n2 <<= 1;
        n4 = n2 >> 2;
        n8 = n2 >> 3;
        is = 0;
        id = n2 << 1;
        do {
            for (i = is; i < n; i += id) {
                i1 = i + 1;
                i2 = i1 + n4;
                i3 = i2 + n4;
                i4 = i3 + n4;
                t1 = x[i4] + x[i3];
                x[i4] -= x[i3];
                x[i3] = x[i1] - t1;
                x[i1] += t1;
                if (n4 == 1)
                    continue;
                i1 += n8;
                i2 += n8;
                i3 += n8;
                i4 += n8;
                t1 = (x[i3] + x[i4]) * SQRTHALF;
                t2 = (x[i3] - x[i4]) * SQRTHALF;
                x[i4] = x[i2] - t1;
                x[i3] = -x[i2] - t1;
                x[i2] = x[i1] - t2;
                x[i1] += t2;
            }
            is = (id << 1) - n2;
            id <<= 2;
        } while (is < n);
        dil = n / n2;
        a = dil;
        for (j = 2; j <= n8; j++) {
            a3 = (a + (a << 1)) & nminus;
            b = a * expand;
            b3 = a3 * expand;
            cc1 = s_cos(b);
            ss1 = s_sin(b);
            cc3 = s_cos(b3);
            ss3 = s_sin(b3);
            a = (a + dil) & nminus;
            is = 0;
            id = n2 << 1;
            do {
                for (i = is; i < n; i += id) {
                    i1 = i + j;
                    i2 = i1 + n4;
                    i3 = i2 + n4;
                    i4 = i3 + n4;
                    i5 = i + n4 - j + 2;
```

```
                    i6 = i5 + n4;
                    i7 = i6 + n4;
                    i8 = i7 + n4;
                    t1 = x[i3] * cc1 + x[i7] * ss1;
                    t2 = x[i7] * cc1 - x[i3] * ss1;
                    t3 = x[i4] * cc3 + x[i8] * ss3;
                    t4 = x[i8] * cc3 - x[i4] * ss3;
                    t5 = t1 + t3;
                    t6 = t2 + t4;
                    t3 = t1 - t3;
                    t4 = t2 - t4;
                    t2 = x[i6] + t6;
                    x[i3] = t6 - x[i6];
                    x[i8] = t2;
                    t2 = x[i2] - t3;
                    x[i7] = -x[i2] - t3;
                    x[i4] = t2;
                    t1 = x[i1] + t5;
                    x[i6] = x[i1] - t5;
                    x[i1] = t1;
                    t1 = x[i5] + t4;
                    x[i5] -= t4;
                    x[i2] = t1;
                }
                is = (id << 1) - n2;
                id <<= 2;
            } while (is < n);
        }
    }
}

void fftinv_hermitian_to_real(double *z, int n)
/*
 * Input is {Re(z^[0]),...,Re(z^[n/2),Im(z^[n/2-1]),...,Im(z^[1]).
 * This is a decimation-in-frequency, split-radix algorithm.
 */
{
    register double cc1, ss1, cc3, ss3;
    register int is, id, i0, i1, i2, i3, i4, i5, i6, i7, i8, a, a3, b,
b3, nminus = n - 1, dil, expand;
    register double *x, c;
    int nn = n >> 1;
    double t1, t2, t3, t4, t5;
    int n2, n4, n8, i, j;

    init_sincos(n);
    expand = cur_run / n;
    x = z - 1;
    n2 = n << 1;
    while (nn >>= 1) {
        is = 0;
        id = n2;
        n2 >>= 1;
        n4 = n2 >> 2;
        n8 = n4 >> 1;
        do {
            for (i = is; i < n; i += id) {
```

```
                i1 = i + 1;
                i2 = i1 + n4;
                i3 = i2 + n4;
                i4 = i3 + n4;
                t1 = x[i1] - x[i3];
                x[i1] += x[i3];
                x[i2] += x[i2];
                x[i3] = t1 - 2.0 * x[i4];
                x[i4] = t1 + 2.0 * x[i4];
                if (n4 == 1)
                    continue;
                i1 += n8;
                i2 += n8;
                i3 += n8;
                i4 += n8;
                t1 = (x[i2] - x[i1]) * SQRTHALF;
                t2 = (x[i4] + x[i3]) * SQRTHALF;
                x[i1] += x[i2];
                x[i2] = x[i4] - x[i3];
                x[i3] = -2.0 * (t2 + t1);
                x[i4] = 2.0 * (t1 - t2);
            }
            is = (id << 1) - n2;
            id <<= 2;
        } while (is < n - 1);
        dil = n / n2;
        a = dil;
        for (j = 2; j <= n8; j++) {
            a3 = (a + (a << 1)) & nminus;
            b = a * expand;
            b3 = a3 * expand;
            cc1 = s_cos(b);
            ss1 = s_sin(b);
            cc3 = s_cos(b3);
            ss3 = s_sin(b3);
            a = (a + dil) & nminus;
            is = 0;
            id = n2 << 1;
            do {
                for (i = is; i < n; i += id) {
                    i1 = i + j;
                    i2 = i1 + n4;
                    i3 = i2 + n4;
                    i4 = i3 + n4;
                    i5 = i + n4 - j + 2;
                    i6 = i5 + n4;
                    i7 = i6 + n4;
                    i8 = i7 + n4;
                    t1 = x[i1] - x[i6];
                    x[i1] += x[i6];
                    t2 = x[i5] - x[i2];
                    x[i5] += x[i2];
                    t3 = x[i8] + x[i3];
                    x[i6] = x[i8] - x[i3];
                    t4 = x[i4] + x[i7];
                    x[i2] = x[i4] - x[i7];
                    t5 = t1 - t4;
```

```
                    t1 += t4;
                    t4 = t2 - t3;
                    t2 += t3;
                    x[i3] = t5 * cc1 + t4 * ss1;
                    x[i7] = -t4 * cc1 + t5 * ss1;
                    x[i4] = t1 * cc3 - t2 * ss3;
                    x[i8] = t2 * cc3 + t1 * ss3;
                }
                is = (id << 1) - n2;
                id <<= 2;
            } while (is < n - 1);
        }
    }
    is = 1;
    id = 4;
    do {
        for (i0 = is; i0 <= n; i0 += id) {
            i1 = i0 + 1;
            e = x[i0];
            x[i0] = e + x[i1];
            x[i1] = e - x[i1];
        }
        is = (id << 1) - 1;
        id <<= 2;
    } while (is < n);
    scramble_real(z, n);
    c = 1 / (double)n;
    for (i = 0; i < n; i++)
        z[i] *= e;
}

void mul_hermitian(double *a, double *b, int n)
/* b becomes b*a in Hermitian representation. */
{
    int k, half = n >> 1;
    register double c, d, e, f;

    b[0] *= a[0];
    b[half] *= a[half];
    for (k = 1; k < half; k++) {
        c = a[k];
        d = b[k];
        e = a[n - k];
        f = b[n - k];
        b[n - k] = c * f + d * e;
        b[k] = c * d - e * f;
    }
}
```

fractalsound.c

```
/*
 * fractalsound.c
 *
 * [Requires NEXTSTEP sound utilities.]
 * Program to generate fractal sound...it sounds essentially the
 * same when played 1.40254 times slower!!
```

```
 *
 * Compile with:
 *
 * % cc fractalsound.c -o fractalsound
 *
 * Run with:
 *
 * % fractalsound file.snd
 *
 * whence file.snd will be the NEXTSTEP sound file to be played.
 */

#include <stdio.h>
#include <sound/sound.h>
#include <math.h>

#define NUM 32
#define F (8000.0)                /* NUM=32, F=8000 Hz. gives a=1.40254
                                   * invariant scale. */
#define SIZE (1<<21)
#define PI (3.14159265358979323846)
#define A exp(log(2*PI*F)/(double)NUM)

void main(int argc, char **argv)
{
    SNDSoundStruct *sound;
    short *sig;
    double tau = 1.0 / SND_RATE_LOW;
    int j, k, ndata;
    double arg, sum;

    SNDAlloc(&sound, SIZE, SND_FORMAT_LINEAR_16, SND_RATE_LOW, 1, 0);
    sig = (short *)sound;
    sig += (sound->dataLocation) / sizeof(short);
    ndata = (sound->dataSize) / sizeof(short);
    for (j = 0; j < ndata; j++) {
        sum = 0;
        arg = j * tau;
        for (k = 0; k < NUM; k++) {
            arg *= A;
            sum += sin(arg);
        }
        sum *= 10000 / (double)NUM;  /* Give signal some strength. */
        sig[j] = sum;

        /* Next, option to print out plottable data.
         *  if(j<2048) {
         *      printf("%d,",(int)sig[j]);
         *      if(j%8==0) printf("\n");
         *   }
         */

        /* Next, option to change sampling rate on-the-fly:
         * if(j%(ndata/4) == (ndata/4)-1) tau *= 0.707;
         */
    }
    SNDWriteSoundfile(argv[1], sound);
```

```
}
```

FSWT.ma

```
(* Fast Square Wave Transform.
   The returned data differ from [Pender and Covey 1992] by simple
   powers of 2.
 *)

fswt[x_] :=
    Block[{n, y, m, j, jj, jm, tmp},
        y = x;
        m = n = Length[x];
        While[(m = Floor[m/2]) > 0,
            For[j = n-2m, j < n-m, j++,
                jj = j + 1; jm = j + m + 1;
               {y[[jj]], y[[jm]]} = {y[[jj]]-y[[jm]],y[[jj]]+y[[jm]]};
            ];
            If[m<n/2,tmp = -y[[n-4m+1]];
               For[j = n-4m, j<n-2m-1, j++,y[[j+1]] -= y[[j+2]]];
               y[[n-2m]] -= tmp;
            ];
         ];
         Return[y];
    ];

x = {1,2,3,-4,5,6,7,-8,7,6,5,4,3,2,1,0};
Print[fswt[x]];
```

FWTInPlace.ma

```
(* Fast wavelet transform, in place.
 *)

hh[n_] := Block[{ },
             If[n==0,Return[(1+Sqrt[3])/(4*Sqrt[2])]];
             If[n==1,Return[(3+Sqrt[3])/(4*Sqrt[2])]];
             If[n==2,Return[(3-Sqrt[3])/(4*Sqrt[2])]];
             If[n==3,Return[(1-Sqrt[3])/(4*Sqrt[2])]];
             Return[0];
         ]

h[n_] := N[hh[n]]
g[n_] := (-1)^n h[3-n]
p[i_] := Mod[i,2]
q[i_] := Floor[(i+1)/2]

m = 4
n = 2^m

(* First, create a signal. *)
Do[s[q] = N[Sin[q/3] * Exp[-3*q/n]], {q,0,n-1}]

Print["signal = ",Table[s[q],{q,0,n-1}]]
```

```
(* Now do recursion, each time splitting the active array
into "s" and "d" parts. *)

For[r=0, r<m, r++,
        c = s[0]; d = s[2^r];
        For[k=0, k < n/2^(r+1)-1, k++,
                a = Sum[h[q] s[(q+2k) 2^r], {q,0,3}];
                b = Sum[g[q] s[(q+2k) 2^r], {q,0,3}];
                s[k 2^(r+1)] = a;
                s[k 2^(r+1) + 2^r] = b;
        ];
        k = n/2^(r+1)-1;
        a = Sum[h[q] s[(q+2k) 2^r], {q,0,1}] +
                h[2] c + h[3] d;
        b = Sum[g[q] s[(q+2k) 2^r], {q,0,1}] +
                g[2] c + g[3] d;
        s[k 2^(r+1)] = a;
        s[k 2^(r+1) + 2^r] = b;
]


Print["wtransform = ",Table[s[q],{q,0,n-1}]]


(* Now perform the inverse FWT. *)
For[r=m, r>=1, r--,
        c = s[n-2^r]; d = s[n-2^r+2^(r-1)];
        For[k=n/2^r-1, k > 0, k--,
          a = Sum[h[2q] s[(k-q) 2^r] +
                        g[2q] s[(k-q) 2^r + 2^(r-1)], {q,0,1}];
          b = Sum[h[2q+1] s[(k-q) 2^r] +
                        g[2q+1] s[(k-q) 2^r + 2^(r-1)], {q,0,1}];
          s[k 2^r] = a;
          s[k 2^r + 2^(r-1)] = b;
        ];
        a = h[0] s[0] + h[2] c + g[0] s[2^(r-1)] + g[2] d;
        b = h[1] s[0] + h[3] c + g[1] s[2^(r-1)] + g[3] d;
        s[0] = a;
        s[2^(r-1)] = b;
]

Print["reconstruction = ",Table[s[q],{q,0,n-1}]]
```

FWTNotInPlace.ma

```
(* Fast wavelet transform, not in place. *)

hh2[n_] := Block[{ },
           If[n==0,Return[3/(5 Sqrt[2])]];
           If[n==1,Return[6/(5 Sqrt[2])]];
           If[n==2,Return[2/(5 Sqrt[2])]];
           If[n==3,Return[-1/(5 Sqrt[2])]];
           Return[0];
        ]
h[n_] := N[hh2[n]]
g[n_] := (-1)^n h[3-n]
```

```
m = 5;
n = 2^m;

Do[s[q] = (-1)^Floor[q/2]; t[q] = 0, {q,0,n-1}];
tt = Table[s[q],{q,0,n-1}];
Print[tt]
(* ListPlot[tt,  PlotJoined->True, PlotRange->{-5,5}];   *)
dep = m;
len = n;
For[r=0, r<dep, r++,
        len = len/2;
        c = s[0]; d = s[1];
        For[j=0, j<len-1,j++,
          k = 2*j;
          a = h[0]*s[k]+h[1]*s[k+1]+h[2]*s[k+2]+h[3]*s[k+3];
          b = g[0]*s[k]+g[1]*s[k+1]+g[2]*s[k+2]+g[3]*s[k+3];
          s[j] = a;
          t[j] = b;
        ];
          j = len-1; k = 2*j;
          a = h[0]*s[k]+h[1]*s[k+1]+h[2]*c+h[3]*d;
          b = g[0]*s[k]+g[1]*s[k+1]+g[2]*c+g[3]*d;
          s[len-1] = a;
          t[len-1] = b;

        For[j=0, j<len, j++,
          s[j+len] = t[j];
        ];
        tt = Table[s[q],{q,0,n-1}];
        (* ListPlot[tt,  PlotJoined->True, PlotRange->{-5,5}];
        *)
]
Print[tt]
(* Next, inverse wavelet transform, starting from depth "dep." *)

len = n/(2^dep);
For[r=dep, r>0, r--,
        len = 2*len;
        For[j=0, j<len, j++,
            t[j] = s[j];
        ];
        a = s[0];
        b = s[len/2];
        c = s[len/2-1];
        d = s[len-1];
        s[1] = h[1]*a - h[2]*b + h[3]*c - h[0]*d;
        s[0] = h[0]*a + h[3]*b + h[2]*c + h[1]*d;
        For[j=len-1; k = len/2-1; i = j, j>1,j -= 2,
          a = t[k];
          b = t[i];
          c = t[k-1];
          d - t[i-1];
          --i; --k;
          s[j] = h[1]*a - h[2]*b + h[3]*c - h[0]*d;
          s[j-1] = h[0]*a + h[3]*b + h[2]*c + h[1]*d;
        ];
```

```
        tt = Table[s[q],{q,0,n-1}];
        (*ListPlot[tt,  PlotJoined->True, PlotRange->{-5,5}];
        *)
]
Print[tt];
```

fwt.c

```
/*
 * fwt.c
 *
 * Routines for fast wavelet transform.
 *
 * FWVT_2(x, n, dn): Order-2 (Haar) wavelet transform, on array x,
 *                   length n, through dn recursive levels.
 * IFWVT_2(): Inverse Haar.
 * FWVT_4(): Order-4 transform, using coefficients H0-H3 from fwt.h.
 * IFWVT_4(): Inverse order-4.
 * FWVT_6(): Order-6 transform, using coefficients A0-A5 from fwt.h.
 * IFWVT_6(): Inverse order-6.
 *
 * FWVT2D_2(x,w,h,dn): Order-2, 2-dimensional transform, on array x,
 *                     width w by height h, through dn levels.
 * IFWVT2D_2(x,w,h,dn): Inverse order-2, 2-dimensional
 * FWVT2D_4(x,w,h,dn): Order-4, 2-dimensional
 * IFWVT2D_4(x,w,h,dn): Inverse order-4, 2-dimensional
 * FWVT2D_6(x,w,h,dn): Order-6, 2-dimensional
 * IFWVT2D_6(x,w,h,dn): Inverse order-6, 2-dimensional
 *
 * The inverse 2-dimensional transforms have some optimizations,
 * namely pointer arithmetic and loop unrolling; though more
 * enhancements are possible.  One major optimization is to normalize
 * out the HSUM by shifting, or by redefining the transform itself.
 */

#include "fwt.h"
#include <stdlib.h>
#include <stdio.h>
#include <math.h>

static DATA_TYPE *aux = NULL;
static DATA_TYPE *column = NULL;
int curaux = 0;


void init_aux(w, h)
    int w, h;
{
    int max = (w > h) ? w : h;

    if (max > curaux) {
        curaux = max;
        if (!aux)
            aux = (DATA_TYPE *) malloc(sizeof(DATA_TYPE) * max / 2);
```

```
        else
            aux = (DATA_TYPE *) realloc(aux,
                                        sizeof(DATA_TYPE) * max / 2);
        if (!column)
            column = (DATA_TYPE *) malloc(sizeof(DATA_TYPE) * max);
        else
            column = (DATA_TYPE *) realloc(column,
                                        sizeof(DATA_TYPE) * max);
    }
}

void FWVT_2(x, n, dn)
    DATA_TYPE *x;
    int n, dn;
{
    int len = n, j, k;
    DATA_TYPE a, b;
    short r;

    init_aux(n, n);
    for (r = 0; r < dn; r++) {
        len >>= 1;
        for (j = 0; j < len; j++) {
            k = j << 1;
            a = rint((x[k] + x[k + 1]) * 0.5);
            b = rint((x[k] - x[k + 1]) * 0.5);
            x[j] = a;
            aux[j] = b;
        }
        for (j = 0; j < len; j++)
            x[j + len] = aux[j];
    }
}

void IFWVT_2(x, n, dn)
    DATA_TYPE *x;
    int n, dn;
{
    DATA_TYPE *u, *v, d, k;
    int len = n >> dn;
    short r;
    int i;

    init_aux(n, n);
    for (r = dn; --r != -1;) {
        len <<= 1;
        bcopy(x, aux, len << 1);
        i = len >> 1;
        u = aux;
        v = x + i;
        for (; --i != -1;) {
            k = *u++;
            d = *v++;
            *x++ = k + d;
            *x++ = k - d;
        }
        x -= len;
    }
```

```
}

void FWVT_4(x, n, dn)
    DATA_TYPE *x;
    int n, dn;
{
    int len = n, a, b, c, d, j, k, r;

    init_aux(n, n);
    for (r = 0; r < dn; r++) {  /* for each recursive depth */
        len >>= 1;
        /* store first 2 values for matrix wraparound */
        c = x[0];
        d = x[1];
        for (j = 0; j < len - 1; j++) {
            k = j << 1;

            /* do matrix multiplication */
            a = rint(H0 * x[k] + H1 * x[k + 1] +
                     H2 * x[k + 2] + H3 * x[k + 3]);
            b = rint(H3 * x[k] - H2 * x[k + 1] +
                     H1 * x[k + 2] - H0 * x[k + 3]);
            x[j] = a;
            aux[j] = b;
        }

        j = len - 1;
        k = j << 1;

        /* do wraparound calculation */
        a = rint(H0 * x[k] + H1 * x[k + 1] +
                 H2 * c + H3 * d);
        b = rint(H3 * x[k] - H2 * x[k + 1] +
                 H1 * c - H0 * d);
        x[len - 1] = a;
        aux[len - 1] = b;

        /* tack second half onto vector */
        for (j = 0; j < len; j++)
            x[j + len] = aux[j];
    }
}

void IFWVT_4(x, n, dn)      /* 4th-order inverse wavelet transform */
    int *x;
    int n, dn;
{
    int k, i;
    int x0, x1;
    int len = n >> (dn - 1), r, hlen;
    int a, b, c, d;
    int *u, *v;


    init_aux(n, n);
    for (r = dn; --r != -1;) {
        hlen = len >> 1;        /* hlen = len/2 */
```

```
        bcopy(x, aux, len * sizeof(DATA_TYPE)); /* copy x to aux */
        k = hlen - 1;
        i = len - 1;
        a = aux[0];
        b = aux[k];
        c = aux[hlen];
        d = aux[i];
        x1 = rint((H1 * a + H3 * b - H2 * c - H0 * d) / HSUM);
        x0 = rint((H0 * a + H2 * b + H3 * c + H1 * d) / HSUM);
        /* increment pointer to array to last element */
        x += i;
        u = aux + k;            /* set up pointers to copy of data */
        v = aux + i;
        --i;
        for (i >>= 1; --i != -1;) {
            b = *u;             /* get values from copy */
            a = *--u;
            d = *v;
            c = *--v;
            /* calc. values */
            *x-- = rint((H3 * a + H1 * b - H0 * c - H2 * d) / HSUM);
            *x-- = rint((H2 * a + H0 * b + H1 * c + H3 * d) / HSUM);
        }
        *x-- = x1;
        *x = x0;
        /* len = len*2 => expand len for next level */
        len <<= 1;

    }
}

void FWVT_6(x, n, dn)           /* 6th-order wavelet transform */
    int *x, n, dn;
{
    int a, b, c, d, e, f;
    int r, len, j, k;


    init_aux(n, n);

    len = n;

    for (r = 0; r < dn; r++) {
        /* save first four values for wraparound */
        c = x[0];
        d = x[1];
        e = x[2];
        f = x[3];

        len >>= 1;

        for (j = 0; j < len - 2; j++) {
            k = j << 1;
            a = rint(A0 * x[k] + A1 * x[k + 1] + A2 * x[k + 2] +
                     A3 * x[k + 3] + A4 * x[k + 4] + A5 * x[k + 5]);
            b = rint(A5 * x[k] - A4 * x[k + 1] + A3 * x[k + 2] -
                     A2 * x[k + 3] + A1 * x[k + 4] - A0 * x[k + 5]);
```

```
            x[j] = a;
            aux[j] = b;
        }

        j = len - 2;
        k = j << 1;
        a = rint(A0 * x[k] + A1 * x[k + 1] + A2 * x[k + 2] +
                 A3 * x[k + 3] + A4 * c + A5 * d);
        b = rint(A5 * x[k] - A4 * x[k + 1] + A3 * x[k + 2] -
                 A2 * x[k + 3] + A1 * c - A0 * d);
        x[j] = a;
        aux[j] = b;

        j = len - 1;
        k = j << 1;
        a = rint(A0 * x[k] + A1 * x[k + 1] + A2 * c + A3 * d +
                 A4 * e + A5 * f);
        b = rint(A5 * x[k] - A4 * x[k + 1] + A3 * c - A2 * d +
                 A1 * e - A0 * f);
        x[j] = a;
        aux[j] = b;

        for (j = 0; j < len; j++)
            x[j + len] = aux[j];
    }
}


void IFWVT_6(x, n, dn)       /* 6th-order inverse wavelet transform */
    int *x, n, dn;
{
    int k, i;
    int x0, x1, x2, x3;
    int len, r, hlen;
    int a, b, c, d, e, f;
    int *u, *v, *tmp;


    init_aux(n, n);
    tmp = x;

    len = n >> (dn - 1);
    for (r = dn; --r != -1;) {
        hlen = len >> 1;
        bcopy(x, aux, len * sizeof(int));

        k = hlen - 1;
        i = len - 1;

        a = aux[0];
        b = aux[k - 1];
        c = aux[k];
        d = aux[hlen];
        e = aux[i - 1];
        f = aux[i];
        x1 = rint((A1 * a + A5 * b + A3 * c - A4 * d - A0 * e -
                   A2 * f) / ASUM);
```

```
        x0 = rint((A0 * a + A4 * b + A2 * c + A5 * d + A1 * e +
                   A3 * f) / ASUM);

        b = aux[1];
        e = aux[hlen + 1];
        x3 = rint((A3 * a + A1 * b + A5 * c - A2 * d - A4 * e -
                   A0 * f) / ASUM);
        x2 = rint((A2 * a + A0 * b + A4 * c + A3 * d + A5 * e +
                   A1 * f) / ASUM);

        x += i;
        u = aux + k;
        v = aux + i;
        i -= 2;
        for (i >>= 1; --i != -1;) {
            a = *(u - 2);
            b = *(u - 1);
            c = *u--;
            d = *(v - 2);
            e = *(v - 1);
            f = *v--;

            *x-- = rint((A5 * a + A3 * b + A1 * c - A0 * d -
                         A2 * e - A4 * f) / ASUM);
            *x-- = rint((A4 * a + A2 * b + A0 * c + A1 * d +
                         A3 * e + A5 * f) / ASUM);
        }

        *x   = x3;
        *x-- = x2;
        *x-- = x1;
        *x = x0;
        len <<= 1;
    }
}

void
FWVT2D_2(x, w, h, dn)
    DATA_TYPE *x;
    int w, h, dn;
{   int j,k,ct;
    int hh = h, ww = w;
    init_aux(w, h);
    for(ct=0; ct<dn; ct++) {
        for(j=0; j<hh; j++) {
                FWVT_2(x+w*j, ww, 1);
        }
        for(j=0; j<ww; j++) {
            for(k=0; k<hh; k++) column[k] = x[j+w*k];
            FWVT_2(column, hh, 1);
            for(k=0; k<hh; k++) x[j+w*k] = column[k];
        }
        ww >>= 1; hh >>= 1;
    }
}

void
```

```
IFWVT2D_2(x, w, h, dn)
/* Showing some pointer speedups and loop unrolling. */
    DATA_TYPE *x;
    int w, h, dn;
{   int j, ct;
    DATA_TYPE *xx, *cc;
    short k;
    int ww = (w>>dn), hh = (h>>dn);
    init_aux(w, h);
    for(ct = 0; ct < dn; ct++) {
        for(j=0; j<ww; j++) {
            xx = x+j;
            cc = column;
            for(k=hh>>1; --k != -1;) {
                *cc++ = *xx;
                xx += w;
                *cc++ = *xx;
                xx += w;
            }
            IFWVT_2(column, hh, 1);
            xx = x+j;
            cc = column;
            for(k=hh>>1; --k != -1;) {
                *xx = *cc++;
                xx += w;
                *xx = *cc++;
                xx += w;
            }
        }
        for(j=0; j<hh; j++) {
            IFWVT_2(x+w*j, ww, 1);
        }
        ww <<= 1;
        hh <<= 1;
    }
}

void
FWVT2D_4(x, w, h, dn)
    DATA_TYPE *x;
    int w, h, dn;
{   int j,k,ct;
    int hh = h, ww = w;
    init_aux(w, h);
    for(ct=0; ct<dn; ct++) {
        for(j=0; j<hh; j++) {
            FWVT_4(x+w*j, ww, 1);
        }
        for(j=0; j<ww; j++) {
            for(k=0; k<hh; k++) column[k] = x[j+w*k];
            FWVT_4(column, hh, 1);
            for(k=0; k<hh; k++) x[j+w*k] = column[k];
        }
        ww >>= 1; hh >>= 1;
    }
}
```

```
void
IFWVT2D_4(x, w, h, dn)
/* Showing some pointer speedups and loop unrolling. */
    DATA_TYPE *x;
    int w, h, dn;
{   int j, ct;
    DATA_TYPE *xx, *cc;
    short k;
    int ww = (w>>dn), hh = (h>>dn);
    init_aux(w, h);
    for(ct = 0; ct < dn; ct++) {
        for(j=0; j<ww; j++) {
            xx = x+j;
            cc = column;
            for(k=hh>>1; --k != -1;) {
                *cc++ = *xx;
                xx += w;
                *cc++ = *xx;
                xx += w;
            }
            IFWVT_4(column, hh, 1);
            xx = x+j;
            cc = column;
            for(k=hh>>1; --k != -1;) {
                *xx = *cc++;
                xx += w;
                *xx = *cc++;
                xx += w;
            }
        }
        for(j=0; j<hh; j++) {
            IFWVT_4(x+w*j, ww, 1);
        }
        ww <<= 1;
        hh <<= 1;
    }
}

void
FWVT2D_6(x, w, h, dn)
    DATA_TYPE *x;
    int w, h, dn;
{   int j,k,ct;
    int hh = h, ww = w;
    init_aux(w, h);
    for(ct=0; ct<dn; ct++) {
        for(j=0; j<hh; j++) {
                FWVT_6(x+w*j, ww, 1);
        }
        for(j=0; j<ww; j++) {
            for(k=0; k<hh; k++) column[k] = x[j+w*k];
            FWVT_6(column, hh, 1);
            for(k=0; k<hh; k++) x[j+w*k] = column[k];
        }
        ww >>= 1; hh >>= 1;
   }
}
```

```
void
IFWVT2D_6(x, w, h, dn)
/* Showing some pointer speedups and loop unrolling. */
    DATA_TYPE *x;
    int w, h, dn;
{   int j, ct;
    DATA_TYPE *xx, *cc;
    short k;
    int ww = (w>>dn), hh = (h>>dn);
    init_aux(w, h);
    for(ct = 0; ct < dn; ct++) {
        for(j=0; j<ww; j++) {
            xx = x+j;
            cc = column;
            for(k=hh>>1; --k != -1;) {
                *cc++ = *xx;
                xx += w;
                *cc++ = *xx;
                xx += w;
            }
            IFWVT_6(column, hh, 1);
            xx = x+j;
            cc = column;
            for(k=hh>>1; --k != -1;) {
                *xx = *cc++;
                xx += w;
                *xx = *cc++;
                xx += w;
            }
        }
        for(j=0; j<hh; j++) {
            IFWVT_6(x+w*j, ww, 1);
        }
        ww <<= 1;
        hh <<= 1;
    }
}
```

fwt.h

```
/*
 * fwt.h
 *
 * Fast wavelet transform header file.
 *
 *
 *
 */

#define DATA_TYPE       int

/* Daubechies 4th order Wavelet */
#define H0 ((1+sqrt(3.0))/4)
#define H1 ((3+sqrt(3.0))/4)
#define H2 ((3-sqrt(3.0))/4)
```

```
#define H3 ((1-sqrt(3.0))/4)


/* Daubechies 6th order wavelet - evaluated by [Russell 1992] */
#define A0      0.3326705529500825
#define A1      0.806891509311092
#define A2      0.4598775021184915
#define A3      (-0.1350110200102545)
#define A4      (-0.0854412738820267)
#define A5      0.03522629188570953

#define HSUM (H0*H0 + H1*H1 + H2*H2 + H3*H3)
#define ASUM (A0*A0 + A1*A1 + A2*A2 + A3*A3 + A4*A4 + A5*A5)
```

Genetic.ma

```
(* Genetic algorithm experiment.

   First we establish the array "path" such that

   path[[q]]

   is the triple {x,y,z} at Hilbert curve coordinate q.
   This first part of the code is lifted from "Hilbert.ma"
   which also shows how to graph the curve if desired.
 *)

my[t_] := {{Cos[t],0,Sin[t]},{0,1,0},{-Sin[t],0,Cos[t]}};
mz[t_] := {{Cos[t],-Sin[t],0},{Sin[t],Cos[t],0},{0,0,1}};
roty[p_, t_] := Table[my[t] . p[[q]],{q,1,Length[p]}];
rotz[p_, t_] := Table[mz[t] . p[[q]],{q,1,Length[p]}];
dil[pat_,d_] := d * pat;
tra[pat_,vec_] := Table[pat[[q]] + vec, {q,1,Length[pat]}];

path =
{{0,0,0},{1,0,0},{1,0,1},{0,0,1},{0,1,1},{1,1,1},{1,1,0},{0,1,0}};

maxdepth = 3;  (* Curve coords will run from 0 through
                  2^(3*(maxdepth+1)-1
                  inclusive. *)
Do[
        small = (2^depth-1)/(2^(depth+1)-1);
        large = (2^depth)/(2^(depth+1)-1);
        piece = dil[path, small];
        path = rotz[roty[piece,-Pi/2],-Pi/2];
        path = Join[path,
tra[roty[rotz[piece,Pi/2],Pi/2],{large,0,0}]];
        path = Join[path,
tra[roty[rotz[piece,Pi/2],Pi/2],{large,0,large}]];
        path = Join[path, tra[roty[piece,Pi],{small,0,1}]];
        path = Join[path, tra[roty[piece,Pi],{small,large,1}]];
        path = Join[path, tra[roty[rotz[piece,-
Pi/2],Pi/2],{large,1,1}]];
        path = Join[path, tra[roty[rotz[piece,-
Pi/2],Pi/2],{large,1,small}]];
```

```
        path = Join[path, tra[rotz[roty[piece,-
Pi/2],Pi/2],{small,1,0}]];
        ,{depth,1,maxdepth}
]

(* Next, the function to be maximized via genetic algorithm. *)
fitness[xyz_List] := Block[{x,y,z},
                x = xyz[[1]]; y = xyz[[2]]; z = xyz[[3]];
                Return[N[Exp[-(x-1/3)^2 - (y-1/5)^2 - (z-1/3)^2]]];
                ];
maxgene = 2^(3*(maxdepth+1)-1;  (* Maximum coordinate of space-filling
curve. *)
mutationrate = 32;
adults = 32;
survived = 8;
offspring = adults/survived;  (* Retain 8, each of which will have 4
offspring.  *)
genecoords = Table[Random[Integer,maxgene],{q,1,adults}];
Do[
  geneexpressions = Table[path[[genecoords[[q]]]],{q,1,adults}];
  fitnesses = Map[fitness, geneexpressions];
  parents = Take[Sort[fitnesses], {Length[fitnesses]-survived+1,
Length[fitnesses]}];
  Print[N[parents,3]];  (* Fitnesses of current parents. *)
  parentgenes = Table[genecoords[[
                    Position[fitnesses,parents[[q]]][[1]][[1]]
                  ]],
                  {q,1,survived}
              ];
  genecoords = {};
  Do[
       newgene = parentgenes[[Floor[1+(q-1)/offspring]]] +
               Random[Integer,{-mutationrate, mutationrate}];
       newgene = Max[0,Min[newgene,maxgene]];
       genecoords = Append[genecoords, newgene],
       {q,1,adults}

  ],
  {generation,1,6}
];
(* Next, output the xyz locations of the current adults. *)
MatrixForm[geneexpressions];
```

giants.c - on companion disk only

giants.h

```
#define MAX_SHORTS ((1<<20)+(1<<8))
#define INFINITY (-1)
#define FALSE 0
#define TRUE 1
#define COLUMNWIDTH 64
#define DIVIDEBYZERO  1         /* error codes */
#define OVERFLOW      2
#define SIGN          3
```

```
#define OVERRANGE     4
#define AUTO_MUL 0
#define GRAMMAR_MUL 1
#define FFT_MUL 2
#define USE_ASSEMBLER_MUL 1
#define TWOPI (2 * 3.14159265358979323846)
#define SQRT2 1.41421356237309504880
#define SQRTHALF 0.70710678118654752440
#define TWO16 65536.0
#define TWOM16 0.0000152587890625
#define MAX_DIGITS 10000          /* Decimal digit ceiling in I/O
                                   * routines. */
#define BREAK_SHORTS 400          /* Number of shorts at which FFT
                                   * breaks over. */
#define min(a,b) ((a)<(b)? (a) : (b))
#define max(a,b) ((a)>(b)? (a) : (b))
#define STEPS 32                  /* Maximum number of recursive steps
                                   * needed to calculate gcds of
                                   * integers */
#define GCDLIMIT 5000             /* The limit below which hgcd is too
                                   * ponderous */
#define INTLIMIT  31              /* The limit below which ordinary
                                   * ints will be used */
#define gin(x)   gread(x,stdin)
#define gout(x)  gwriteln(x,stdout)

typedef struct _giant {
    int sign;                     /* number of shorts = abs(sign) */
    unsigned short n[1];
} *giant;

typedef struct _gmatrix {
    giant ul, ur, ll, lr;         /* upper left, upper right, etc. */
} *gmatrix;

typedef struct {
    double re, im;
} complex;

/*
 * Creates a new giant, numshorts = INFINITY invokes the maximum
 * MAX_SHORTS.
 */
giant newgiant(int numshorts);

/* Returns the bit-length n; e.g. n=7 returns 3. */
int bitlen(giant n);

/* Returns the value of the pos bit of n. */
int bitval(giant n, int pos);

int iszero(giant g);              /* Returns whether g is zero. */
int isone(giant g);               /* Returns whether g is one. */

/* Copies one giant to another. */
void gtog(giant src, giant dest);
```

```
/* Gives a giant an int value. */
void itog(int n, giant g);

/* Returns the sign of g: -1, 0, 1. */
int gsign(giant g);

/* Returns 1, 0, -1 as a>b, a=b, a<b. */
int gcompg(giant a, giant b);

/* g := g/n, and (g mod n) is returned. */
int idivg(int n, giant g);

/*
 * Output the giant in decimal, with neither '\'-newline notation at
 * right terminal margin, nor a final newline.
 */
void gwrite(giant g, FILE * fp, int newlines);

/*
 * Output the giant in decimal, with both '\'-newline
 * notation and a final newline.
 */
void gwriteln(giant g, FILE * fp);

/*
 * Input the giant in decimal, assuming the formatting of 'gwriteln'.
 */
void gread(giant g, FILE * fp);

void negg(giant g);              /* g := -g. */
void addg(giant a, giant b);     /* b += a. */
void subg(giant a, giant b);     /* b -= a. */

/* g += i, with i non-negative and < 2^16. */
void iaddg(int i, giant g);

/*
 * If 1/x exists (mod n), 1 is returned and x := 1/x.  If inverse
 * does not exist, 0 is returned and x := GCD(n, x).
 */
int invg(giant n, giant x);

/*
 * Stable-divide version of invg, init_divide() should be called
 * first if the denominator is new.
 */
int stableinvg(giant n, giant x);
/* Same as stableinvg, except uses binary method. */
int stablebinvg(giant p, giant x);

/* u becomes greatest power of two not exceeding u/v. */
void bdivg(giant v, giant u);
/* Same as invg, but uses bdivg. */
int binvg(giant n, giant x);

void cgcdg(giant n, giant x);    /* Classical GCD, x:= GCD(n, x). */
void gcdg(giant n, giant x);     /* General GCD, x:= GCD(n, x). */
```

```
void bgcdg(giant n, giant x);    /* Binary GCD, x:= GCD(n, x). */

/* g := m^n, no mod is performed. */
void powerg(int a, int b, giant g);

/* num := num mod den, any positive den. */
void modg(giant den, giant num);

/* num := |num|/den, any positive den. */
void divg(giant den, giant num);

/* same as modg but  reciprocal storage is invoked for den. */
void stablemodg(giant den, giant num);

/* same as divg but reciprocal storage is  invoked for den. */
void stabledivg(giant den, giant num);

/* x := x^n (mod z). */
void powermod(giant x, int n, giant z);
/* x := x^n (mod z). */
void powermodg(giant x, giant n, giant z);
/* x := x^n (mod 2^q+1). */
void fermatpowermod(giant x, int n, int q);
/* x := x^n (mod 2^q+1). */
void fermatpowermodg(giant x, giant n, int q);
/* x := x^n (mod 2^q-1). */
void mersennepowermod(giant x, int n, int q);
/* x := x^n (mod 2^q-1). */
void mersennepowermodg(giant x, giant n, int q);
/* g := g (mod 2^n+1). */
void fermatmod(int n, giant g);
/* g := g (mod 2^n-1). */
void mersennemod(int n, giant g);

void smulg(unsigned short s, giant g);  /* g *= s. */
void absg(giant g);                     /* g := |g|. */
void squareg(giant g);                  /* g *= g. */
void mulg(giant a, giant b);            /* b *= a. */

/* Shift g left by bits, introducing zeros on the right. */
void gshiftleft(int bits, giant g);
/* Shift g right by bits, losing bits on the right. */
void gshiftright(int bits, giant g);

/*
 * Set AUTO_MUL for automatic FFT crossover (this is the default),
 * set FFT_MUL for forced FFT multiply, set GRAMMAR_MUL for forced
 * grammar school multiply.
 */
void setmulmode(int mode);
```

Henon.ma

```
(* Henon attractor.  *)

a = 1.4; b = 0.3;
```

```
x = 1; y = 1;
(* Next, reject transients. *)
Do[{x,y} = {1 - a x^2 + y, b x}, {q,1,50}];
attractor = {{x,y}};
(* Next, fill an attractor list. *)
Do[
    {x,y} = {1 - a x^2 + y, b x};
    attractor = Append[attractor, {x,y}],
    {q,1,8000}
]
ListPlot[attractor]
```

Hilbert.ma

```
(* Hilbert fractals, plane- and space-filling;
   including option to draw book's frontispiece
   fractal structure.
 *)

(* Next, make a planar Hilbert curve. *)
m[t_] := {{Cos[t],-Sin[t]},{Sin[t],Cos[t]}};
rot[pat_, mat_] := Table[mat . pat[[q]],{q,1,Length[pat]}];
dil[pat_,d_] := d * pat;
tra[pat_,vec_] := Table[pat[[q]] + vec, {q,1,Length[pat]}];
rev[pat_] := Table[pat[[Length[pat]-q+1]],{q,1,Length[pat]}];

path = {{0,0},{1,0},{1,1},{0,1}};
Do[
        Show[Graphics[Line[N[path]]], AspectRatio->1];
        piece = dil[path, (2^depth-1)/(2^(depth+1)-1)];
        path = tra[rev[rot[piece, m[Pi/2]]],
                {(2^depth-1)/(2^(depth+1)-1),0}];
        path = Join[path, tra[rot[piece,m[0]],
                        {(2^depth)/(2^(depth+1)-1),0}]];
        path = Join[path, tra[rot[piece,m[0]],
                {(2^depth)/(2^(depth+1)-1),
                (2^depth)/(2^(depth+1)-1)}]];
        path = Join[path, tra[rev[rot[piece,m[-Pi/2]]],
                {0,1}]];
        ,{depth,1,6}
]

(* Next, generalize to 3 dimensions. *)
my[t_] := {{Cos[t],0,Sin[t]},{0,1,0},{-Sin[t],0,Cos[t]}};
mz[t_] := {{Cos[t],-Sin[t],0},{Sin[t],Cos[t],0},{0,0,1}};

roty[p_, t_] := Table[my[t] . p[[q]],{q,1,Length[p]}];
rotz[p_, t_] := Table[mz[t] . p[[q]],{q,1,Length[p]}];

dil[pat_,d_] := d * pat;
tra[pat_,vec_] := Table[pat[[q]] + vec, {q,1,Length[pat]}];

path =
{{0,0,0},{1,0,0},{1,0,1},{0,0,1},{0,1,1},{1,1,1},{1,1,0},{0,1,0}};
Show[Graphics3D[Line[N[path]]], AspectRatio->1, Boxed->False,
ViewPoint->{-2,-10,6},
```

```
Lighting->True];

Do[
        small = (2^depth-1)/(2^(depth+1)-1);
        large = (2^depth)/(2^(depth+1)-1);
        piece = dil[path, small];
        path = rotz[roty[piece,-Pi/2],-Pi/2];
        path = Join[path,
                    tra[roty[rotz[piece,Pi/2],Pi/2],{large,0,0}]];
        path = Join[path,
                    tra[roty[rotz[piece,Pi/2],Pi/2],{large,0,large}]];
        path = Join[path, tra[roty[piece,Pi],{small,0,1}]];
        path = Join[path, tra[roty[piece,Pi],{small,large,1}]];
        path = Join[path,
                    tra[roty[rotz[piece,-Pi/2],Pi/2],{large,1,1}]];
        path = Join[path,
                   tra[roty[rotz[piece,-Pi/2],Pi/2],{large,1,small}]];
        path = Join[path,
                    tra[rotz[roty[piece,-Pi/2],Pi/2],{small,1,0}]];
        Show[Graphics3D[Line[N[path]]],
              AspectRatio->1, Boxed->False, ViewPoint->{-2,-10,6}];
        ,{depth,1,2}
];

(* Next, make a Hilbert 'sausage'. *)
rad = 0.02; rad2 = 2*rad;
norm[t_List] := Sqrt[t[[1]]^2 + t[[2]]^2 + t[[3]]^2];
point[t_] := Block[{p,q,f},
               p = Floor[t] + 1;
               q = Ceiling[t] + 1;
               f = t-Floor[t]; f *= 4/3;
               Return[((1-f)(1+rad2)-rad2*f)* path[[p]] +
                (f*(1+rad2)-rad2*(1-f))*path[[q]]];
              ];
vw[t_] := Block[{low, high, tmp, vec, cross},
              low = Floor[t];
              high = Ceiling[t];
              If[high+1 > Length[path], high = Length[path]-1];
              vec = path[[high+1]]   path[[low+1]];
              vec /= norm[vec];
              tmp = {vec[[2]], -vec[[1]], 0};
              If[norm[tmp] == 0,
                  tmp = {vec[[3]], 0, -vec[[1]]}];
              tmp /= norm[tmp];
              cross = {vec[[2]] * tmp[[3]] - tmp[[2]]*vec[[3]],
                       vec[[3]] * tmp[[1]] - tmp[[3]]*vec[[1]],
                       vec[[1]] * tmp[[2]] - tmp[[1]]*vec[[2]]};
              Return[Join[{tmp}, {cross}]];
         ];
surf[t_, u_] := Block[{f,ve},
               f = point[t];
               ve = rad* vw[t];
               Return[f + ve[[1]]*Cos[u-Pi/4] + ve[[2]]*Sin[u-Pi/4]];
               ];

ParametricPlot3D[surf[w,u]   ,
```

```
                            {w,0.0001,Length[path]-1-0.0001,1/4},
                            {u,0,2 Pi,Pi/2}, Boxed->False,
                            ViewPoint->{1.5,-1.6,3},
                            LightSources->{
                                  {{1., 1., 1.}, RGBColor[0, 1, 0]},
                                  {{0., 1., 1.}, RGBColor[0, 0, 1]}}
];
```

huffman.c

```
/*
 * huffman.c
 *
 * Generator for Huffman codes of the command line arguments.
 * Courtesy [Buhler 1993].
 *
 * Compile with:
 *
 *.% cc huffman.c -o huffman
 *
 * % huffman p1 p2 p3 ...
 *
 * where the p_i are the symbol relative probabilities (whose sum
 * is normalized to unity herein).  The symbol-code allocation is
 * then printed out.
 *
 */

#include <stdio.h>
#include <stdlib.h>
#include <math.h>

#define MAXN   128
#define EMPTY  (-1)

struct node {
    double weight;              /* total weight of all leaves */
    int left;                   /* indices of children */
    int right;
} a[2 * MAXN];

void huff();                    /* merge the two smallest nodes */
void print(int, char *);        /* print nodes */

int ntrees;                     /* number of trees in the forest */
int avail;                      /* next available node */
char code[MAXN + 1];            /* the codeword; used by print */
double sum = 0.0;               /* the sum of initial weights; used
                                 * in print */

int main(int argc, char **argv)
{
    int i;

    if (argc == 1) {
        fprintf(stderr, "Usage: %s p1 p2 ...\n", argv[0]);
        fprintf(stderr, "to see Huffman codes for p1, p2, ...\n");
```

```
        return 1;
    }
    avail = ntrees = argc - 1;
    for (i = 0; i < ntrees; i++) {
        sum += a[i].weight = atof(argv[i + 1]);
        a[i].left = a[i].right = EMPTY;
    }
    for (; ntrees > 1; --ntrees, avail++)
        huff();
    print(0, code);
    return 0;
}

/*
 * Next, Merge the two smallest nodes, leaving the current ntrees
 * trees in the first ntrees positions of the table
 */

#define swap(i,j) t = a[i], a[i] = a[j], a[j] = t

void huff(void)
{
    int i, m1, m2;
    double min1, min2, x;
    struct node t;

    if (a[0].weight > a[1].weight)
        swap(0, 1);
    m1 = 0, min1 = a[0].weight;
    m2 = 1, min2 = a[1].weight;
    for (i = 2; i < ntrees; i++) {        /* find the two smallest
                                           * nodes */
        x = a[i].weight;
        if (x < min1) {
            min2 = min1, min1 = x, m2 = m1, m1 = i;
        } else if (x < min2) {
            min2 = x, m2 = i;
        }
    }
    if (m1 == ntrees - 1) {         /* ensure that m1 is in
                                     * a[0...ntrees-2] */
        i = m1, m1 = m2, m2 = i;
    }
    swap(avail, m1);
    swap(ntrees - 1, m2);
    a[m1].weight = a[avail].weight + a[ntrees - 1].weight;
    a[m1].left = avail;
    a[m1].right = ntrees - 1;
}

/*
 * Next, recursively print the codewords of the nodes of the tree
 * whose root is in a[n]; the second argument is a pointer to the end
 * of the code word so far.
 */

void print(int n, char *end)
```

```
{
    if (a[n].left == EMPTY) {    /* we've hit a leaf */
        printf("%8g: %.*s\n", a[n].weight / sum, end - code, code);
    } else {
        *end = '0';
        print(a[n].left, end + 1);
        *end = '1';
        print(a[n].right, end + 1);
    }
}
```

IFS.ma

```
(* Fractal fern and fractal tree via
   Iterated Function System (IFS).
 *)

(* Next, an IFS fractal fern. *)

dil = {{0.0, 0.0, 0.0, 0.16},
        {0.85, -0.04, 0.04, 0.85},
        {0.2, -0.26, 0.23, 0.22},
        {-0.15, 0.28, 0.26, 0.24}};
tra = {{ 0.0, 0.0},
         { 0.0, 1.6},
         { 0.0, 1.6},
         { 0.0, 0.44}};
prob  = {0.01, 0.65, 0.07, 0.27};
cum = Table[Sum[prob[[q]],{q,1,s}],{s,1,Length[prob]-1}];

fern = {};
{x,y} = {-1,-1};
Do[
        ran = Random[Real,1];
        Which[ran<cum[[1]], ind=1,
              ran<cum[[2]], ind=2,
              ran<cum[[3]], ind=3,
              ran<1, ind=4];
        mm = dil[[ind]];
        {x,y} = {mm[[1]]*x + mm[[2]]*y, mm[[3]]*x + mm[[4]]*y};
        {x,y} += tra[[ind]];
        If[q>10,fern = Append[fern, {x,y}]],
        {q,1,8000}
]
ListPlot[fern, PlotRange->{0,10}, AspectRatio->Automatic,
        Axes->None]

(* Fractal fern and fractal tree via
   Iterated Function System (IFS).
 *)

(* Next, an IFS fractal fern. *)

dil = {{0.0, 0.0, 0.0, 0.16},
        {0.85, -0.04, 0.04, 0.85},
        {0.2, -0.26, 0.23, 0.22},
```

```
        {-0.15, 0.28, 0.26, 0.24}};
tra = {{ 0.0, 0.0},
         { 0.0, 1.6},
         { 0.0, 1.6},
         { 0.0, 0.44}};
prob  = {0.01, 0.65, 0.07, 0.27};
cum = Table[Sum[prob[[q]],{q,1,s}],{s,1,Length[prob]-1}];

fern = {};
{x,y} = {-1,-1};
Do[
        ran = Random[Real,1];
        Which[ran<cum[[1]], ind=1,
              ran<cum[[2]], ind=2,
              ran<cum[[3]], ind=3,
              ran<1, ind=4];
        mm = dil[[ind]];
        {x,y} = {mm[[1]]*x + mm[[2]]*y, mm[[3]]*x + mm[[4]]*y};
        {x,y} += tra[[ind]];
        If[q>10,fern = Append[fern, {x,y}]],
        {q,1,8000}
]
ListPlot[fern, PlotRange->{0,10}, AspectRatio->Automatic,
        Axes->None]
```

Invariant.ma

```
(* Plot of the invariant set of the extremal quadratic
   map.
 *)

num = 256;
dil = 32;
a = 3.4;
f[x_] := 4.0 x (1-x);

particles = Table[1.0*q/(dil*num),{q,0,dil*num}];
Do[
        hist[q] = Table[0,{r,0,num-1}];
        his = hist[q];
        Do[i = Floor[particles[[r]]*(num-1)+0.5];
                his[[i+1]] += 1,
                {r,1,dil*num+1}
        ];
        hist[q] = his;
        particles = Map[f, particles];
        ,
        {q,0,7}
];

f[x_,y_] := hist[Floor[y]][[Floor[(num-1)*x]+1]];
DensityPlot[f[x,y],{x,1/(2*num),1},{y,0,7.999},
        Axes->Automatic, Mesh->False,PlotPoints->num];
```

inverse.c

```
/*
 * inverse.c
 *
 * Matrix inversion test using Gauss pivoting, adapted from
 * [Press et. al. 1988]
 *
 * Compile with:
 *
 * % cc inverse.c -o inverse
 *
 * Run with:
 *
 * % inverse N <print>
 *
 * which creates then inverts a random N-by-N matrix.
 * If a second arg (any int) is entered, matrices will be output
 * in Mathematica format.
 *
 */

#include <stdlib.h>
#include <stdio.h>
#include <math.h>

#define DEN (2147483648.0)
#define SWAP(a,b) {double temp=(a);(a)=(b);(b)=temp;}
#define MAX 1250

static double *a[MAX + 1], b[MAX + 1];

void solve(a, b, n)
    double **a;
    double *b;
    int n;
{
    register int *indxc, *indxr, *ipiv;
    int i, icol, irow, j, k, l, ll;
    register double tabs, big, dum, pivinv;

    indxc = (int *)malloc(sizeof(int) * (n + 1));
    indxr = (int *)malloc(sizeof(int) * (n + 1));
    ipiv = (int *)malloc(sizeof(int) * (n + 1));
    for (j = 1; j <= n; j++)
        ipiv[j] = 0;
    for (i = 1; i <= n; i++) {
        big = 0.0;
        for (j = 1; j <= n; j++)
            if (ipiv[j] != 1)
                for (k = 1; k <= n; k++) {
                    if (ipiv[k] == 0) {
                        tabs = fabs(a[j][k]);
                        if (tabs >= big) {
                            big = tabs;
                            irow = j;
                            icol = k;
```

```
                        }
                    } else if (ipiv[k] > 1) {
                        fprintf(stderr, "Sing. mat.\n");
                        exit(1);
                    }
                }
        ++(ipiv[icol]);
        if (irow != icol) {
            for (l = 1; l <= n; l++)
                SWAP(a[irow][l], a[icol][l]);
            SWAP(b[irow], b[icol]);
        }
        indxr[i] = irow;
        indxc[i] = icol;
        if (a[icol][icol] == 0) {
            fprintf(stderr, "Sing. mat.\n");
            exit(2);
        }
        pivinv = 1.0 / a[icol][icol];
        a[icol][icol] = 1.0;
        for (l = 1; l <= n; l++)
            a[icol][l] *= pivinv;
        b[icol] *= pivinv;
        for (ll = 1; ll <= n; ll++)
            if (ll != icol) {
                dum = a[ll][icol];
                a[ll][icol] = 0.0;
                for (l = 1; l <= n; l++)
                    a[ll][l] -= a[icol][l] * dum;
                b[ll] -= b[icol] * dum;
            }
    }
    for (l = n; l >= 1; l--) {
        if (indxr[l] != indxc[l])
            for (k = 1; k <= n; k++)
                SWAP(a[k][indxr[l]], a[k][indxc[l]]);
    }
    free(indxc);
    free(indxr);
    free(ipiv);
}

void printmat(a, n)
    double **a;
    int n;
{
    int j, k;

    printf("{");
    for (j = 1; j <= n; j++) {
        printf("{");
        for (k = 1; k <= n; k++) {
            printf("%f", a[j][k]);
            if (k < n)
                printf(",");
            else
                printf("}");
```

```
        }
        if (j < n)
            printf(",");
        else
            printf("}");
    }
    printf("\n");
}


void main(argc, argv)
    int argc;
    char **argv;
{
    int n = atoi(argv[1]), j, k;

    for (j = 1; j <= n; j++)
        a[j] = (double *)malloc(sizeof(double) * (1 + MAX));
    for (j = 1; j <= n; j++) {
        b[j] = 1.0;
        for (k = 1; k <= n; k++) {
            a[j][k] = random() / DEN;
        }
    }
    if (argc > 2)
        printmat(a, n);
    solve(a, b, n);
    if (argc > 2) {
        printf("\n\n");
        printmat(a, n);
    }
}
```

Julia.ma

```
(* Filled Julia set plot.
   The Breaking criterion is not exact, but
   the checking of Re and Im parts is relatively
   fast.
 *)

c = 0.3 - 0.5I;
escape = 80;
julia[z_, c_] := Block[{w, q},
                 w = N[z];
                 q = 1;
                 While[q < escape,
                         w = w^2 + c;
                         If[Re[w]>2, Break[],
                              If[Im[w]>2, Break[]] ];
                         ++q;
                 ];
                 Return[q];
                ];
DensityPlot[-julia[x + I y,c],{x,-0.9,0.9},{y,-1.1,1.1},
        PlotPoints->128, Mesh->False, Axes->None]
```

Lorenz.ma

```
(* Lorenz attractor, by simple Eulerian
   iteration.  *)

s = 10.0; b = 8/3.0; r = 28.0;
dt = 0.02;
x = 0; y = 3; z = 0;
attractor = {};
Do[
        vx = s (y-x);
        vy = r x - y - x z;
        vz = x y - b z;
        x += vx*dt;
        y += vy*dt;
        z += vz*dt;

        If[q>100,attractor = Append[attractor, {x,y}]],
        {q,1,800}
]
ListPlot[attractor, PlotJoined->True, Ticks->None]
```

LSystem.ma

```
(* Fractal construction via Lindenmayer systems. *)

turt = {y,f};  (* Initial turtle instruction list. *)
depth = 6;
delta = Pi/3;
Do[
        len = Length[turt];
        Do[rep = Switch[turt[[q]],
                        (* f,{f,f,cc,f,cc,f,cc,f,cc,f,f},
                           Tablecloth!
                           Initial turt = {f,cc,f,cc,f,cc,f},
                           delta = Pi/2.
                         *)
                        (* x,{cw,y,f,cc,x,f,x,cc,f,y,cw},
                           y,{cc,x,f,cw,y,f,y,cw,f,x,cc},
                           f,f,
                           Hilbert curve! Initial turt = {x} above,
                           delta = Pi/2.
                         *)
                        (*
                  x,{x,f,y,f,x,cc,f,cc,y,f,x,f,y,cw,f,cw,x,f,y,f,x},
                  y,{y,f,x,f,y,cw,f,cw,x,f,y,f,x,cc,f,cc,y,f,x,f,y},
                           f,f,
                           Peano curve!  Initial turt = {x} above,
                           delta = Pi/2.
                         *)
                        (* l,{l,cc,r,f,cc},
```

```
                          r,{cw,f,l,cw,r},
                          Dragon curve! Initial turt = {f,l} above,
                          delta = Pi/2.
                        *)
                        (* f,{f,cc,f,cw,cw,f,cc,f},
                          Koch curve!  Initial turt = {f} above,
                          delta = Pi/3.
                        *)
                        x,{y,f,cw,x,f,cw,y},
                        y,{x,f,cc,y,f,cc,x},
                        (* Sierpinski gasket!
                           Initial turt = {f} above,
                           delta = Pi/3.
                         *)
                       f,f,
                       cc,cc,
                       cw,cw
                 ];
                 turt = Insert[Drop[turt,{q,q}],rep,q],
                {q,1,len}
        ];
        turt = Flatten[turt];
        state = {{0,0},0,1}; (* Initial state (x,y,a,r). *)
        pts = {state[[1]]};
        len = Length[turt];
    Do[
          sym = turt[[q]];
          Which[sym == f, state[[1]] += {state[[3]] Cos[state[[2]]],
state[[3]] Sin[state[[2]]]};
                         pts = Append[pts,state[[1]]],
               sym == cc, state[[2]] += delta,
               sym == cw, state[[2]] -= delta
          ]
          ,
          {q,1,len}
    ];
    Show[Graphics[Line[pts]], AspectRatio->Automatic],
        {dep,1,depth}
];
```

Mountains.ma

```
(* Fractal mountain generation.
  The dimension can be selected, below;
  dimension near 3 means jagged; near 2 means
  gentle hills.
*)

size = 64;
height[x_, y_] :=
       Block[{a,b,c},
(* Next, force a circular lake for effect! *)
              If[y^2 + (1-x)^2 < 0.22, Return[0]];
                     c = (size-1)/2;
                     a = Round[1 + c x];
```

```
                            b = Round[1 + c y];
                            Return[Re[flis[a][[b]]]];
       ];

Do[phase[u] = Table[Exp[ 1.0 I Random[Real, N[2 Pi]]],
                {v,1,size}],{u,1,size}];
dimension = 2.85;
H = 3-dimension;
pow = (H+1)/2;
  Do[jlis[u] = Table[ If[(u>0) && (w>0),1,0] *
      (u^2+w^2)^(-pow) * phase[w][[u]],
      {w,1,size}],{u,1,size}];
  Do[flis[u] = Fourier[jlis[u]],{u,1,size}];
  Do[col[u] = Table[flis[v][[u]],{v,1,size}],{u,1,size}];
  Do[flis[u] = Fourier[col[u]],{u,1,size}];

ParametricPlot3D[{x,y,10*height[x,y]}  ,
                            {x,0,1, 0.02},
                            {y,0,1,0.02}, Boxed->False,
                            ViewPoint->{1.5,-1.6,3},
                            LightSources->{
                                  {{1., 1., 1.}, RGBColor[0, 1, 0]},
                                  {{0., 1., 1.}, RGBColor[0, 0, 1]}}
];
```

MultiNewton.ma

```
(* Multidimensional Newton method for solving non-linear
   systems.

 *)

(* Next, enter the expressions to be forced to be
   zero.
 *)
dim = 3;
f[1][x_] := x[[1]]^3 - x[[2]]^2 - 1;
f[2][x_] := x[[1]]^2 + x[[3]]^4 - 4;
f[3][x_] := x[[3]] x[[2]] + 2;

Clear[u];
jac[x_] := N[Table[
              D[f[m][u],u[[n]]] /. u->x,{m,1,dim},{n,1,dim}
             ],20];

x = {2,2,2}; (* Here is the initial guess. *)
Do[
    x = N[x - Inverse[jac[x]] . Table[f[q][x],{q,1,dim}],20];
    Print[x],
    {e,1,16}
]
```

Neural.ma

```
(* Neural network experiment; network
   mimics "exclsuive-or" logic.
 *)

gain = 0.75;

sigmoid[z_] := 1/(1+Exp[-gain*z]);

nodeout[input_,weight_] :=
        sigmoid[input . weight[[1]] + weight[[2]]];

response[x_,y_] :=
   nodeout[{nodeout[{x,y},w1],
        nodeout[{x,y},w2]},w3];

(* Next, give the weights and offsets for each perceptron. *)
w1 = {{-2.69, -2.80}, -2.21};
w2 = {{-3.39,-4.56},4.76};
w3 = {{-4.91,4.95},-2.28};

Plot3D[response[x,y],{x,-1,1},{y,-1,1},
   PlotPoints->32,Axes->None,Boxed->False, Mesh->False,
   ViewPoint->{1,2,4}]
```

NewtonMatrixInverse.ma

```
(* Newton method for matrix inversion. *)

(* Next, the matrix to be inverted. *)
a = {{4,3,0},
     {0,3,0},
     {1,1,2}}/4;

x = a/2;  (* The initial guess matrix. *)
Do[
    x = N[2 x - x . a . x],
    {q,1,6}
];
Print[MatrixForm[x]]
Inverse[N[a]]  (* Test. *)
```

NewtonTrefoil.ma

```
(* Newton trefoil fractal plot.
   A point of the complex plane is colored
   according to which of the three roots of unity
   is its destiny under Newton iteration.
 *)

escape = 30;
r1 = 1.0;
r2 = -0.5 + N[I Sqrt[3]/2];
r3 = Conjugate[r2];
newton3[z_] := Block[{w, q, d1, d2, d3, p},
```

```
                w = N[z];
                If[Abs[w] < 0.02, Return[4]];
                q = 1;
                While[q < escape,
                        w -= (w^3-1)/(3w^2);
                        If[Re[w]>2, Break[],
                             If[Im[w]>2, Break[]] ];
                        ++q;
                ];
                d1 = Abs[w-r1];
                d2 = Abs[w-r2];
                d3 = Abs[w-r3];
                If[d1<d2, p = 1, p = 2];
                If[p==1, If[d1<d3, p=1, p=3],
                         If[d2<d3, p=2, p=3]
                ];
                Return[p]
               ];
DensityPlot[newton3[x + I y],{x,-1, 1},{y,-1,1},
        PlotPoints->128, Mesh->False, Axes->None
]
```

PhiHat.ma

```
(* Plotting of wavelet function via resolution of
   the Fourier transform of the mother function.
   This method amounts to iteration of the dilation
   equation but now in frequency space.
 *)

(* Next, the Daubechies order-4 wavelet coefficients. *)

c0 = (1+Sqrt[3])/4;
c1 = (3+Sqrt[3])/4;
c2 = (3-Sqrt[3])/4;
c3 = (1-Sqrt[3])/4;

(* Next, generate the transform of the scalar
   mother function.
 *)

f[k_] := Product[N[(c0 + c1 Exp[-I k/2^q] +
     c2 Exp[-2 I k/2^q] + c3 Exp[-3 I k/2^q])/2],{q,1,24}];
Plot[Abs[f[w]],{w,-30,30}]

(* Next, make a discrete table of transform values. *)
n = 512;
trans = Table[N[f[2 r]],{r,-n/2, n/2-1}];
ListPlot[Abs[trans], PlotJoined->True, PlotRange->{0,1}]

(* Next, plot the wavelet by transforming back. *)
four = Fourier[trans];
phase = Table[Exp[I Pi w],{w,0,Length[four]-1}];
```

```
four *= phase;
ListPlot[Re[four], PlotJoined->True]
```

Pi.ma

```
(* Fast algorithms for computation of Pi. *)

(* Next, arithmetic-geometric mean (AGM) sequence for Pi.
   This is asymptotically faster than the Ramanujan
   sequence below, but of course this AGM scheme has
   a square root in the loop.
 *)

x = N[Sqrt[2],200];
y = Sqrt[x];
s = y;
p = 2 + x;
true = N[Pi,200];
Do[
        x = (s + 1/s)/2;
        p *= (1+x)/(1+y);
        s = Sqrt[x];
        y = (y*s + 1/s)/(y+1);
        Print[p - true],
        {u,1,5}
];

(* Next, a Ramanujan formula, claimed to yield 14 decimal
   digits per summand!  The implicit square root can of course
   be taken out and performed just once.
 *)

w = 12 Sum[(-1)^k (6k)! /((k!)^3 (3k)!) *
        (13591409 + k*545140134)/(640320^3)^(k+1/2),{k,0,10}];
Print[N[1/w,200] - N[Pi,200]];
```

PolynomialSolve.ma

```
(* Polynomial solver for all roots,
   assuming all roots are real.
   The loop ends when a List of all roots is full.
   Uses Maehly algorithm based on Newton iteration.
   [Stoer and Bulirsch 1993].
 *)

(* Next, specify the polynomial to be solved. *)
deg = 8;
p[x_] := (x-1)(x-2)(x-5)(x-6)(x+2)(x+4)(x-3)(x+1);

eps = 10^(-10);
z0 = 10.0;
roots = {};
Do[
        m = 2;
```

```
        zs = z0;
        z = zs+eps;
        While[zs < z,
          s = 0;
          z = zs-eps;
          (* Print[z]; *)
          Do[
            s = s + 1.0/(z-roots[[i]]);
            ,{i,1,j-1}
          ];
          zs = p[z];
          zs = z - m zs/((D[p[x],x] /. x->z) - zs s);
          If[m==2,zs = z-eps; m = 1];
        ];

          roots = Append[roots,z];
          z0 = z
        ,{j,1,deg}
];

roots
```

quasi.c

```
/*
 * quasi.c
 *
 * Routines for quasi-Monte Carlo discrepancy integration for
 * multidimensional integrals.
 */

#include <math.h>
#include <time.h>

#define MAX_DIM 32
#define DEPTH 64
#define DEN 2147483648.0

int pt[MAX_DIM][DEPTH];
double pc[MAX_DIM][DEPTH];
int pr[MAX_DIM] = {2, 3, 5, 7, 11, 13, 17, 19, 23, 29, 31, 37, 41, 43,
47, 53, 59, 61, 67, 71, 73, 79, 83, 89, 97, 101, 103, 107, 109, 113,
127, 131};

void set_random_seed()
/* Start the random number generator at a new position */
{
    srandom(time(NULL));
}

void new_point(double *x, int n)
/* Generate n new coordinates. */
{
    int m, j, p;
```

```
    for (m = 0; m < n; m++) {
        p = pr[m];
        for (j = 0; j < DEPTH; j++) {
            ++pt[m][j];
            x[m] += pc[m][j];
            if (pt[m][j] == p) {
                pt[m][j] = 0;
                x[m] -= p * pc[m][j];
                continue;
            }
            break;
        }
    }
}

void new_random_point(double *x, int n)
/* Generate n random coordinates. */
{
    int m;

    for (m = 0; m < n; m++) {
        x[m] = random() / DEN;
    }
}

void seed_coordinate(double *x, int m, unsigned int s)
/* Start the Wozniakowski process at a chosen s value. */
{
    int j, a, p;

    p = pr[m];
    pc[m][0] = 1.0 / p;
    for (j = 1; j < DEPTH; j++)
        pc[m][j] = pc[m][j - 1] / p;
    x[m] = 0;
    for (j = 0; j < DEPTH; j++) {
        a = s % p;
        pt[m][j] = a;
        s = (s - a) / p;
        x[m] += a * pc[m][j];
    }
}

void seed_point(double *x, int n, unsigned int s)
/*
 * Start the Wozniakowski process at a chosen s value for all n
 * coordinates.
 */
{
    int j;

    for (j = 0; j < n; j++) {
        seed_coordinate(x, j, s);
    }
}

void random_seed_point(double *x, int n)
```

```
/*
 * Start the Wozniakowski process at a random value for all n
 * coordinates.
 */
{
    int j;

    for (j = 0; j < n; j++) {
        seed_coordinate(x, j, random());
    }
}
```

RecursiveFHT.ma

```
(* Recursive Fast Hartley Transform. *)

fht[x_] :-
   Block[{n, xeven, xodd, xrev, xsplit, c, s},
        If[(n=Length[x])==1,Return[x]];
        xsplit = Transpose[Partition[x,2]];
        xeven = fht[xsplit[[1]]]; xodd = fht[xsplit[[2]]];
        c = Table[Cos[2 Pi m/n],{m,0,n-1}];
        s = Table[Sin[2 Pi m/n],{m,0,n-1}];
        xrev = Reverse[Append[Take[xodd,{2,n/2}],xodd[[1]]]];
        Return[Join[xeven,xeven] +
                        c * Join[xodd,xodd] +
                               s * Join[xrev, xrev]];
   ]

(* Next, definition of transform for testing
   purposes.
 *)

dht[x_] :- Table[Sum[x[[j+1]] cas[2 Pi j k/Length[x]],
                 {j,0,Length[x]-1}
                 ] ,
                 {k,0,Length[x]-1}
            ]

z = {1,3,2,4,5,6,7,9};
Print[fht[z]]
Print[dht[z]]
```

RombergLimits.ma

```
(* Romberg tableaux for computation of
   Euler's constant, Log[], and Pi.
*)

(* Next, do the Euler constant. *)
lim = 9;
x = N[1/4.0,80];
q = Table[0,{i,1,lim},{j,1,lim}];
Do[
        q[[n+1,1]] = N[-n Log[2] + 1/(2^(n+1)) +
                          Sum[1/k,{k,1,2^n-1}],80],
```

```
        {n,0,lim-1}

];
Do[
  Do[
   q[[n,m]] = N[(q[[n,m-1]] - x^(m-1) q[[n-1,m-1]])/
                (1-x^(m-1)),80],
   {m,2,n}
  ],
  {n,2,lim}
];
Print[MatrixForm[q]]

(* Next, do Log[y]. *)
lim = 9;
x = N[1/4.0,80];
y = 2.0;
q = Table[0,{i,1,lim},{j,1,lim}];
q[[1,1]] = (y^2-1)/(2y); q[[2,1]] = N[(y-1)/Sqrt[y]];
Do[
        q[[n+1,1]] =
          q[[n,1]] Sqrt[2q[[n,1]]/(q[[n,1]]+q[[n-1,1]])];
        ,
        {n,2,lim-1}

];
Do[
  Do[
   q[[n,m]] = N[(q[[n,m-1]] - x^(m-1) q[[n-1,m-1]])/
                (1-x^(m-1)),80],
   {m,2,n}
  ],
  {n,2,lim}
];

Print[MatrixForm[q]]

(* Next, do Pi. *)
lim = 9;
x = N[1/4.0,80];
y = 2.0;
q = Table[0,{i,1,lim},{j,1,lim}];
q[[1,1]] = 1.0; q[[2,1]] = N[Sqrt[2]];
Do[
        q[[n+1,1]] =
          q[[n,1]] Sqrt[2q[[n,1]]/(q[[n,1]]+q[[n-1,1]])];
        ,
        {n,2,lim-1}

];
Do[
  Do[
   q[[n,m]] = N[(q[[n,m-1]] - x^(m-1) q[[n-1,m-1]])/
                (1-x^(m-1)),80],
   {m,2,n}
  ],
  {n,2,lim}
```

```
];

Print[MatrixForm[N[2.0 * q,80]]]
```

Sessile.ma

```
(* Symbolic solution of the sessile drop
differential equation. *)

prec = 4
f = Sum[a[i] r^(2i),{i,1,prec}]
ex = Normal[Series[(1+h^2)^(3/2),{h,0,3}]]
g = ex /. h->D[f,r]
g = Expand[- g*(1-f) k]
m = D[f,{r,2}] + (1+D[f,r]^2) D[f,r]/r
m = Expand[m]
log = True
Do[
   log = log && LogicalExpand[(D[m,{r,q}] /. r->0) ==
                (D[g,{r,q}] /. r->0)],
   {q,0,2*(prec-1)}
]
soln = Solve[log]
f /. soln[[1]]
fr = (f /. soln[[1]]) /. k->0.1;
Plot3D[If[x^2+y^2<9^2,fr /. r->Sqrt[x^2+y^2],
            fr /. r->9], {x,-12,12},{y,-12,12},
            Axes->None, Boxed->False, Mesh->False,
            PlotRange->{-4,1}, Lighting->True,
            PlotPoints->64, ViewPoint->{0,-10,0}]
Plot[If[Abs[s]<9,fr /. r->s ,
            fr /. r->9], {s,-12,12},
            Axes->None,                PlotPoints->64]
```

SloshingPacket.ma

```
(* Animated plot of sloshing quantum packet.
 *)

aa = 3
w = 0.5
d[x_, t_] := Cos[w t]^2 + 1/w^2 Sin[w t]^2
a[t_] := aa Cos[w t]
p[x_, t_] := 1/Sqrt[d[x,t]] Exp[-(x-a[t])^2/(2 d[x,t]) ]
Do[
    Plot[p[x,t],{x,-9,9}, PlotRange->{0,1.2 Max[w,1]},Ticks->None],
    {t,0, Pi/w,Pi/(8 w)}
]
```

SoundEntropy.ma

```
(* Plots of statistics and entropy calculation
   for sound data.
 *)

lis = {};  (* Create this data as a set of histogram
              bins for the quantized sound. For example
              lis = {20,100,100,30} might be the occurrences
              of four sound data values. *)
lis2 = lis;
(* Next, convert if CODEC sound *)
Do[
  lis2[[q]] = If[q<128, lis[[q-128]],
                          lis[[256-q]]], {q,1,Length[lis]-1}
];
ListPlot[lis2, PlotRange->{0,3000}]

tot = Sum[N[lis2[[q]]],{q,1,Length[lis]}]

lis2 /= tot;
entropy = Sum[ -N[lis2[[q]] Log[2, lis2[[q]]+0.0001]],
                        {q,1,Length[lis]}]

twodlist = {};  (* Create this 2-dimensional histogram
                   bin, arranged lexicographically. *)
side = Sqrt[Length[twodlist]];
f[x_, y_] := twodlist[[side*Floor[x] + Floor[y] + 1]];
DensityPlot[f[x,y],{x,0,30},{y,0,30}, PlotPoints->32]

tot = Sum[N[twodlist[[q]]],{q,1,Length[twodlist]}]

twodlist /= tot;
entropy = Sum[ -N[twodlist[[q]] Log[2, twodlist[[q]]+0.0001]],
                        {q,1,Length[twodlist]}]/2
```

SoundFileReader.ma

```
(* NEXTSTEP sound file format reader.
   From [Chandra A 1992]
 *)

(* usage:

        mylist = readSF["/me/mysound.snd"] ;

This will read in the file mysound.snd, and if the soundfile is 16-bit
linear, convert all the samples to signed integers between -2^15 and
+2^15,
then return those values to the list "mylist".

"mylist" can then be manipulated, as any other Mathematica list.

If "mysound.snd" is stereo, the list "mylist" will consist of two
lists, the first Left, the second Right.
```

```
If you want to play this sound from within Mathematica, the options
for
ListPlay must be set:

        SetOptions[ListPlay, PlayRange->{-2^15,2^15}, SampleDepth->16,
                          SampleRate->44100]

        ListPlay[mylist]

*)

readSF[name_String] := Block[

        {dataLocation, dataSize, dataFormat, samplingRate,
         channelCount, samples, sizeofheader, binaryData, shortData,
         sf, header },
        sizeofheader = 24 ; (* minimum size of header *)
        sf = OpenRead[name] ;
        header = ReadList[sf,Byte,sizeofheader];
        (* skip the first four bytes (magic number) *)
        (* location of data *)
        dataLocation = ((((header[[5]] * 256) +
                              header[[6]]) * 256) +
                              header[[7]]) * 256 + header[[8]] ;
        (* data size in bytes *)
        dataSize - ((((header[[9]] * 256) +
                              header[[10]]) * 256) +
                                                   header[[11]]) * 256 +
                                                          header[[12]] ;

        (* NeXT format of data *)
        dataFormat = ((((header[[13]] * 256) +
                              header[[14]]) * 256) +
                              header[[15]]) * 256 + header[[16]] ;


        (* sampling rate *)
        samplingRate = ((((header[[17]] * 256) +
                              header[[18]]) * 256) +
                              header[[19]]) * 256 + header[[20]] ;
        (* number of channels *)
        channelCount = ((((header[[21]] * 256) +
                              header[[22]]) * 256) +
                              header[[23]]) * 256 + header[[24]] ;

        (* calculate number of samples, assume 16-bit data *)
        samples = dataSize / 2 / channelCount ;

        (* display information *)
        Print["Soundfile: ", name];
        Print["Size: ",dataSize," bytes"];
        Print["Samples: ", samples];
        Print["Duration: ",N[samples/samplingRate]," seconds"];
        Print["Format: ",dataFormat];
        Print["SamplingRate: ",samplingRate];
        Print["Channels: ",channelCount];
```

```
        (* Stop here if the format is not 16-bit linear *)

        If[ dataFormat != 3,(Close[sf]; Return[0];)];
        (* Otherwise, the format IS 16-bit linear *)
        (* skip over Info string *)
        If[ dataLocation > sizeofheader,
                Skip[sf,Byte,(dataLocation - sizeofheader)]];
        Print["reading"];
        binaryData = ReadList[sf,Byte];
        Print["converting to signed integers"];
        (* high-order byte: #1, low-order byte: #2 *)
        shortData = Apply[ If[ #1 > 127,
                            ( #1 - 256 ) * 256 + #2,
                            ( #1 * 256 ) + #2]&,
                            Partition[binaryData,2],{1}];

        Print["done"];
        Close[sf];

        (* if the soundfile is stereo, return two lists *)

        If[ channelCount == 2,
                ( Print["returning stereo data"];
                Return[Transpose[Partition[shortData,2]]]),
                ( Print["returning mono data"];
                 Return[shortData])
        ];
]
```

sphere.c

```
/*
 * sphere.c
 *
 * Discrepancy integration test.
 * Compile with:
 *
 * % cc -O sphere.c quasi.c -o sphere
 *
 * Run with
 *
 * % sphere n k
 *
 * to approximate the volume of the unit n-ball by dropping k points.
 *
 */

#include <stdio.h>
#include <stdlib.h>
#include <math.h>

void main(int argc, char **argv)
{
    int num = atoi(argv[2]), dim = atoi(argv[1]), j, k;
```

```
    double x[32], y, integ, sum;

    seed_point(x, dim, 0);
    integ = 0.0;
    for (j = 0; j < num; j++) {
        new_point(x, dim);      /* Right here the array x is a new
                                 * point in the cloud. */
        sum = 0;
        for (k = 0; k < dim; k++) {
            y = 2 * x[k] - 1;
            sum += y * y;
        }
        if (sum <= 1.0)
            integ += 1.0;

    }
    printf("%d %3.20f\n", dim, (1 << dim) * integ / num);
}
```

strassentest.c

```
 * Run with:
 *
 * % strassen N
 *
 * to multiply two N-by-N matrices and report times.
 * The random number generator can be seeded if desired.
 *
 */

#include <stdio.h>
#include <stdlib.h>
#include "strassen.h"

matrix newmatrix(int);              /* allocate storage */
void randomfill(int, matrix);       /* fill with random values in the
                                     * the range [0,1) */
void mult(int, matrix, matrix, matrix); /* classical algorithm */
void discrepancy(int, matrix, matrix);  /* print places where
                                         * matrices differ */
int diff(double, double);           /* do two doubles differ? */
void check(int, char *);            /* check for error conditions */

int main(int argc, char **argv)
{
    int n;
    matrix a, b, c, d;

    check(argc == 2, "main: Need matrix size on command line");
    n = atoi(argv[1]);
    a = newmatrix(n);
    b = newmatrix(n);
    c = newmatrix(n);
    d = newmatrix(n);
    randomfill(n, a);
```

```
    randomfill(n, b);
    start_timer();
    multiply(n, a, b, c, d);     /* strassen algorithm */
    end_timer();
    timer_summary(1);
    start_timer();
    mult(n, a, b, d);            /* classical algorithm */
    end_timer();
    timer_summary(1);
    discrepancy(n, c, d);
    return 0;
}

/* Fill the matrix a with random values between 0 and 1 */
void randomfill(int n, matrix a)
{
    if (n <= BREAK) {
        int i, j;
        double **p = a->d, T = -(double)(1 << 31);

        for (i = 0; i < n; i++)
            for (j = 0; j < n; j++)
                p[i][j] = random() / T;
    } else {
        n /= 2;
        randomfill(n, a11);
        randomfill(n, a12);
        randomfill(n, a21);
        randomfill(n, a22);
    }
}

void mult(int n, matrix a, matrix b, matrix c)   /* c = a*b */
{
    matrix scratch;

    if (n <= BREAK) {
        double sum, **p = a->d, **q = b->d, **r = c->d;
        int i, j, k;

        for (i = 0; i < n; i++) {
            for (j = 0; j < n; j++) {
                for (sum = 0., k = 0; k < n; k++)
                    sum += p[i][k] * q[k][j];
                r[i][j] = sum;
            }
        }
    } else {
        n /= 2;
        scratch = newmatrix(n);
        mult(n, a11, b11, scratch);
        mult(n, a12, b21, c11);
        add(n, scratch, c11, c11);
        mult(n, a11, b12, scratch);
        mult(n, a12, b22, c12);
        add(n, scratch, c12, c12);
        mult(n, a21, b11, scratch);
```

```
        mult(n, a22, b21, c21);
        add(n, scratch, c21, c21);
        mult(n, a21, b12, scratch);
        mult(n, a22, b22, c22);
        add(n, scratch, c22, c22);
        /* we should free scratch at this point but we are lazy */
    }
}

/* return new square n by n matrix */
matrix newmatrix(int n)
{
    matrix a;

    a = (matrix)malloc(sizeof(*a));
    check(a != NULL, "newmatrix: out of space for matrix");
    if (n <= BREAK) {
        int i;

        a->d = (double **)calloc(n, sizeof(double *));
        check(a->d != NULL,
                     "newmatrix: out of space for row pointers");
        for (i = 0; i < n; i++) {
            a->d[i] = (double *)calloc(n, sizeof(double));
            check(a != NULL, "newmatrix: out of space for rows");
        }
    } else {
        n /= 2;
        a->p = (matrix *)calloc(4, sizeof(matrix));
        check(a->p != NULL,
                     "newmatrix: out of space for submatrices");
        a11 = newmatrix(n);
        a12 = newmatrix(n);
        a21 = newmatrix(n);
        a22 = newmatrix(n);
    }
    return a;
}

/*
 * Print the first MAXERRS locations where the two matrices differ
 */

#define MAXERRS 100

void discrepancy(int n, matrix a, matrix b)
{
    static int errors = 0;

    if (errors >= MAXERRS)
        return;
    if (n <= BREAK) {
        int i, j;
        double **p = a->d, **q = b->d;

        for (i = 0; i < n; i++) {
            for (j = 0; j < n; j++) {
```

```
                if (diff(p[i][j], q[i][j])) {
                    if (errors++ >= MAXERRS)
                        return;
                    else {
                        printf("strassen[%d][%d] = %.12f, ",
                                                    i, j, p[i][j]);
                        printf("classical = %.12f\n", q[i][j]);
                    }
                }
            }
        }
    } else {
        n /= 2;
        discrepancy(n, a11, b11);
        discrepancy(n, a12, b12);
        discrepancy(n, a21, b21);
        discrepancy(n, a22, b22);
    }
}

/*
 * Are two doubles significantly different?  Since the original
 * numbers are between 0 and 1, we arbitrarily decide they are
 * different if their difference is at most 1e-12.  This is somewhat
 * fragile, and could be replaced by a consideration of their
 * relative difference.
 */

int diff(double x, double y)
{
    x -= y;
    if (x < 0)
        x = -x;
    return x > 1e-12;
}

/*
 * If the expression e is false print the error message s and quit.
 */

void check(int e, char *s)
{
    if (!e) {
        fprintf(stderr, "Fatal error -> %s\n", s);
        exit(1);
    }
}
```

strassen.c

```
/*
 * strassen.c
 *
 * Courtesy [Buhler 1993].
 * Routines to realize the Strassen recursive matrix multiplication.
 * Multiply n by n matrices a and b, putting the result in c, and
 * using the matrix d as scratch space.  The Strassen algorithm is:
```

```
 *
 *      q7 = (a12-a22)(b21+b22)
 *      q6 = (-a11+a21)(b11+b12)
 *      q5 = (a11+a12)b22
 *      q4 = a22(-b11+b21)
 *      q3 = a11(b12-b22)
 *      q2 = (a21+a22)b11
 *      q1 = (a11+a22)(b11+b22)
 *      c11 = q1+q4-q5+q7
 *      c12 = q3+q5
 *      c21 = q2+q4
 *      c22 = q1+q3-q2+q6
 *
 * where the double indices refer to submatrices in an obvious way.
 * Each line of multiply() that recursively calls itself computes one
 * of the q's.  Four scratch half-size matrices are required by the
 * sequence of computations here; with some rearrangement this
 * storage requirement can be reduced to three half-size matrices.
 *
 * The small matrix computations (i.e., for n <= BREAK) can be
 * optimized considerably from those given here; in particular, this
 * is important to do before the value of BREAK is chosen optimally.
 *
 */

#include "strassen.h"

/* c = a*b */
void multiply(int n, matrix a, matrix b, matrix c, matrix d)
{
    if (n <= BREAK) {
        double sum, **p = a->d, **q = b->d, **r = c->d;
        int i, j, k;

        for (i = 0; i < n; i++) {
            for (j = 0; j < n; j++) {
                for (sum = 0., k = 0; k < n; k++)
                    sum += p[i][k] * q[k][j];
                r[i][j] = sum;
            }
        }
    } else {
        n /= 2;
        sub(n, a12, a22, d11);
        add(n, b21, b22, d12);
        multiply(n, d11, d12, c11, d21);
        sub(n, a21, a11, d11);
        add(n, b11, b12, d12);
        multiply(n, d11, d12, c22, d21);
        add(n, a11, a12, d11);
        multiply(n, d11, b22, c12, d12);
        sub(n, c11, c12, c11);
        sub(n, b21, b11, d11);
        multiply(n, a22, d11, c21, d12);
        add(n, c21, c11, c11);
        sub(n, b12, b22, d11);
        multiply(n, a11, d11, d12, d21);
```

```
        add(n, d12, c12, c12);
        add(n, d12, c22, c22);
        add(n, a21, a22, d11);
        multiply(n, d11, b11, d12, d21);
        add(n, d12, c21, c21);
        sub(n, c22, d12, c22);
        add(n, a11, a22, d11);
        add(n, b11, b22, d12);
        multiply(n, d11, d12, d21, d22);
        add(n, d21, c11, c11);
        add(n, d21, c22, c22);
    }
}

/* c = a+b */
void add(int n, matrix a, matrix b, matrix c)
{
    if (n <= BREAK) {
        double **p = a->d, **q = b->d, **r = c->d;
        int i, j;

        for (i = 0; i < n; i++) {
            for (j = 0; j < n; j++) {
                r[i][j] = p[i][j] + q[i][j];
            }
        }
    } else {
        n /= 2;
        add(n, a11, b11, c11);
        add(n, a12, b12, c12);
        add(n, a21, b21, c21);
        add(n, a22, b22, c22);
    }
}

/* c = a-b */
void sub(int n, matrix a, matrix b, matrix c)
{
    if (n <= BREAK) {
        double **p = a->d, **q = b->d, **r = c->d;
        int i, j;

        for (i = 0; i < n; i++) {
            for (j = 0; j < n; j++) {
                r[i][j] = p[i][j] - q[i][j];
            }
        }
    } else {
        n /= 2;
        sub(n, a11, b11, c11);
        sub(n, a12, b12, c12);
        sub(n, a21, b21, c21);
        sub(n, a22, b22, c22);
    }
}
```

strassen.h

```
/*
 * strassen.h
 *
 * Copurtesy [Buhler 1993].
 * Header file for strassen matrix multiplication functions.
 */

/*
 * Square matrices of size <= BREAK are handled with the ''classical
 * algorithms.  The shape of almost all functions is therefore
 * something like
 *
 *      if ( n <= BREAK )
 *          classical algorithms
 *      else
 *          n/= 2
 *          recursive call for 4 half-size submatrices
 */

#define BREAK 8

/*
 * A matrix is defined to be a pointer to a ''union _matrix'', which
 * is (if the size is <= BREAK) a matrix of numbers, or else is an
 * array of four submatrices of half size.
 */

typedef union _matrix {
    double **d;
    union _matrix **p;
} *matrix;

void multiply(int, matrix, matrix, matrix, matrix);
void add(int, matrix, matrix, matrix);
void sub(int, matrix, matrix, matrix);

/*
 * Notational shorthand to access submatrices for matrices named
 * a,b,c,d
 */

#define a11 a->p[0]
#define a12 a->p[1]
#define a21 a->p[2]
#define a22 a->p[3]
#define b11 b->p[0]
#define b12 b->p[1]
#define b21 b->p[2]
#define b22 b->p[3]
#define c11 c->p[0]
#define c12 c->p[1]
#define c21 c->p[2]
#define c22 c->p[3]
#define d11 d->p[0]
#define d12 d->p[1]
#define d21 d->p[2]
```

```
#define d22 d->p[3]
```

Strassen.ma

```
(* Strassen recursive matrix multiplication.
   This is an O(n^Log[2,7]) algorithm.
 *)

(* Next, return a List of
4 quadrant matrices that comprise matrix a. *)
quads[a_List] :=
        Block[{c,d,k,f},
              d = Length[a]/2;
              c = Table[
                    Table[Take[Take[a,{k,k}][[1]],
                      {1+d Mod[(f-1),2],d+d Mod[(f-1),2]}],
                      {k,1+d Floor[(f-1)/2],d+d Floor[(f-1)/2]}
                    ],
                    {f,1,4}
                  ];
             Return[c];
        ];

(* Next, return reassembly of a matrix
from its list of 4 quadrants. *)
glue[c_List] :=
        Block[{a,d,k},
              d = Length[c[[1]]];
              a = Join[
                    Table[Join[c[[1]][[k]],c[[2]][[k]]],
                        {k,1,d}
                    ],
                    Table[Join[c[[3]][[k]],c[[4]][[k]]],
                        {k,1,d}
                    ]
                  ];
              Return[a];
        ];

strassen[a_List, b_List] :=
      Block[{c,d,q1,q2,q3,q4,q5,q6,q7},
                If[Length[a]==2, Return[a . b]];
                c = quads[a]; d = quads[b];
                q1 = strassen[c[[1]]+c[[4]],d[[1]]+d[[4]]];
                q2 = strassen[c[[3]]+c[[4]],d[[1]]];
                q3 = strassen[c[[1]],d[[2]]-d[[4]]];
                q4 = strassen[c[[4]],-d[[1]]+d[[3]]];
                q5 = strassen[c[[1]]+c[[2]],d[[4]]];
                q6 = strassen[-c[[1]]+c[[3]],d[[1]]+d[[2]]];
                q7 = strassen[c[[2]]-c[[4]],d[[3]]+d[[4]]];
                Return[glue[{q1+q4-q5+q7,
                            q3+q5,
                            q2+q4,
                            q1+q3-q2+q6}
                        ]
                ]
```

```
    ];

a = Table[i+j,{i,1,8},{j,1,8}];
b = Table[i^2-j^2,{i,1,8},{j,1,8}];
Print[MatrixForm[strassen[a,b] - (a . b)]];
```

threebody.c

```
/*
 * 3body.c
 * C program for 3-body model.
 * Three bodies have respective coordinates:
 *      (x0, y0), (x1, y1), (x2, y2)
 * and corresponding initial velocities:
 *      (vx0, vy0), (vx1, vy1), (vx2, vy2)
 *
 * Typical command line:
 *      > a.out 100  0   0 100 -50 87 -87 -50 -50 -87  87 -50
 *               x0 y0 vx0 vy0  x1 y1 vx1 vy1  x2  y2 vx2 vy2
 *
 * The output is formatted:
 *      x0 y0 z0
 *      x1 y1 z1
 *      x2 y2 z2
 *      x0 y0 z0 ... etc.
 * where every triple of rows is a new DT increment.
 */

#include <stdlib.h>
#include <stdio.h>
#include <math.h>

#define DT 0.01                     /* The time increment. */
#define M 1                         /* The mass constant. */
#define G 1200000                   /* Constant of Gravitation. */

double x[3][3], dx[3][3], v[3][3], a[3][3], r[3], f[3], mass[3];

void main(int argc, char **argv)
{
    int n, m, j, k, ct;

    ct = 1;
    for (n = 0; n < 3; n++) {
        for (j = 0; j < 2; j++)
            x[n][j] = atoi(argv[ct++]);
        x[n][2] = 0;
        for (j = 0; j < 2; j++)
            v[n][j] = atoi(argv[ct++]);
        v[n][2] = 0;
        mass[n] = M;
    }
    for (ct = 0; ct < 1000; ct++) {
        for (n = 0; n < 3; n++) {
            m = (n + 1) % 3;
            r[n] = 0;
```

```
            for (k = 0; k < 3; k++) {
                dx[n][k] = x[m][k] - x[n][k];
                r[n] += dx[n][k] * dx[n][k];
            }
            f[n] = G * mass[m] * mass[n] / (r[n] * sqrt(r[n]));
        }
        for (n = 0; n < 3; n++) {
            m = (n + 2) % 3;
            for (k = 0; k < 3; k++) {
                a[n][k] = f[n] * dx[n][k] - f[m] * dx[m][k];
                v[n][k] += a[n][k] * DT;
                x[n][k] += v[n][k] * DT;
                printf("%f ", x[n][k]);
            }
            printf("\n");
        }
    }
}
```

timing.c

```
/*
 * timing.c
 *
 * Courtesy [Tevanian 1992].
 * Timer facilities.  These are Unix BSD (POSIX) dependent.
 */

#include <sys/time.h>
#include <sys/resource.h>

#include <stdio.h>


static struct rusage rusage_start, rusage_end;
static struct timeval time_start, time_end;

void start_timer()
{
    getrusage(RUSAGE_SELF, &rusage_start);
    gettimeofday(&time_start, NULL);
}

void end_timer()
{
    getrusage(RUSAGE_SELF, &rusage_end);
    gettimeofday(&time_end, NULL);
}

double diff_time(struct timeval end, struct timeval start)
{
    double e, s;

    e = ((double)end.tv_sec + ((double)end.tv_usec / 1000000.0));
    s = ((double)start.tv_sec + ((double)start.tv_usec / 1000000.0));
    return (e - s);
}
```

```
/*
 * Return times in milliseconds (floating point).
 */
void timer_summary(int count)
{
    double user, system, elapsed;
    double user_time, system_time, elapsed_time;

    user_time = diff_time(rusage_end.ru_utime,
                          rusage_start.ru_utime);
    system_time = diff_time(rusage_end.ru_stime,
                            rusage_start.ru_stime);
    elapsed_time = diff_time(time_end, time_start);

    user = 1000 * user_time / count;
    system = 1000 * system_time / count;
    elapsed = 1000 * elapsed_time / count;

    printf("CPU usage (ms/op): %.1f total (%.1f user, %.1f system)",
           user + system, user, system);
    printf(", %.1f elapsed.\n", elapsed);
}
```

TwoCycle.ma

```
(* Symbolic solution for 2-cycles of quadratic map

   x := a x (1-x)

   to show that sometimes such exact solutions can
   be found.
 *)

sol = Solve[ x == a y (1-y) && y == a x (1-x),{x,y}]
```

upsampler.c

```
/*
 * upsampler.c
 *
 * Routines for upsampling from a "down" frequency to an "up"
 * frequency, using sinc interpolation as in [Smith and Gossett
 * 1984].
 *
 * The routines are slow but demonstrative.  One should, for
 * efficiency, turn sinc() into a table lookup, and optimize
 * pointer arithmetic in the loop convolutions.
 *
 */

#define PI 3.14159265358979323846

#include <math.h>
```

```
double sincpi(double x)
/*
 * This atrociously slow function call should eventually be replaced
 * by a tabular approach.
 */
{
    if (x == 0.0)
        return (1.0);
    return (sin(PI * x) / (PI * x));
}

int upsample(short *x, short *y, int xsamps, double up, double down)
/*
 * Convert low frequency data x, having xsamps 16-bit samples, to a
 * certain number ysamps of high-frequency y samples. The exact
 * number ysamps is the return value of this function, said value
 * determined by a formula in the declarations below. The high and
 * low frequencies are up, down respectively.
 */
{
    double down_ratio = down / up;
    double a, b;
    int ysamps = 1 + (int)((xsamps - 6) / down_ratio);
    int i, j;

    for (j = 0; j < ysamps; j++) {
        b = j * down_ratio;
        i = (int)b;
        a = b - i;
        y[j] = 0.5 * (
                    x[i]   * sincpi(2+a) + x[i+1] * sincpi(1+a) +
                    x[i+2] * sincpi(a)   + x[i+3] * sincpi(1-a) +
                    x[i+4] * sincpi(2-a) + x[i+5] * sincpi(3-a));
    }
    return (ysamps);
}
```

walshfile.c

```
/*
 * walshfile.c
 *
 * Requires NEXTSTEP sound utilities.
 * Program to transform sound (with compression intended)
 * via Walsh-Hadamard transform.
 *
 * Compile with:
 *
 * % cc -O walshfile.c walsh.c -o walshfile
 *
 * Run with:
 *
 * % walshfile a.and b.snd dir
 *
 * where a.snd is converted into b.snd.  The direction "dir"
 * says whether we are converting to transform, or inverse
```

```
 * transforming to original format.  The only difference is
 * in the normalization because thw Walsh-Hadamard is its
 * own inverse.
 *
 * Compression experiments can performed on the transformed files
 * by such as:
 *
 * % compress b.snd
 *
 * or via some other compressor than LZW.
 */

#include <stdio.h>
#include <sound/sound.h>

#define N 256

void cohere(s, n, nd)
    short *s;
    int n, nd;
{
    int j, k, b = nd / n;
    short *t = (short *)malloc(nd * sizeof(short));

    bcopy(s, t, nd * sizeof(short));
    for (j = 0; j < n; j++) {
        for (k = 0; k < b; k++) {
            s[j * b + k] = t[k * n + j];
        }
    }
    free(t);
}

void uncohere(s, n, nd)
    short *s;
    int n, nd;
{
    int j, k, b = nd / n;
    short *t = (short *)malloc(nd * sizeof(short));

    bcopy(s, t, nd * sizeof(short));
    for (j = 0; j < n; j++) {
        for (k = 0; k < b; k++) {
            s[k * n + j] = t[j * b + k];
        }
    }
    free(t);
}

void main(int argc, char **argv)
{
    int ibuf[N];
    SNDSoundStruct *sound;
    short *sig;
    int j, k, ndata;
    int dir = atoi(argv[3]);
    double norm;
```

```
    if (dir < 0)
        norm = 16.0;
    else
        norm = (1 / (16.0 * N));
    SNDReadSoundfile(argv[1], &sound);
    sig = (short *)sound;
    sig += (sound->dataLocation) / sizeof(short);
    ndata = (sound->dataSize) / sizeof(short);
    ndata = N * (ndata / N);
    /* if(dir<0) uncohere(sig,N,ndata); */
    for (j = 0; j < ndata; j += N) {
        for (k = 0; k < N; k++)
            ibuf[k] = sig[j + k];
        FWTraw(ibuf, N);
        for (k = 0; k < N; k++)
            sig[j + k] = ibuf[k] * norm;
    }
    /* if(dir>0) cohere(sig,N,ndata); */
    sound->dataSize = ndata * sizeof(short);
    SNDWriteSoundfile(argv[2], sound);
}
```

walsh.c

```
/*
 * walsh.c
 *
 * Walsh-Hadamard transform library.
 */

#include <stdio.h>

int sequencytoindex(j, N)
    int j, N;
{
    register int i = j ^ (j >> 1), x, s;

    for (x = 0, s = 1; s < N; s <<= 1) {
        x <<= 1;
        if (i & s)
            x++;
    }
    return (x);
}

int walsh(m, t)
    int m, t;
{
    register int k = m & t, s, par = 1;

    for (s = 1; s <= k; s <<= 1)
        if (s & k)
            par = -par;
    return (par);
}
```

```
int scanintdata(x)
    int x[];
{
    int n = 0;

    while (scanf("%d\n", x + n++) != EOF);
    return (n - 1);
}

void putseqdata(x, n)
    int x[];
    int n;
{
    int j;

    for (j = 0; j < n; j++)
        printf("%d\n", x[sequencytoindex(j, n)]);
}

void putintdata(x, n)
    int x[];
    int n;
{
    int j;

    for (j = 0; j < n; j++)
        printf("%d\n", x[j]);
}

void FWTraw(x, n)
    int x[];
    int n;
{
    register int s, j, tmp;

    for (s = 1; s < n; s <<= 1) {
        for (j = 0; j < n; j++)
            if ((j & s) == 0) {
                tmp = x[j];
                x[j] += x[j + s];
                x[j + s] = tmp - x[j + s];
            }
    }
}

void FWTskip(x, n, skip)
    int x[];
    int n, skip;
{
    register int s, j, ss, jj, tmp;

    for (s = 1; s < n; s <<= 1) {
        for (j = 0; j < n; j++)
            if ((j & s) == 0) {
                jj = j * skip;
                ss = (j + s) * skip;
                tmp = x[jj];
```

```
                x[jj] += x[ss];
                x[ss] = tmp - x[ss];
            }
    }
}
void FWT(x, y, n)
    int x[], y[];
    int n;
{
    register int j;

    FWTraw(x, n);
    for (j = 0; j < n; j++)
        y[j] = x[sequencytoindex(j, n)];
}

void FWT2dimensional(x, w, h)
    int *x;
    int w, h;
{
    register int j;

    for (j = 0; j < h; j++)
        FWTskip(x + j * w, w, 1);
    for (j = 0; j < w; j++)
        FWTskip(x + j, h, w);
}
```

wavdim.c

```
/*
 * wavdim.c
 *
 * Experimental program for estimating fractal dimension via
 * discrete wavelet transform
 *
 * Compile with:
 *
 * % cc -O wavdim.c fwt.c -o wavdim
 *
 * Run with:
 *
 * % wavdim N
 *
 * which creates an N-sample piece of a Weierstrass function
 * which has known dimensional properties; and reports the
 * wavelet-based estimate.
 */

#include <stdlib.h>
#include <stdio.h>
#include <math.h>
#include "fwt.h"

#define PI 3.14159265358979323846
#define MAX_RUN 65536
```

```
void create_weierstrass(DATA_TYPE *x, int len, double a, double dim)
{
    int j;
    double sum, arg;
    double fac, fac0 = exp((dim - 2) * log(a));

    for (j = 0; j < len; j++) {
        arg = j / ((double)len);
        sum = 0.0;
        fac = fac0;
        while (fac > 0.0001) {
            arg *= a;
            sum += sin(arg) * fac;
            fac *= fac0;
        }
        x[j] = sum;
    }
}

double i_a(DATA_TYPE *x, int a, int b)
{
    double sum = 0.0;
    int j;

    for (j = a; j < b; j++) {
        sum += x[j] * x[j];
    }
    return (sum / (b - a));
}

double fdim(double u)
{
    return ((5 - log(u) / log(2.0)) / 2.0);
}

void main(int argc, char **argv)
{
    DATA_TYPE x[MAX_RUN];
    int n = atoi(argv[1]), co;
    double dim = 0, t, u, v, w;

    for (co = 1; co < 10; co++) {
        dim = 1.0 + co / 10.0;
        create_weierstrass(x, n, 2.0, dim);

        FWVT_6(x, n, 3);
        t = i_a(x, n / 16, n / 8);
        u = i_a(x, n / 8, n / 4);
        v = i_a(x, n / 4, n / 2);
        w = i_a(x, n / 2, n);
        printf("%d %f %f %f %f\n", n, fdim(t / u), fdim(u / v), fdim(v
/ w), dim);
    }
}
```

WaveletMatrix.ma

```
(* Generate full wavelet matrix from T matrix decomposition.

   These matrices (when applied to vectors of signal data)
   are precisely equivalent to the fast wavelet transforms.

   Remove N[] functions to obtain exact symbolic matrices.
 *)

hh[n_] := Block[{ },
             If[n==0,Return[(1+Sqrt[3])/(4)]];
             If[n==1,Return[(3+Sqrt[3])/(4)]];
             If[n==2,Return[(3-Sqrt[3])/(4)]];
             If[n==3,Return[(1-Sqrt[3])/(4)]];
             Return[0];
         ];

h[i_] := N[hh[i]];
t[n_] := 1/Sqrt[2] *
         Table[If[i<n/2,Sum[h[j-b n - 2i],{b,-1,1}],
                        Sum[(-1)^(j) h[n-1-(j-b n - 2 i)],{b,-10,10}]
                  ],{i,0,n},{j,0,n-1}];

ord = 32;  (* Global data size. *)
tp[n_] := Block[{tt, u},
                  tt = N[t[n]];
                  u = Table[If[(i<=n) &&(j<=n),
                                tt[[i,j]],
                                If[i==j,1,0]
                                ],{i,1,ord},{j,1,ord}];
                    Return[u];
                ];
(* Next, here comes the full wavelet matrix for signals
of length "ord." *)
w = tp[ord]; q = ord/2; While[q>1, w = tp[q] . w; q /= 2];
MatrixForm[w]
```

WaveletOnSignal.ma

```
(* Wavelet transform is applied to a synthetic signal
   of choice.

   Then wavelet components are removed (usually to allow
   high compression), and thresholded
   so the reconstruction is optionally imperfect.
 *)

(* Next, define various wavelet systems. *)

hhs[n_] := Block[{ },
             If[n==0,Return[1/2]];
             If[n==1,Return[-1/2]];
             Return[0];
         ]
hh0[n_] := Block[{ },
             If[n==0,Return[1/Sqrt[2]]];
```

```
            If[n==1,Return[-1/Sqrt[2]]];
            Return[0];
        ]
hh[n_] := Block[{ },
            If[n==0,Return[(1+Sqrt[3])/(4*Sqrt[2])]];
            If[n==1,Return[(3+Sqrt[3])/(4*Sqrt[2])]];
            If[n==2,Return[(3-Sqrt[3])/(4*Sqrt[2])]];
            If[n==3,Return[(1-Sqrt[3])/(4*Sqrt[2])]];
            Return[0];
        ]

hh2[n_] := Block[{ },
            If[n==0,Return[3/Sqrt[50]]];
            If[n==1,Return[6/Sqrt[50]]];
            If[n==2,Return[2/Sqrt[50]]];
            If[n==3,Return[-1/Sqrt[50]]];
            Return[0];
        ]

hh3[n_] := Block[{ },
            If[n==0,Return[ 5/Sqrt[578]]];
            If[n==1,Return[20/Sqrt[578]]];
            If[n==2,Return[12/Sqrt[578]]];
            If[n==3,Return[-3/Sqrt[578]]];
            Return[0];
        ]

h[n_] := N[hh[n]];
g[n_] := (-1)^n h[3-n];
p[i_] := Mod[i,2];
q[i_] := Floor[(i+1)/2];


m = 8;
n = 2^m;
(* Next, create a signal having n data. *)
Do[s[q] = N[Sin[q/3] * Exp[-3*q/n]], {q,0,n-1}];

test = Table[s[q],{q,0,n-1}];
ListPlot[test,  PlotJoined->True];
For[r=0, r<m, r++,
        c = s[0]; d = s[2^r];
        For[k=0, k < n/2^(r+1)-1, k++,
                a = Sum[h[q] s[(q+2k) 2^r], {q,0,3}];
                b = Sum[g[q] s[(q+2k) 2^r], {q,0,3}];
                s[k 2^(r+1)] = a;
                s[k 2^(r+1) + 2^r] = b;
        ];
        k = n/2^(r+1)-1;
        a = Sum[h[q] s[(q+2k) 2^r], {q,0,1}] +
                h[2] c + h[3] d;
        b = Sum[g[q] s[(q+2k) 2^r], {q,0,1}] +
                g[2] c + g[3] d;
        s[k 2^(r+1)] = a;
        s[k 2^(r+1) + 2^r] = b;
];
(* Next, reorder the wavelet components so that
   detail sectors are sorted in order.
```

```
 *)
u[0] = s[0];
ct = 1;
For[r = 0, r < m, r++,
        For[w = 0, w< 2^r, w++,
                u[ct] = s[2^(m-r-1) + w*2^(m-r)];
                ++ct;
        ];
];

(* Next, plot the in-order transform. *)
t = Table[u[q],{q,0,n-1}];
ListPlot[t,  PlotJoined->True, PlotRange->{-3,3}];

remain = n;
(* Next, zero those components below some threshold. *)
Do[If[Abs[s[q]]<0.3, s[q] = 0; --remain],{q,0,n-1}]
(* Or quantize the components:
    Do[s[q] = (Round[s[q]*128/3])/128,{q,0,n-1}] *)

u[0] = s[0];
ct = 1;
For[r = 0, r < m, r++,
        For[w = 0, w< 2^r, w++,
                u[ct] = s[2^(m-r-1) + w*2^(m-r)];
                ++ct;
        ];
];
(* Next, plot the thresholded components in order. *)
t = Table[u[q],{q,0,n-1}];
ListPlot[t,  PlotJoined->True, PlotRange->{-3,3}];
For[r=m, r>=1, r--,
        c = s[n-2^r]; d = s[n-2^r+2^(r-1)];
        For[k=n/2^r-1, k > 0, k--,
          a = Sum[h[2q] s[(k-q) 2^r] +
                        g[2q] s[(k-q) 2^r + 2^(r-1)], {q,0,1}];
          b = Sum[h[2q+1] s[(k-q) 2^r] +
                      g[2q+1] s[(k-q) 2^r + 2^(r-1)], {q,0,1}];
          s[k 2^r] = a;
          s[k 2^r + 2^(r-1)] = b;
        ];
        a = h[0] s[0] + h[2] c + g[0] s[2^(r-1)] + g[2] d;
        b = h[1] s[0] + h[3] c + g[1] s[2^(r-1)] + g[3] d;
        s[0] = a;
        s[2^(r-1)] = b;
];

(* Next, print how many components were non-zero; this
   gives an idea of compressibility. *)

Print[remain];

(* Next, plot the imperfect
   reconstruction of the original signal. *)
t = Table[s[q],{q,0,n-1}];
ListPlot[t,  PlotJoined->True, PlotRange->{-1,1}];
```

waveletsound.c

```
/*
 * waveletsound.c
 *
 * Requires NEXTSTEP sound utilities.
 * Courtesy Russell [1992].
 * Experimental program for sound compression via
 * wavelet transform.
 *
 * Compile with:
 *
 * % cc -O waveletsound.c fwt.c -o waveletsound
 *
 * Run with:
 *
 * % waveletsound orig.snd final.snd f
 *
 * where orig.snd is the original sound, final.snd is the
 * reconstructed sound, and f is the (floating point)
 * fraction of the maximum wavelet component value below
 * which fraction components are zeroed.
 * The report of "bytes in compressed version" simply estimates
 * compression possibilities by reporting the number of non-zero
 * bytes.  It turns out that compressors such as "compress" (LZW)
 * perform fairly closely to this measure of non-zero bytes,
 * especially for sound-based transform data.  Of course one can
 * always create the byte file and try actual compressors.
 */

#include <stdio.h>
#include <sound/sound.h>

#define N 1024
#define DEPTH 6

short muLaw[256] = { /* This is a complete mu-law conversion table. */
    0x8284, 0x8684, 0x8a84, 0x8c84, 0x9284, 0x9684, 0x9a84, 0x9e84,
    0xa284, 0xa684, 0xaa84, 0xae84, 0xb284, 0xb684, 0xba84, 0xbe84,
    0xc184, 0xc384, 0xc584, 0xc784, 0xc984, 0xcb84, 0xcd84, 0xcf84,
    0xd184, 0xd384, 0xd584, 0xd784, 0xd984, 0xdb84, 0xdd84, 0xdf84,
    0xe104, 0xe204, 0xe304, 0xe404, 0xe504, 0xe604, 0xe704, 0xe804,
    0xe904, 0xea04, 0xeb04, 0xec04, 0xed04, 0xee04, 0xef04, 0xf004,
    0xf0c4, 0xf144, 0xf1c4, 0xf244, 0xf2c4, 0xf344, 0xf3c4, 0xf444,
    0xf4c4, 0xf544, 0xf5c4, 0xf644, 0xf6c4, 0xf744, 0xf7c4, 0xf844,
    0xf8a4, 0xf8e4, 0xf924, 0xf964, 0xf9a4, 0xf9e4, 0xfa24, 0xfa64,
    0xfaa4, 0xfae4, 0xfb24, 0xfb64, 0xfba4, 0xfbe4, 0xfc24, 0xfc64,
    0xfc94, 0xfcb4, 0xfcd4, 0xfcf4, 0xfd14, 0xfd34, 0xfd54, 0xfd74,
    0xfd94, 0xfdb4, 0xfdd4, 0xfdf4, 0xfe14, 0xfe34, 0xfe54, 0xfe74,
    0xfe8c, 0xfe9c, 0xfeac, 0xfebc, 0xfecc, 0xfedc, 0xfeec, 0xfefc,
    0xff0c, 0xff1c, 0xff2c, 0xff3c, 0xff4c, 0xff5c, 0xff6c, 0xff7c,
    0xff88, 0xff90, 0xff98, 0xffa0, 0xffa8, 0xffb0, 0xffb8, 0xffc0,
    0xffc8, 0xffd0, 0xffd8, 0xffe0, 0xffe8, 0xfff0, 0xfff8, 0x0,
    0x7d7c, 0x797c, 0x757c, 0x717c, 0x6d7c, 0x697c, 0x657c, 0x617c,
    0x5d7c, 0x597c, 0x557c, 0x517c, 0x4d7c, 0x497c, 0x457c, 0x417c,
```

```
    0x3e7c, 0x3c7c, 0x3a7c, 0x387c, 0x367c, 0x347c, 0x327c, 0x307c,
    0x2e7c, 0x2c7c, 0x2a7c, 0x287c, 0x267c, 0x247c, 0x227c, 0x207c,
    0x1efc, 0x1dfc, 0x1cfc, 0x1bfc, 0x1afc, 0x19fc, 0x18fc, 0x17fc,
    0x16fc, 0x15fc, 0x14fc, 0x13fc, 0x12fc, 0x11fc, 0x10fc, 0xffc,
    0xf3c, 0xebc, 0xe3c, 0xdbc, 0xd3c, 0xcbc, 0xc3c, 0xbbc,
    0xb3c, 0xabc, 0xa3c, 0x9bc, 0x93c, 0x8bc, 0x83c, 0x7bc,
    0x75c, 0x71c, 0x6dc, 0x69c, 0x65c, 0x61c, 0x5dc, 0x59c,
    0x55c, 0x51c, 0x4dc, 0x49c, 0x45c, 0x41c, 0x3dc, 0x39c,
    0x36c, 0x34c, 0x32c, 0x30c, 0x2ec, 0x2cc, 0x2ac, 0x28c,
    0x26c, 0x24c, 0x22c, 0x20c, 0x1ec, 0x1cc, 0x1ac, 0x18c,
    0x174, 0x164, 0x154, 0x144, 0x134, 0x124, 0x114, 0x104,
    0xf4, 0xe4, 0xd4, 0xc4, 0xb4, 0xa4, 0x94, 0x84,
    0x78, 0x70, 0x68, 0x60, 0x58, 0x50, 0x48, 0x40,
    0x38, 0x30, 0x28, 0x20, 0x18, 0x10, 0x8, 0x0
};

void main(int argc, char **argv)
{
    int ibuf[N];
    SNDSoundStruct *sound;
    unsigned char *sig;
    int j, k, origdata, ndata, remain, total, m, mag, sign;
    double f = atof(argv[3]), max, fract, threshold;

    SNDReadSoundfile(argv[1], &sound);
    sig = (unsigned char *)sound;
    sig += (sound->dataLocation) / sizeof(char);
    origdata = (sound->dataSize) / sizeof(char);
    ndata = N * (origdata / N);
    total = 0;

    /* Next, loop over blocks of N sound samples each. */
    for (j = 0; j < ndata; j += N) {

        /* Next, get the data into an integer array. */
        for (k = 0; k < N; k++)
            ibuf[k] = muLaw[sig[j + k]];

        /*
         * Next, take the wavelet transform, down to level DEPTH, of
         * the block.
         */
        FWVT_4(ibuf, N, DEPTH);
        max = 0.0;

        /* Next, find the maximum component. */
        for (k = 0; k < N; k++) {
            if (abs(ibuf[k]) > max)
                max = abs(ibuf[k]);
        }
        threshold = max * f;
        remain = 0;
        for (k = 0; k < N; k++) {
            mag = abs(ibuf[k]);

            /* Next, cut off the smallest components. */
            if (mag < threshold)
                ibuf[k] = 0;
```

```
            else {
                ++remain;

                /* Next, set up 8-bit quantization. */
                sign = (ibuf[k] < 0) ? -1 : 1;
                fract = (mag - threshold) / (max - threshold);
                m = (int)(1.5 + fract * 254.0);
                ibuf[k] = sign * (((m - 1) / 254.0) *
                                (max - threshold) + threshold);
            }
        }
        total += remain;

        /*
         * Next, print out for each block the number of non-zero
         * components.
         */
        printf("%d %d\n", j / N, remain);
        IFWVT_4(ibuf, N, DEPTH);
        for (k = 0; k < N; k++)
            sig[j + k] = SNDMulaw((short)ibuf[k]);
    }
    printf("Bytes in compressed version: %d\n", total);
    printf("Bytes in original soundfile: %d\n", origdata);
    sound->dataSize = ndata * sizeof(short);
    SNDWriteSoundfile(argv[2], sound);
}
```

ZetaGamma.ma

```
(* Computation of Riemann Zeta function and Euler's
   constant via functional incomplete gamma
   expansions.
 *)

zeta[s_]:= Block[{n,q,lim},
                lim = 5; (* Depends on desired error bound. *)
                g = NSum[Gamma[s/2,Pi n^2]/n^s +
                    Pi^(s-1/2)*
                        Gamma[(1-s)/2,Pi n^2]/n^(1-s),
                                {n,1,lim}];
                Return[N[(Pi^(s/2)/(s*(s-1)) + g)/Gamma[s/2],20]];
        ];
gamma = -2 + N[Log[4 Pi],20] +
        2 NSum[Gamma[1/2,Pi n^2]/(n Sqrt[Pi]) +
                Gamma[0,Pi n^2],{n,1,5}]
```

References

Aho A, Hopcroft J, and Ullman J 1974, *The Design and Analysis of Computer Algorithms*, Addison Wesley Publishing Company

Aho A, Hopcroft J, and Ullman J 1983, *Data Structures and Algorithms*, Addison-Wesley Publishing Company

Akansu A N et al. 1993, *IEEE Trans. Sig. Proc.*, 41, 1, 13-19

Arnott S, Dover S D, and Wonacott A J 1969, *Acta Cryst.*, B25, 2192

Arnott S and Hukins D W L 1972, *Biochem. and Biophys. Res. Comm.*, 47, 6, 1504

Atken A O L and Morain F 1993, *Math. Comp.*, 60, 399-405

Bahn A 1992, *Galactic Dynamics: Computationand Theory*, Thesis, Reed College

Bailey D H and Swarztrauber P N 1991, *SIAM Review*, 33, 3, 389-404

Baker G L and Golub J P 1990, *Chaotic Dynamics: an Introduction*, Cambridge University Press

Barlow R B Jr., Prakash R, and Solessio E 1993, *Amer. Zool.*, 33:66-78

Barnes J and Hut P 1986, *Nature*, Dec, 324, 446-449

Barnsley M 1988, *Fractals Everywhere*, Academic Press, Inc.

Bedau M 1993, *private comm.*

Bedau M and Packard N 1992, in *Artificial Life II, Santa Fe Institute Studies in the Sciences of Complexity*, Vol. X, Addison-Wesley

Bedau M and Seymour R 1993, *private comm.*

Bedford T, Keane M, and Series C 1991, *Ergodic Theory, Symbolic Dynamics, and Hyperbolic Spaces*, Oxford University Press

Beier T and Neely S 1992, *SIGGRAPH Proc. Comp. Graphics*, 26, 2, 35

Berry M V 1988, *Nonlinearity*, 1, 399-407

Beyer R P Jr. 1992, *J. Comp. Phys.*, 98, 145-162

Bingham 1992, *Phys. Lett. A*, 163, 419-424

Black S C and Kennedy A D 1989, *Computers in Physics*, May/Jun 59-68

Borwein J M and Borwein P B 1987, *Pi and the AGM*, John Wiley & Sons, Inc.

Bosma W 1993, *Math. Comp.*, 61, 203, 97

Bracewell R N 1986, *The Hartley Transform*, Oxford University Press, New York

Briggs K 1991, *Math. Comp.*, 57, 195, 435-439

Brown L, et al. 1989, *private comm. from* Caldwell C. 1992

Buchanan J L and Turner P R 1992, *Numerical Methods and Analysis*, McGraw-Hill

Buchler et al. eds. 1985, *Chaos in Astrophysics*, D. Reidel Publishing Company

Buhler 1990, 1993, *private comm.*

Buhler J P and Crandall R E 1990, *J. Phys. A. Math. Gen.*, 23, 2523-2528

Buhler J P, Crandall R E, and Sompolski R W 1992, *Math. Comp.*, 59, 200, 717-722

Buhler J P, Crandall R E, Ernvall R and Metsankyla 1993, *Math. Comp.*, 61, 203, 151

Buhler J P and Gleason A 1993, *private comm.*

Burrus C S 1988, "Efficient Fourier Transform and Convolution Algorithms," in [Lim and Oppenheim 1988]

Burrus C S 1990, "Notes on the FFT", preprint, Department of Electrical and Computer Engineering, Rice University

Cantor C R 1990, *Science*, 247, 6.

Carlson D A 1992, *J. Supercomputing*, 6, 107-116

Chakrabarti and JaJa 1990, *IEEE Transactions on Computers*, 39, 11, 1359-1368

Chandra A 1992, *private comm.*

Chudnovksy D V and Chudnovsky G 1987, in *Ramanujan Revistited*, Academic Press

Chui C K 1992 I,II, *Wavelets: Introduction (Vol. I) and Tutorial (Vol. II)*, Academic Press

Cohen H 1993, *A Course in Computational Algebraic Number Theory*, Springer-Verlag, Berlin

Colquitt W N and Welsh L Jr. 1991, *Math. Comp.*, 56, 194, 867-870

Collett G H and Germain L 1993, *Math. Comp.*, 61, 204, 659-673

Combes J M et al., *Wavelets: Time-Frequency Methods in Phase Space*, Springer-Verlag

Cooley J W and Tukey J W 1965 *Math. Comp.* 19, 297-301

Corless R M 1992, *Amer. Mathematical Monthly*, 99, 3, 203

Crandall R E 1991, *Mathematica for the Sciences*, Addison-Wesley

Crandall R E 1993, *J. Phys. A. Math. Gen.*, 26, 14, 3627-3648

Crandall R E and Buhler J P 1987, *J. Phys. A. Math. Gen.*, 20, 5497-5510

Crandall R E and Fagin B F 1994, "Discrete Weighted Transforms and large-integer arithmetic," *Math. Comp.* [to appear]

Crandall R E and Litt B R 1983, *Ann. Phys.*,146,2,458-469

Crandall R, McClellan M, Arch S, Doenias J, and Piper R, *Analytical Biochemistry*, 167, 15-22, 15-22

Crilly E et al. eds 1991, *Fractals and Chaos*, Springer-Verlag, New York

Crochiere R E and Rabiner L R 1988, "Multirate Processing of Digital Signals," in [Lim and Oppenheim 1988]

Daubechies I 1988, *Comm. Pure and Applied Math.*, 41, 909-996

Daubechies I 1992, *Ten Lectures on Wavelets.*, SIAM, Philadelphia PA

Dorren H J S and Tip A, *J. Math. Phys.*, 32 (11), 3060-3070

Edwards H M 1974, *Riemann's Zeta Function*, Academic Press

Evans D 1987, "Bit-reversed permutation," *IEEE ASSP* Aug.

Evans D 1989, "Improved bit-reversed permutation," *IEEE ASSP* Aug.

Falconer K 1990, *Fractal Geometry: Mathematical Foundations and Applications*, John Wiley & Sons, Ltd.

Feigenbaum M J 1978, *J. Stat. Phys.*, 19, 25-32

Fiala E R and Greene D H 1989, *Comm. ACM*, 32, 4, April, 490

Fleischmann et al. eds. 1989, *Fractals in the Natural Sciences*, The Royal Society of London

Frame M and Erdman L 1990, *Computers in Physics*, Sep/Oct, 500-505

Freeman J A and Skapura D M 1992, *Neural Networks: Algorithms, Applications and Programming Techniques*, Addison-Wesley Publishing Company

Friedhoff R M and Kiely T 1990, *Computer Graphics World*, Aug.

Froberg C-E 1985, *Numerical Mathematics: Theory and Computer Applications*, Benjamin-Cummings Publishing Company, Inc.

Gash P W 1991, *Amer. J. Phys.*, 59(6), 509-515

Gekelman W, Maggs J, and Xu L 1991, *Computers in Physics*, Jul/Aug, 372-385

Goldberg A, Schey H M and Schwarz J L 1967, *Amer. J. Phys.*, 35, 177-186

Golub G H and Van Loan C F 1989, *Matrix Computations,* 2nd ed, The Johns Hopkins University Press

Graffagnino P 1993, *private comm.*

Graham R L, Knuth D E, and Patashnik O 1989, *Concrete Mathematics*, Addison-Wesley

Grassberger P and Procaccia I 1983, *Physica*, 9D 189-208

Grassberger P and Procaccia I 1984, *Physica*, 13D 34

Gray A and Knill R J 1992, *Mathematica in Education,* 1,4,12-16

Greenspan D and Casulli V 1988, *Numerical Analysis for Applied Mathematics, Science and Engineering*, Addison-Wesley

Griffiths D 1993, *private comm.*

Guy R K 1980, *Unsolved Problems in Number Theory*, Springer-Verlag, New York

Hardy G H and Wright E M 1978, *An Introduction to the Theory of Numbers*, 5th. ed., Clarendon Press, Oxford

Hautot A 1975, *J. Phys. A. Math. Gen.*, 8, 853-62

He B and Cohen R J 1992, *IEEE Trans. Biomed. Eng.*, 39,11,1179-1191

Hejhal D A 1987, *Zeros of Epstein Zeta Functions and Supercomputers*, Univ. Minn. Supercomp. Inst., January

Henrici P 1977, *Applied and Computational Complex Analysis, Vol. 2*, John Wiley & Sons, Inc.

Hertz P 1990. *Computers in Physics*, Jan/Feb, 86-90

Hillis W D and Boghosian B M 1993, *Science*, 261, 13 Aug., 856-863

Hocking WH 1989, *Computers in Physics*, Jan/Feb, 59-65

Holden ed 1986, *Chaos*, Princeton University Press, Princeton New Jersey

Hoppensteadt F C and Peskin C S 1992, *Mathematica in Medicine and the Life Sciences*, Springer-Verlag, New York

Horowitz E J 1990, *Computers in Physics*, Jul/Aug, 418-422

Hsiung P-K, Thibadeau R H, and Dunn R H P 1990, *Pixel*, Jan/Feb, 10-18

Hu N G et al. 1992, *IEEE Trans. Signal Proc.* , 40, 12

Hush D R and Horne B G 1993, *IEEE Sig. Proc. Mag.*, January, 8-39

Jacobs E W, Fisher Y, and Boss R D 1992, *Signal Processing*, 29, 251-263

Jagers P 1975, *Branching Processes with Biological Applications*, John Wiley & Sons

Kahn P 1990, *Mathematical Methods for Scientists and Engineers: Linear and Non-linear*

Systems, John-Wiley & Sons, Inc.
Karin S and Smith N P 1987, *The SuperComputer Era*, Harcourt Brace Jovanovich, Inc.
Katzenelson J 1989, *SIAM J. Stat. Sci. Comp.*, 10,4,787-815
Keiper J 1992, *private comm.*
Keller W 1992, *Math. Comp.*
Kim R C and Lee S U 1992, *Signal Processing*, 28, 77-90
Knuth D 1981, *The Art of Computer Programming*, Vols I,II, 2nd Ed., Addison-Wesley
Koblitz N 1984, *Introduction to Elliptic Curves and Modular Forms*, Springer-Verlag, New York
Koblitz N 1987, *Course in Number Theory and Cryptography*, Springer-Verlag, New York
Koonin S E 1986, *Computational Physics*, Benjamin/Cummings Publishing Co., Inc.
Kosko B 1992, *Neural Networks and Fuzzy Systems: a Dynamical Approach to Machine Intelligence*, Prentice Hall, Englewood Cliffs, N.J.
Johnson J and Tolimieri R 1989, *FFT algorithms for parallel computers*, Springer-Verlag, New York
Keiper J 1992, *Math. Comp.*, 58, 198, 755-764
Keller H B and Swenson J R 1963, *Math. Comp.*, 17, 223-230
Lagarias J C, Miller V S, and Odlyzko A M 1985, *Math. Comp*, 44, 537-560.
Lagarias J C and Odlyzko A M 1987, *J. Algorithms*, 8, 173-191.
Le-Ngoc T and Vo M T 1989, *IEEE Micro*, 9, 5, 20-27
Leddy J 1990, *private comm.*
Lee M A and Schmidt K E 1992, *Computers in Physics*, 6, 2, 192-196
Lelewer D A and Hirschberg D S, *ACM Computing Surveys*, 19, 3, Sept.,261
Lenstra A K and Lenstra, H W Jr., eds 1993, *The Development of the Number Field Sieve*, Springer Lecture Notes 1554
Lenstra A, Lenstra H W Jr., Manasse M, and Pollard J M 1993, *Math. Comp.*, 61, 203, 319
Lim J S and Oppenheim A V 1988, *Advanced Topics in Signal Processing*, Prentice-Hall, New Jersey
Ling F H and Schmidt G 1992, *J. Comp. Phys.*, 99, 196-202
Lippman R P 1987, *IEEE Acoust.. Sp. Sig. Proc. Mag.*, April, 4(2), 4-22
Longfellow H W 1874, *The Theologian's Tale, Elizabeth, iv*
Lorenz E N 1963, *J. Atmos. Sci.*, 20, 130-41
Lu M and Chiang J-S 1992, *IEEE Trans. Comp.*, 41,8,1026

Lune van de, te Riele H J J, and Winter D T 1986, *Math. Comp.*, 46, 667-681

Mallat S G 1989, *IEEE Trans. Patt. Rec. Mach. Intell.*, 11, 7, 674-693

Malvar H S 1987, *IEEE Trans. Acous. Sp. and Sig. Proc.*, ASSP-35, 10, 1484-1485

Mann P J 1987, *J. Comp. Phys.*, 72, 467-485

Marsaglia G 1991, *J. Supercomp.*, 5, 49-55

May M and White R H 1967, *Math. Comp. Phys.*, 7, 219

Mayer R and Buhler J 1993, *private comm.*

McCammon J A and Harvey S C 1987, *Dynamics of Proteins and Nucleic Acids*, Cambridge University Press, Cambridge

McClellan J H and Rader C M 1979, *Number Theory in Digital Signal Processing*, Prentice-Hall, Englewood Cliffs, NJ

Meyer Y 1991 ed., *Wavelets and Applications*, Springer-Verlag, Masson

Montgomery P 1987, *Math. Comp.*, 48, 177, 243-264

Montgomery P 1992, *Ph.D. Thesis*, UCLA

Montgomery P and Silverman R 1990, *Math. Comp.*, 54, 839-854

Moorer A 1979, *Audio Eng. Soc. preprint* 1443, 62nd AES Convention, Brussels, Belg.

Morain F 1990, *Advances in Cryptology*-EUROCRYPT '90, 110-123

Morales J J 1993, *J. Comp. Chem.*, 14, 6, 728-735

Morlet J et al. 1982, *Geophysics*, 47, 203-236

Niedereiter H 1992, *Random Number Generation and Quasi-Monte Carlo Methods*, SIAM, Philadelphia, PA.

Nussbaumer H J 1981, *Fast Fourier Transform and Convolution Algorithms*, Springer-Verlag, New York

Odlyzko A M 1993, *Math. Comp. Lehmer Memorial Issue*

Odlyzko A M 1993b, *private comm.*

Olson C L and Olsson M G 1991, *Amer. J. Phys.*, 59 (10), 907-911

Pao Y 1989, *Adaptive Pattern Recognition and Neural Networks*, Addison-Wesley

Pender J and Covey D 1992, *IEEE Trans. Sig. Proc.*, 40, 8, 2095-2097

Peskin C S 1977, *J. Comp. Phys.*, 25, 220

Pinkerton B 1993, *private comm.*

Pomerance C, Smith J W and Tuler R 1988, *SIAM J. Comp.*, 17, 2, 1988

Potvin J 1991, *Computers in Physics*, May/Jun, 333-338

Powell J W and Crandall R E 1993, *Mathematica in Education*, Spring, 15-20

Press W H et al. 1988, *Numerical Recipes in C*, Cambridge University Press

Press W H and Teukolsky S A 1989, *Computers in Physics*, Jan/Feb, 91-94

Press W H and Teukolsky S A 1989b, *Computers in Physics*, Sep/Oct, 84-86

Press W H and Teukolsky S A 1992, *Computers in Physics*, Jan/Feb, 82-83

Press W H and Teukolsky S A 1992b, *Computers in Physics*, Sep/Oct, 522-524

Prusinkiewicz P and Hanan J 1989, *Lindenmayer Systems, Fractals, and Plants, Lecture Notes in Biomathematics*, Springer-Verlag, New York

Rabbani M and Jones P W 1991, *Digital Image Compression Techniques*, SPIE Optical Engineering Press, Bellingham, WA

Rabiner L R 1979, "FFT subroutines for sequences with special properties," in *IEEE ASSP Programs for Digital Signal Processing*, IEEE Press, New York

Resnikoff H L 1989, *Aware preprint* AD890507.1, Aware, Inc., Cambrdige MA

Resnikoff H L and Burrus C S 1990, *SPIE Adv. Sig. Proc. Alg., Arch., Imp.*, 1348, 291-300

Ribenboim P 1988, *The Book of Prime Number Records*, Springer-Verlag New York, Inc.

Riesel H 1985, *Prime Numbers and Computer Methods for Factorization*, Birkhauser

Roche C 1992, *IEEE Trans. Sig. Proc.*, 40, 5, 1273-1276

Rosen J and Shamir J 1989, *Applied Optics*, 28,2,240-244

Russell S 1992, *Thesis*, Reed College

Schiff 1992, *Optical Engineering*, 31, 11, 2492-2495

Schroeder M 1991, *Fractals, Chaos, Power Laws*, W. H. Freeman and Company, New York

Senoo T and Girod B 1992, *IEEE Trans. Image Proc.*, 1,4,526

Shiu P 1986, *Math. Comp.*, 47, 175, 351-360

Silverman R 1991, *private comm.*

Sinisalo M K 1993, *Math. Comp.*, 61, 204, 931

Skeel R D and Keiper J B 1993, *Elementary Numerical Computing with Mathematica*, McGraw-Hill, Inc.

Smith J O and Gossett P 1984, *Proc. IEEE Int'l. Conf. Acoust. Sp. Sig. Proc.*, San Diego, March, 19.4.4

Snyder G 1993, *private comm.*

Sorenson et al. 1987, *IEEE Trans. Acous. Sp. and Sig. Proc.*, ASSP-35, 6, 849-863

Stein A D, Wang Z, and Leigh J S Jr. 1992, *Computers in Physics*, 6,4,Jul/Aug, 389-392

Stoer J and Bulirsch R 1993, *Introduction to Numerical Analysis*, Springer-Verlag

Strang G 1989, *SIAM Review*, 31, 4, 614-627

Stump D 1986, *Amer. J. Phys.*, December 54(12), 1096-1100

Sturtevant M and Newlin J 1989, *private comm.*

Tevanian A 1991, *private comm.*

Tobochnik J, Gould H and Mulder K 1990, *Computational Physics*, Jul/Aug, 431-435

Traub J F and Wozniakowski H 1992, *Bull. Amer. Math. Soc.*, 26, 3-52

Trefethen L N 1992, *SIAM News*, November, 6

Truhlar D G ed. 1988, *Mathematical Frontiers in Computational Chemical Physics*, Springer-Verlag, New York

Tufillaro N B, Abbott T, and Reilly J 1992, *An Experimental Approach to Nonlinear Dynamics and Chaos*, Addison-Wesley Publishing Company

Vardi I 1991, *Computational Recreations in Mathematica*, Addison-Wesley

Vardi I 1993, *private comm.*

Varga R S 1990, *Scientific Computation on Mathematical Problems and Conjectures,* SIAM, Philadelphia, PA.

Verhulst P 1845, *Nouv. Mem. de l'Academia Roy, des Sciences et Belles-Lettres de Bruxelles,* XVIII.8,1-38

Vetterli M 1988, "Trade-offs in the computation of mono- and multidimensional DCT's.," Center for Telecom. Res. Tech., Report CU/CTR/TR-090-88-18, Columbia University, NY

Wagon S 1991, *Mathematica in Action*, W.H. Freeman & Co., New York

Wagstaff S 1993, *Factoring Newsletter*, from Purdue Univ., e-mail: ssw@cs.purdue.edu

Wang Y, Twizell E H, and Price W G 1991, *Comm. Appl. Numer. Meth.*, 8, 511-518

Wang Z Q et al. 1993, *Applied Optics*, 32, 2, 184-189

Watson J D 1990, *Science*, 247, April 6.

Wegner T et al. 1992, *Fractals for Windows*, Waite Group Press, Corte Madera CA

Whittaker E T and Watson G N 1972, *A Course of Modern Analysis*, 4th ed., Cambridge Univ. Press, New York

Winfree E, Smith B and Gayley T, *private comm.*

Wozniakowski H 1991, *Bull. Amer. Math. Soc.*, 24, 185-194

Index

A

Abbott T 229, 454
AC pulse 201-202
acoustics 198
Action potential 87
Adams J xiv
adaptive filters 74
adaptive resonance systems 74
ADCT 328-329 (see also DCT)
adenine 89
Adleman-Pomerance-Rumely-Cohen-Lenstra (see APRCL test)
affine transformations 329
AGM (see arithmetic-geometric mean)
Aho A 116, 135, 295, 447
Aitken 5
Aitken Δ^2 formula 9
Aitken method 64
AitkenLimits.ma 333, 6, 16
Akansu A N 447
Alford 127

algorithm 1, 2
analog systems 61
analytic number theory 27
angular frequency 244
annealing 74
ansatz 250
Antonini 329
Append[ ] 269
APRCL test 127
Arch S xv, 449
Archimedes 10
area of polygons 54
area-preserving property 231
arithmetic encoders 296, 298
arithmetic-geometric mean 4, 11, 12, 18, 94
Arneodo 288, 290
Arnold tongues 241
Arnold's Cat map 231, 241
Arnott S 89, 447
artificial life 68, 76-78
ascending series 22-27
Asimov I vii
astrophysics 94, 96
asymptotic series 22-27
Atken A O L 143, 447
atomic site 90
atoms 89
attractor 229, 231, 232, 237, 239, 243, 248, 249, 252, 267, 268
Attractor8.ma 333, 236

B

Babylonians 10
backward diffusion 88
Bahm A 96, 447
Bailey D H 179, 182, 447
Baker G L 291, 447
baker's map 231, 238, 241, 245, 252
balanced digit representation 132
bandwidth 293
Barisich A & A xv
Barisich B xv
Barisich L xv
Barlow R B Jr. 89, 447
Barnes J 96, 447
Barnsley M 266, 268, 447
base pairs 89, 90
Bedau M xv, 76-78, 448
Bedford T 254, 448
Beier 323-324, 448
Benes J xv
Bernoulli numbers 4, 123, 124, 136
Berry M V 448
Bernstein conjecture 63-64
Bessel functions 22, 30-31, 67
Beta function 28
Beyer R P 88, 448
bifurcation 236, 238-240, 243, 245, 254
Binet (see Stirling-Binet)
Bingham 282, 284, 287, 448
biology 79
birthday paradox 139
bitstream encoding 300, 301, 304
Bittner A xiv
Black S C 45, 448
Bluestein 182
Boghosian B M 97, 181, 450
Borwein J M and P B 93-94, 4, 8, 10-12, 18, 19, 448
Bosma W 119, 448
Boss R D 450
box dimension 281-282, 284, 287
box/capacity dimension 256
Boynton L xiv
Bracewell R N 183, 448
Bragdon P and N xv
Brent R 4, 125
Briggs K 90, 239, 448
Brillhart 142
Brouncker's formula 11
Brown L 120, 125, 448
Brownian motion/fractals 161, 272-275, 280, 284, 289
Brownian mountains 278
Brownian.ma 273, 334
Brun's constant 125
BSTW compressors 296, 302-303
Buchanan J L 2, 448
Buchler 254, 282, 448
Buffon's needle 52-53

Buhler J P xiv, 56, 58, 107, 108, 116, 123, 132, 133, 135, 448-449, 452
Buhler L xv
Bulirsch R 2, 4, 35, 37, 38, 453
Burrus C S 151, 159, 172, 175-176, 178, 226, 448, 453
butterflies (FFT style) 167
Butz G xv
Byron M xiv

C

Calderon 197
camera 293
Cantor C R 92, 448
Cantor measure 288, 290
Cantor set 241, 254-256, 258-259, 278, 288, 290
CantorGen.ma 224, 258, 260
capacity dimension 256, 281-284
Carlson D A 179, 448
Carmichael number 127
Casabona H xv
Cascade integrals 58
Cassini division (Saturn) 101-102
Casulli V 2, 36, 450
Catalan constant 29, 32
Cayley's operator 104
Chakrabarti 186, 448
Chandra A 448
chaos 82, 229-231, 236, 238, 240, 242, 244-246, 250-251, 253-254
chaotic map 255, 284
Chebyshev polynomial 60
chemistry 92
Chiang J-S 136, 451
Chinese Remainder Theorem 123, 132, 135, 194
chirp wave 160
Christenson S xv
chroma 318, 325
chromatographic models 58
chromosomes 92
Chudnovksy G and D V 448, 11, 12
Chui C K 201, 203, 448
cicle problem (Gauss') 65
circle error 66
circle map 231
classical physics 93
CODEC 312
Cohen H 121, 127, 134, 141, 144, 448
Cohen R J 84, 450
coherent states 198
Coifman 198
Colgrove M xv
Collett G H 449
color reproduction 56
Colquitt W N 118, 449
Combes J M 198, 203, 449
combinatorial analysis 304-305
combinatorial bracket 56, 304
compact support 212
complex analysis 62, 66, 99
complex plane 246-247, 275
complexity 68
compression 198, 224-225, 300, 315
Compressors.ma 297, 298, 335
Compton wavelength 110
congruent numbers 60
conical pendulum 100
ContFract.ma 6, 16, 30, 138, 341
continued fraction 13-14, 16, 62, 254
continuous wavelet transform 202
contour plots 97
convex regions 248
convolution 130, 200
Convolution Theorem 157, 174, 180, 183, 194, 201
Cooley J W 167, 449
Cooley-Tookey 151, 170-172, 175, 178
Corless R M 64, 254, 449
Corliss G xv
coronary arteries 319
corrdim.c 284-286, 342
correlation dimension 282, 286-287, 291
Cottet 319
Coursey J xiv
Covey D 189, 191, 452

CPU xiv
Crandall R E 87-88, 105, 107-110, 118, 123, 129, 130, 132, 134-135, 138, 140, 143, 146, 181, 195, 238, 268, 448-449, 452
Crank-Nickolson methods 104
crilly E 268, 280, 449
Crochiere R E 317, 449
Crow G xiv
crystals 106-108
Cunningham 144
cytosine 89

D

damping 102
Danielson-Lanczos identity 157, 159, 165-166, 168, 171, 184
DanielsonLanczos.ma 166, 344
Daubechies I 198, 202-203, 206, 209, 211, 212, 214, 216, 223, 225, 449
Daubechies.ma 345, 212
DCT 151, 184, 186, 316, 328
decorrelation 201
Deleglise M 120
Delord conjecture 107
Delord J xv
DeMoney M xiv
determinant 225
devil's staircase 241
DFT 158-160, 174, 181, 183, 186, 218, 226, 328
DFTgraphs.ma 160, 346
dielectric breakdown (DBM) fractal 269
differential equation 40-41, 82, 86-87, 93, 95
difficulty level ix
diffusion-limited aggregation (DLA) 269-270
digital filters 210, 281, 317
digital holography 321
dilation equation 207, 210-212, 258-260
dimension (fractal) 256, 260, 266, 271-279, 281
Diophantine relations 61
dipole 85, 96
Dirichlet L-functions 28
discrepancy integration 46-49
discrete cosine transform (see DCT)
    Fourier transform (see DFT)
    wavelet transform 204, 216, 218, 310
divisionless divide 133
DNA 89, 109
Doenias J xiv, 118, 147, 449
Dorren H J S 259, 449
downsampling 312
dragon fractal 262, 264, 290
Dragon.ma 262, 349
DragonRecursion.ps 347, 262
Duffing system 245
Duhamel 159
Duhamel-Hullman 175
Dunn R H 450
Dunne T xv
dynamical systems 281, 283

E

e, method for evaluating 13-15
Earth-moon system 102-103
Edwards H M 63, 449
eigenvalue 36, 59, 63, 106
Eigenvalue.ma 33, 36, 350
Einstein A 110
EKG signals 84
electric dipole 84
    field lines 97
    charge 97, 253, 259
electrocardiography 85
Elias codes 300-302
eliascoder.[ch] 300-302
eliastest.c 297, 299, 353
Elliott G xv
elliptic curve 137, 142, 143, 147
elliptic integral 93
encryption system 253
entropy 294, 305, 312-313, 318, 325

epidemiology 79
equilibrium chemistry 58
equipotential 97-98, 254
Equipotential.ma 93, 355
Erdman L 268, 449
erf() (see error function)
Erfc.ma 355, 30
Erlang-Steffenson extinction 81
Ernvall R 448
error function 22-24, 31, 45
Euclid algorithm 134
Euler constant xiii, 6, 9, 29, 32
Euler method 38-40
Euler numbers 124, 128
Euler-angle rotation 90
Euler-Lotke equation 82-83
Euler-Maclaurin 4, 9, 48
Euler-Mascheroni 6
Evans D 176, 449
evolution 68
Ewald expansion 108
extended Euclid method 134

F

factoring 136
Fagin B xv, 118, 129, 132, 134, 140, 143, 146, 181, 195, 449
Falconer K 256, 272, 276, 449
fast wavelet transform 222, 224, 226
Feigenbaum constant 239-240
Feigenbaum M J 449
Fermat 133, 144
Fermat number transform 195
Fermat numbers 119, 127, 135, 145
Fermat primes 145
Fermat test 122
Fermat's "Last Theorem" 121
Fermat's little theorem 116
FermatConvolution.ma 145, 356
Fermi's Zero-th Rule 215
FFT xii, xiii, 96, 129-130, 132-133, 136, 140, 142, 151, 155, 159, 163, 165, 174, 176, 201, 214, 223, 226, 260, (cont.) 278, 280, 310, 316, 318-319, 324, 328-329
fft.c 162, 173-174, 177-180, 367
fft2D_real.c 179-180, 359
FFTs.ma 173-174, 194, 357
fft_real.c 173, 177, 375
Fiala E R 296, 449
Fibonacci codes 302
Fibonacci numbers 79, 128
Fisher Y 450
Fleischmann 270, 449
Floyd cycle-finding algorithm 138
fluid flow 88, 97
food function 78
Forrest 74
FORTRAN 173, 175
Fourier transform 130, 199, 202, 216, 259, 272, 275 (see also FFT, DFT)
fractal 216, 229-230, 242, 255, 257-258, 261, 264, 266, 269, 291
fractal dimension 226, 251-252, 256, 262, 270, 272, 280, 284, 287-290
fractal mountain surface 279-280
fractalsound.c 307, 310, 381
Frame M 268, 449
Freeman J A 74, 449
frequency modulated signal 201
frequency spectrum 253
Friedhoff R M 293, 449
Froberg C-E 1, 2, 4, 15, 449
Frye R xiv
FSWT 191
FSWT.ma 191, 383
functional spaces 198
fuzzy logic 73
fwt.[ch] 222, 224, 314, 326, 386
FWTInPlace.ma 383, 222
FWTNotInPlace.ma 384, 222

G

Gabor 197
Gabor transform 201
Gage 118

galaxies 274
Galois transform 132, 195
gamma function 31, 122 (see also incomplete gamma function)
Gash P W 99, 450
Gauss 4, 11, 66, 151
Gauss circle problem 65
Gauss map 254
Gauss quadrature 47
Gaussian 44, 45, 88, 106, 102, 104, 195, 201-202
  packet 103
  random variable 161, 273
  reduction technique 142
  statistical ensembles 51
  unitary ensembles 63
Gayley T xiv, 454
Gekelman W 110, 450
genes 89
genetic algorithms 74, 77
genetic code 75, 91
Genetic.ma 77, 395
Gentleman-Sande 175
geometric mean 4, 15
Germain L 319, 449
giants.[ch] 114, 115, 121, 129, 132-135, 396
Gibreath conjecture 121
Gillespie S xiv
gingerbread map 232, 241
Girod B 453
Gleason A 56, 448
global truncation error 39
Goertzel 163
Goldbach conjecture 128
Goldberg A 104, 105, 450
golden ratio 79
golf xiii, xiv
Golub G H 2, 447
Golub J P 238, 245, 252, 291, 447
Gossett P 311, 312, 453
Gostin 140
Gould H 454
Graffagnino P xiv, 56, 450
Graham R L 4, 16, 450
Granville 127
grapestake problem 53
Grassberger P 282, 284, 291, 450
gravitation 94, 96
Gray A 450
Gray codes 75, 78, 263
Gray T xiv
Green P xv
Greene D H 296, 449
Greenspan D 2, 36, 450
Griffiths D xv, 86, 450
Grossman 198
ground state 106
guanine 89
Guy R K 60, 64, 124, 450
Gwilliam G xv

H

Haar wavelet system 197, 207, 210, 219, 204, 206, 223
Hadamard J 315
Hadamard matrices 189
Haden W xv
Hajhal 62
Hallstrom C xiv
Hamilton's equations 42
Hamiltonian 42, 63, 104, 244
Hamming Error 305
Hanan J 264, 265, 453
Hankel function 22, 30, 105
Hardy G H 66, 118, 119, 127, 450
harmonic oscillator potential 102
Hartley transform 151, 177, 182-184, 186
Harvey S C 109, 452
Hausdorff dimension 256, 272
Hautot A 108, 450
He B 84, 450
Heaviside unit step function 282
Hejhal D A 450
Henon attractor 231, 241-242, 253, 285
Henon.ma 240, 399
Henrici P 450, 7, 8, 10
Hermite polynomials 106

Hermitian symmetry 155, 156, 159, 163-164, 274, 280
Hertz P 181, 450
Hilbert 263, 310
Hilbert curve 75, 77, 263-264, 290
Hilbert.ma 400, 262-263, 400
Hildreth 319
Hillis W D 97, 181, 450
Hindus 1
Hirschberg D S 296, 298, 302-303, 451
Hocking W H 181, 450
Hodgkin-Huxley oscillator 85-87
Holden 234, 243, 245, 450
holography 321
Hopcroft J 447
Hoppensteadt F C 82, 85, 450
Horne B G 69, 71-72, 450
Horowitz E J 97, 450
horseshoe map 231
Hourvitz L xiv
Hsiung P-K 110, 450
Hu N G 184, 450
Huffman codes 295-297, 300, 302, 304, 314
huffman.c 297, 302, 402
Hukins D W L 447
Hullman 159
Hullot J-M xiv
Human Genome project 91
Hunter W xiv
Hurst exponent 272, 280
Hush D R 69, 71, 450
Huskins 89
Hut P 96, 447
hypergeometric function 24-26

I

IFS fractal, fern, tree 266-268, 329
IFS.ma 262, 267-268, 404
image compression 326
image processing 225, 290, 318
incomplete gamma function 26-27, 31
information dimension 282
information theory 176, 281
integration (numerical) 59, 94, 104, 252
  one dimensional 47-48
  higher-dimensional 46-48
invariant distribution 239-246, 251
Invariant.ma 405, 232
inverse wavelet transform 208, 222-223
  continuous version 202
  Fourier transform 199
inverse matrix 60
inverse.c 33, 405
ion channel action potentials 69
iterated function system (see IFS)
iterative map 251

J

Jacobi theta functions 109
Jacobian matrix 37-38, 72
Jacobs C xv
Jacobs E W 329, 450
Jagers P 81, 450
Jaja 186, 448
Jobs L xv
Jobs S xiv
Johnson J 181, 451
Jones A xv
Jones P W 185-186, 295, 326, 328, 453
JPEG 328, 329 (see also DCT)
Julia set 245-249
Julia.ma 240, 246, 408

K

Kahn P xv, 450
Kaplan R xv
Kaplan-Yorke relation 291
Karatsuba algorithm 129
Karin S 51, 451
Karman vortex sheets 99
Katzenelson J 96, 451
Keane M 448
Keiper J xiv, 2, 22, 32, 239-240, 451
Keller H B 68, 451

Keller W 146, 451
Kennedy A D 45, 448
Khinchin 64
  constant 15
Kim R C 317, 451
Knill R J 55, 450
Knuth D E 2, 43-44, 133-134, 450-451
Koblik S xv
Koblitz N 60-61, 142, 451
Koch fractal 257, 260-261
Koonin S E 196, 451
Kosko B 73, 451
Kuethe D xv

L

L-systems 262, 264-266
Lagarias J C 121, 451
Lagaris-Miller-Odlyzko 121
Lagrange multiplier 306
Lagrange quadratic surd theorem 64
Lagrange's continued fraction 20-21
Laird A xiv
Landau 66
Laplace equation 99
Laplacian operator 84-85
  probability density 45, 318-319, 321
  sampling pulse 201
  signal 206-209
Larson C xv
Le-Ngoc T 183-184, 451
Leach K xv
least-squared-error regression 37
Leddy J 58, 451
Lee M A 106, 451
Lee S U 317, 451
Leeb E xv
Legendre 4, 11, 120
Lehmer 63, 121, 134
Leigh J S Jr. 453
Lelewer D A 296, 298, 302-303, 451
Lempel (see LZW compressors)
Lenstra A K 144, 147, 451
Lenstra, H W Jr. 451
Lerch-Hurwitz Zeta function 28
Levich M xv
Levy flights 274
Levy P 254
lexicographically 327
Lim J S 451
Lindemayer system (see L-system)
line integrals 98
linear congruential method 44
  equation systems 32-34
  filters 314
  intensity gradient 328
  regression 37
Ling F H 287, 451
Lippman R P 72, 451
Litt B R 105, 449
Littlewood-Paley wavelet 203
logarithmic radial transforms 322
Longfellow H W 293, 451
Lorentz 110
Lorenz attractor 242-244, 252, 286
Lorenz E N 451
Lorenz.ma 409, 240
lossless compression 296-297, 304-306
lossy compression 296, 305-306, 314, 325
Love S xv
LSystem.ma 262, 264, 409
Lu M 136, 451
lub norm 35
Lucas numbers 128
Lucas-Lehmer test 117-119, 127, 146
Lune van dc 452
Lupino P xv
Lyapunov dimension 291
  ellipsoid 252
  exponent 250-254
LZW (Lempel-Ziv) compressors 296-297, 303, 316

M

Madelung constant 99, 107-108
Maehly polynomial solver 38

Maggs J 450
magnetic resonance imaging 198
Mallat S G 198, 289, 452
Malthusian growth 79, 81-82, 232
Malvar H S 186, 452
Manasse M 451
Mandelbrot set 229, 245-248, 253
Mann P J 110, 452
Mapes 121
Maple x
Markovian power law 224
Marr-Hildreth 319
Marsaglia G 45, 452
Masheroni (see Euler-Mascheroni)
Mathematica x, xi, 18, 22, 67, 97-99, 114-115, 154, 165, 172-173, 223, 268, 285, 307, 311
MathTensor 110
MatLab x
matrix algebra 33-35, 59, 88, 198, 220-221, 225, 294
  software 33
  inversion 25, 34, 35
  multiplication 34
Matteson T xiv
May M 110, 452
Mayer R xv, 58, 452
McCammon J A 90, 109, 452
McClellan J H 178, 180, 452
McClellan M xv, 449
McCurley K xiv
McFarlane E xv
McNamara B xv
medical imaging 319
medicine 84
Meissel 121
Mellin-Barnes representation 24
Menasse M 147, 451
Mercury 110
Mersenne primes 116-117, 120, 127-128, 133, 144, 194-195
Metsankyla 448
Mexican hat 202, 319
Meyer M xiv
Meyer Y 198, 209, 225, 288, 290, 452
midpoint method 38-39
Miller V S xv, 451
Mimas 101
Minnick M xiv, 315
mode-locking phenomena 231, 241
modified Bessel function 30-31
molecular biology 89
molecules 89-90, 107
Monte Carlo 42, 46, 48-49, 81, 91, 106
Montgomery P 134, 140, 142-143, 452
Moorer A 315, 452
Morain F 127, 143, 447, 452
Morales J J 106, 452
Morlet wavelet 203
morphing 322-323
Morrison 142
Moss D xiv
mother functions for wavelets 206-208, 217
Mountains.ma 273, 280, 410
move-to-front 296, 302
Mulder K 454
multidimensional Newton method 37-38
MultiNewton.ma 38, 411
multiresoluiton analysis 198, 208, 211, 216, 219, 222
music 198, 310
mutation 73-76, 78

N

N-body problem 94, 95
Navier-Stokes 88
Neely S 323-324, 448
negacyclic convolution 184
neural networks 68-71, 73-74
Neural.ma 411, 70
neurobiology 84
neuron propagation model 86
neurophysiology 198
Newlin J xiv, 453
newton attractors 248
Newton fractals 13, 249

Newton method (Newton iteration)
 3, 12, 13, 16, 17, 18, 19, 35, 36, 37, 38
Newton method divide 133
Newton's law 41
NewtonMatrixInverse.ma 33, 412
NewtonTrefoil.ma 412, 240
NeXT, Inc. vii, xiv
NEXTSTEP x, xii, xiv, 89, 307, 310, 324
Niedereiter H 43-44, 46, 49, 452
noise (1/f) 161
 generator 253-254
 spectrum 288
non-linear diffusion 319
 equation systems 37-38
 dynamics 229
Norrie C xiv, 147
NTSC video 325
nucelotides 92
nuclear engineering 198
number field sieve (NFS) 137, 144
number theoretic transform 151
 functions 192-193
numerical integration (see integration)
Nussbaumer convolution 132, 136, 177
Nussbaumer H J 132, 136, 177, 195, 452
Nyquist 165

O

O() 3
object-oriented x
Objective-C x
Odlyzko A M xv, 63, 120-121, 451-452
Olson C L 245, 452
Olsson M G 245, 452
Oppenheim A V 451
optical recognition 322
optics 198
Ormond R xv
orthonormal bases 198, 207

P

Packard N 76, 79, 448
Pade approximants 19-21, 124
Page R xiv
Pao Y 73, 452
Paquette M xiv
parallel inversion scheme 134
 transform compression 314, 317
parallelism 89
Parseval identity 155, 156, 157, 163, 203, 206, 223, 325
partition-of-unity theorem 214
Patashnik O 450
path integration 105
pattern recognition 322
Peano curve 264-265
Pender J 189, 191, 452
Pender-Covey square wave basis functions 191
pendulum 93, 244, 252
Pepin test 119, 127, 135, 146-147
perceptron 69-71
period-doubling 236
Perkinson D xv
Peskin C S 82, 85, 450, 452
phase space analysis 198, 244
PhiHat.ma 413, 212
physics 92
physiology 83-84
π, method for evaluating 10-12, 21
Pi.ma 414, 6
Pinkerton B xiv, 91, 452
Piper R 449
pixel 317
planar geometry 51-52
planimetry equation 54
Poisson identity 64-66
 probability density 44, 299
polar method 44
Pollard J M 143, 451
Pollard Rho factoring algorithm 138-139, 144, 147
polylog function 28

PolynomialSolve.ma 38, 79, 81, 414
polypeptide 90
PolySolver.ma
Pomerance C 127, 143, 452
population biology 79, 230
 dynamics 82, 229
Portland OR vii
PostScript xii
Potts 109
Potvin J 109, 452
Powell J W xv, 109, 452
Powell L xv
power law 288
 method (for eigenvalues) 36
Prakash R 447
precession of perihelion of Mercury 110
predictor-corrector 327-328
Press W H 2, 19-20, 36, 44, 104, 171, 180-181, 452, 453,
Price W G 454
primality proofs 125, 126
prime number xiii
prime-factor FFT algorithms 178
Prinz A xv
probability distribution 42, 43, 54, 81, 233-234
Procaccia I 284, 291, 450
Prusinkiewicz P 264-265, 453
Purdy Pelosi S xv

Q

quadratic convergence 2, 18, 35
 map 232, 240, 252-254, 286
 sieve 137, 142-143
quantum chemistry 106
 theory 51, 59, 102-103
quasi-Monte Carlo integration 46
quasi.c 415, 47

R

Rabbani M 185, 186, 295, 326, 328, 453
rabbit problem 79-80
Rabin-Miller 127
Rabiner L R 158, 317, 449, 453
Rademacher functions 190
Rader C M 179-180, 195, 452
Radon transform 280
Ramanujan 10, 48
random numbers 42-46
random walk (see Brownian motion/fractals)
random() call 43
real signal FFTs 177
real-result inverse FFT 177
real-valued FFT 151
reciprocation (inversion) 17
recursive multiplication 35
RecursiveFHT.ma 183, 184, 417
Redwood City vii
Reed College vii, xv
Reilly J 229, 454
rejection method 43
relativity 110
relaxation 99
renormalization group 198
Renyi entropy 281
repunit numbers 128
Resnikoff H L 206, 226, 453
retina 89, 293
Reynolds R xv
RGB 325, 326
Ribenboim P 113, 118, 122, 125, 453
Riccatti 81
Richardson 5, 63
Riemann 28
 hypothesis 62
 Zeta function 6, 15, 21-22, 27-29, 32, 51, 54, 62-63, 107-108
Riesel H 121, 142-143, 146, 453
Ringle M xv
Ripoll 91
Rivat 120
RKF-45 x
RMSE (RMS error) 163, 325, 329
Roche C 181, 453
Romberg 5
Romberg extrapolation 6, 10
 integration 47-48

RombergLimits.ma 417, 6, 16
Rosen J 322, 453
run-length encoding 304 305
Runge-Kutta method x, 40-42, 94-85, 244
Russell P xv
Russell S xiv, 453

S

Salamin 4
sample rate 307-311
satellite orbits 101
Saturn 101-102
scaling law 288-289
Schey H M 450
Schiff 203, 453
Schlickeiser G xv
Schroeder M 229, 241, 248, 274, 280, 281, 308, 310, 453
Schroedinger equation 59
Schulz' matrix iteration 35
Schwarz J L 450
Seamons J xiv
self-similarity 229, 308, 309, 311
semitone 310, 311
Senoo T 453
Series C 448
sessile drop 100
Sessile.ma 93, 419
Seymour R 77, 448
Shamir J 322, 453
Shiu P 68, 453
Shroedinger equation 103-106
Siegel C L ix, 62, 124
Sierpinski arrowhead 261
Sierpinski gasket 268, 290
signal processing 74, 160, 163, 182, 293-294, 299, 306
Silverman R xiv, 140, 144, 452-453
Simpson's integral approximation 47-48
sinc() 311, 312
Sinisalo M K 128, 453
Skapura D M 74, 449
Skeel R D 2, 453
SloshingPacket.ma 102-103, 419
Slowinski D xiv, 118, 120
Smith B xiv, 454
Smith J O xiv, 311, 312, 453
Smith J W 452
Smith N P 451
Smith-Gossett 311, 312
Smitley D 118-119
smoothness (of integers) 137
Snyder G 324, 453
Solessio E 447
soliton 105
Sompolski R W 448
Sophie Germaine prime 119
Sorenson 177, 453
sound compression 307, 313-316
SoundEntropy.ma 307, 312-313, 419
SoundFileReader.ma 307, 420
SoundKit 307
space-filling fractal 265, 279
spatial derivative 318
spectral power law 260
speech compression 317
    discrimination 198
sphere.c 422, 47
splines 198
split-radix FFT 177, 178
square wave transform 151, 189-191
star distribution 96
statistical fractals 257, 261
Stein A D 321, 453
Stirling-Binet series 122-123
Stoer J 2, 4, 35, 37-38, 453
Strang G 197, 453
strange attractor 229, 243, 281, 283
    (see also attractor)
Strassen algorithm 34-35
Strassen matrix recursion 129
strassen.[ch] 33, 426
Strassen.ma 33-34, 430
strassentest.c 423
Stucki N xv
Stump D 453
Sturtevant M 453
sub-band coding 198, 317
subquadratic convergence 8

Swarztrauber P N 182, 447
Swenson J R 68, 451
SWT (see square-wave transform)

T

tablecloth fractal 266
te Riele H J J 452
television 293
tent map 241, 253, 255-259
Teukolsky S A 19-20, 44, 180, 181, 453
Tevanian A xiv, 453
theta functions (see Jacobi theta functions)
Thibadeau R H 450
Thomas 91
three-body problem 96
threebody.c 93, 95, 431
thymine 89
TIFF xii
timing.c 432
Tip A, J. 259, 449
Titus W xv
Tobochnik J 106, 454
Tolimieri R 181, 451
topological dimension 273, 289
Traub J F 46, 454
Trefethen L N 1, 454
Tribble G xiv
Truhlar D G 106, 454
Tufillaro N B 229, 454
Tukey J W 167, 449
Tuler R 452
Tunnell's theorem 61
Turner P R 2, 448
Twizell E H 454
TwoCycle.ma 433, 232

U

Ulam's number 64
Ullman J 447
Unix 303, 316
upsampler.c 307, 311, 433
upsampling 307

V

Van de Corput 66
van de Lune 62
Van der Pol oscilator 85
Van Loan 2, 170, 172, 450
Vandiver conjecture 124
Vandiver criterion 123, 124
Vardi I xv, 4, 15, 122, 138-139, 454
Varga R S 63, 454
variational theory 286
VCON oscillator 85
vector quantization 306, 316, 317, 328, 329
Verhulst P 81, 230, 454
vertex 268
Vetterli M 187, 454
Vinogradov 128
Vo M T 183-184, 451
vortex sheets (see Karman vortex sheets)

W

Wagon S 63, 454
Wagstaff S xv, 136, 143-144, 454
Walsh functions 187-189
Walsh-Hadamard transform 151, 315-316
walsh.c 188-189, 314, 436
walshfile.c 314, 316, 434
Wang Y 82, 454
Wang Z 453
Wang Z Q 322, 454
Watson G N 48, 109, 454
Watson J D 92, 454
wavdim.c 284, 290, 438
wave packet 105
wavelet compression 316, 329
matrices 223
transform 288-290, 320-321, 325, 329
WaveletMatrix.ma 222, 439
WaveletOnSignal.ma 224, 290, 314, 449
waveletsound.c 314, 316, 443

Wegner T 253, 454
Weierstrass function 277, 280, 288-290, 308
Weisskopf 110
Welch J xv
Welsh L Jr. 118, 449
Wheeler N xv
white noise 274, 275, 280-281
White R H 452
Whittaker E T 48, 98, 109, 452, 454
Whittaker functions 24
Wieferich primes 121
Wieting T xv
Williams primes 128
Wilson primes 122
Wilson's formula 123
windowed Fourier transform 200-201
Winfree E xiv, 454
Winograd Fourier transform 178
Winter D T 452
Wolfram S xiv
Wozniakowski H 46, 454
Wright E M 66, 118, 119, 127, 450
Wylde A xv

X

Xu L 450

Y

Yamamoto B xiv
Young J 147
YUV color matrix 318, 325, 326

Z

z-sorting 90
zeta function (see Riemann Zeta function)
ziggurat method 45
Ziv (see LZW compressors)